Pax Britannica:
British Counterinsurgency in Northern Ireland,
1969-1982

A Dissertation
Presented to the Faculty of the Graduate School
of
Yale University
In Candidacy for the Degree of
Doctor of Philosophy

By
Montgomery Cybele Carlough

December, 1994

Pax Britannica:
British Counterinsurgency in Northern Ireland 1969-1982

Copyright © 1995, 2014 Montgomery McFate
All rights reserved.

Cover photo
Copyright © 2014 Arlene Wege
All rights reserved.

ISBN-13: 978-0-9905-7431-6

Published by Wilberforce Codex
New York, New York

www.paxbritannica.net

table of contents

preface [2014]

In 1990 when I began graduate school at Yale, few anthropologists had any interest in armed conflict. Faculty and fellow students suggested that my research might fit better in the political science department. I found this suggestion disconcerting - why should war be excluded from anthropological inquiry? How human beings fight is as much a matter of culture as table manners, death rituals, or transgendered prostitution. And, after visiting Northern Ireland for the first time, it was clear to me that Republicanism was not just a political philosophy; it was a unique culture with its own norms, narratives, symbols, rituals, and language. Decades of war (or hundreds of years, depending on who is counting) had destroyed neither Republican culture nor Republican political philosophy, but rather had strengthened, clarified and deepened it. It was impossible not to empathize with these self-disciplined, courageous people who continued to fight British forces despite the fact that the consequence was likely to be death or imprisonment; despite the fact that they had neither the manpower nor the firepower to secure a military victory; and despite the fact that their political goal of unification of Ireland's 32 counties was absent from the political agenda of the Republic of Ireland.

This complex mixture of politics, violence and culture was going to be the subject of my dissertation. After a research trip to Northern Ireland in 1991, I returned to Yale. One afternoon browsing in the Sterling Memorial Library, I found Allen Feldman's book, Formations of Violence: Narrative of the Body and Political Terror in Northern Ireland. Holding it in my hands, I had the experience all graduate students dread: somebody else had written what I intended to write, and had done a much better job than I ever could have. That afternoon, I went to see my advisor Professor Hal Scheffler and told him I was dropping out and going to law school. Smiling with bemusement at my predicament, Scheffler told me "you've done the hard part – the fieldwork. Compared with that, the writing will be easy." He asked me to ponder how the topic could be reframed so that it would not duplicate existing work, but still draw upon the research I'd already conducted.

I spent that weekend thinking about the fallacies and assumptions underpinning the conventional academic approach to political violence. At the time in the early 1990s, most journalists and scholars who wrote about Northern Ireland treated it as an ethno-sectarian conflict or as 'terrorism' – spectacular violence decontextualized from history, community and politics. But unlike political science, anthropology usually begins with subjective inquiry into ground truth as perceived by indigenous inhabitants. What do the locals have to say? What are their views about cross-cousin marriage, witchcraft or flint knapping? How do they make sense of the world? During the time I had spent in Ireland, nobody in the Republican community identified their adversary as the Loyalists or Protestants. Instead, in response to any question about sectarian conflict, they would provide a list of historical examples of Protestants who had fought for the Republican cause (such as Wolfe Tone). Likewise, they didn't frame their own violence as terrorism – they defined it as a war, viewed the British security forces as the enemy, and saw themselves as soldiers. The Republican version of reality barely intersected with outsider's interpretations of their world. What if I accepted the Republican version of reality as truth? What if all the poetry, songs, speeches, and books produced by the IRA and its supporters – generally dismissed as self-justifying propaganda – were accepted at face value?

research

If this was indeed a war between the British Army and the Provisional IRA, then most journalists and scholars were asking the wrong questions about the wrong subject. Instead of focusing on the historical origins of the conflict, examining the effect of sectarian divisions on civil society, or positing the IRA as a criminal gang (all topics *du jour* in the early 1990s), the more compelling question was: what social, legal, political, cultural or economic forces sustained the conflict? War is more than just uninhibited violence that expresses some primordial instinct inherited from our club-wielding ancestors; war is a social construct. Ireland was Britain's first colony, and fighting a war against the same enemy for hundreds of years meant that these adversaries knew each other very well; they understood each other's political goals, strategic weaknesses, rhetorical postures, and tactical predilections. Through their long and violent interaction these adversaries had developed unwritten norms, shared patterns, and certain expectations about the other's behavior. In fact, these adversaries were as much opposed as they were bound together; through their relationship as combatants, they had constructed each other. Moreover, any conflict that could not be resolved after roughly 800 years (again, depending on who is counting) had to be viewed as a self-perpetuating system, bounded geographically, with interconnected parts, self-governing rules, and recurrent processes.

With those ideas whirring around in my mind, I went back to the UK. Nothing in my graduate school education or my personal life had prepared me to research military topics. I understood the conflict in Northern Ireland from the Republican point of view, but my knowledge of the British Army had come from WWII movies and popular novels such as Kingsley Amis' fantastic <u>Anti-Death League</u>. So, beginning with Clausewitz, I educated myself on how state forces were organized, how they developed doctrine, how they conceptualized leadership, how they used history as a resource, how the application of force was managed, and so on. Then, I began to read about British military history – from Agincourt to Isandlwana to the Falklands. Only then, with some solid foundation, did I begin to read post-structuralist military theory. Guidance came from many sources: the faculty at Sandhurst, the archivists at the Imperial War Museum, the librarians at the Royal United Services Institute, and the many soldiers with whom I became friends. Hearing their personal stories (and understanding those individual experiences within the larger context of the war) humanized the conflict but also showed me the terrible personal cost paid by soldiers and their families. It was impossible not to admire these self-disciplined, courageous soldiers who continued to fight the IRA despite the fact that the consequence might be death or injury; despite the fact that they were prevented from using the manpower and the firepower necessary to secure a military victory; and despite the fact that their government's political goal contradicted the constitutional provisions of the Republic of Ireland. In the early 1990s, Northern Ireland seemed like war which neither combatant could win and that both were doomed to fight in perpetuity.

objectives/objectivity

During the years I spent conducting research, I had the rare opportunity to see a war from both sides, with access neither combatants nor most journalists have. During the fieldwork I internalized the subjective viewpoints of the people I was interacting with, on both sides of the conflict. Fieldwork – for me anyway – required complete suspension of disbelief, total psychological immersion, and most importantly, an acknowledgement of the truth of each combatant's narrative. Afterwards, while I was writing, holding both perspectives simultaneously in my mind produced some minor cognitive dissonance. Like a Zen *koan* presenting a paradox that cannot be solved through reasoning alone, I knew that both combatant cosmologies were absolutely true, and yet both were grievously false. As an anthropologist, I believed that my role was not to validate, legitimate, advocate or represent the views of either the British Army or the Provisional IRA. They were perfectly capable of doing so without my assistance, and to believe otherwise constitutes the worst type of intellectual paternalism. Rather, my task as an anthropologist was to approach the complexities of the war as a detached observer. By detached observation I mean aspiring to see a phenomenon – in this case the war in Northern Ireland – as a social totality, without interjecting my political views, my ethics, or my emotions into the object of study. As a social scientist, my goal was to understand how these two military organizations interacting over hundreds of years produced a unique war system – self-regulating, autonomous, resilient, and beyond the ability of most participants to apprehend. (Certainly I have read all the critiques of positivism in social science, and concur that all observation already distorts reality through the inclusion of the observer's perspective. Yet if anthropology claims to a science, shouldn't we *aspire* to objectivity? If not, then anthropology is nothing more than a self-referential intellectual cult that should only be taught in universities as a form of creative writing.)

In this dissertation, I aimed to explain the longevity of conflict in Northern Ireland, and approached the war as a self-perpetuating system, in which various elements (such as emergency legislation, geographical containment and the ritualization of martyrdom) maintained equilibrium and prevented resolution. (A later paper published in the <u>Journal of Conflict Studies</u> examined how these same factors hindered the ceasefire process.) As a detached observer – as a social scientist – my goal was to understand and describe how the system actually functioned. My goal was not pass judgment on whether the system, the individuals in the system, or the outcome of the system was good or bad, right or wrong, moral or immoral. Judgment is the prerogative of moral philosophers and religious leaders; not practitioners of the so-called 'science of man.'

At the beginning of <u>Discipline & Punish</u>, Michel Foucault describes the use of torture as a judicial sanction in mid-18th century France; he does not *advocate* the use of torture. Likewise, this book offers description of how a war system works, not a prescription concerning how it *should* work. I offer no suggestions to either paramilitaries or state armed forces for conducting a better military campaign, nor do I indulge in partisan politics. While I certainly do believe that military organizations *should* seek to understand their enemy as a matter of strategy, this book offers no recipes for doing so. Careful readers will thus avoid mistaking description for prescription.

avoiding harm

When I was writing my dissertation in London in 1993, a friend serving as a Member of Parliament asked for a draft copy so that he might share it with the Secretary of State for Northern Ireland. I had lived in Belfast long enough to know that Irish Republicans treated Frank Kitson's book <u>Low Intensity Operations</u> as a reference manual that explained exactly how the British military conceptualized operations. Likewise, I had spent enough time with the defence community in Great Britain to know the value of the apocryphal <u>Green Book</u> in understanding Republican strategy and operations. Unwilling to have my dissertation become the British reference manual for fighting the IRA, or the IRA's reference manual for fighting the British Army, I was determined to craft the manuscript to minimize its utility to either combatant.

Unlike other contemporaneous ethnographies about the IRA (which *inter alia* detailed levels of political support by location, political tactics of nationalist women, and symbol systems of the IRA) this dissertation contained very little information of interest to British intelligence services. In fact, the only chapter about the Republican movement concerns the hungerstrikes, which took place more than a decade earlier. Likewise, the dissertation contains no information with operational relevance to the Republican movement. How can I be sure of that? Not only did they have access to all the books, manuals, newspapers, speeches, magazines and archives which I used in my research, in many cases they told me where to look for source material. Moreover, they sometimes clarified my understanding of British tactics (such as 'framework operations') or explained the significance of particular geographic features (such as a cul-de-sac) within the urban battlespace. Neither the British security forces nor the Republican movement learned anything of interest from my dissertation, unless they had a compelling curiosity about post-structuralist military theory (which I sincerely doubt).

When anthropologists discuss their prime ethical directive – do no harm – they sometimes conflate different types of harm. In the field of public health, practitioners categorize prevention as primary (risk reduction), secondary (treatment) and tertiary (amelioration). This model can also be flipped around and used to describe harm. Harm at the tertiary level would include ethnographic sources being individually targeted as a result of one's research. Secondary harm would mean that belligerent parties derive operationally relevant information about their adversary or adversary's sustaining community to include its political objectives, its internal structure, its organizational weaknesses, and so on. Primary harm – which is the most indirect, most 'upstream' type of harm – would include general information already known to both parties, such as economic indicators, physical geography, or political system. This type of information only has intelligence value in the aggregate.

Proceeding from this model, there are three ways to avoid harm from research results. To avoid harm completely – primary, secondary and tertiary – publish absolutely nothing. Quit and go to law school. Anything in the open source domain automatically becomes a potential link in the so-called 'kill chain.' Regardless of one's personal beliefs about how research results should or should not be utilized or what 'rights' anthropologists have over the information they produce, knowledge always circumvents the creator's intent. It is utter folly to believe that intellectual products can be controlled. (Ironically, however, the more directly you engage with the military, the greater the chance to control the use of ethnographic information.) To avoid secondary harm, ensure that your research results have no military value to the combatants. To determine what has military value, simply ask them what they want to know about their enemy. Generally in my experience, they will tell you exactly what they want to know in the hope that you've got it written in your little notebook. Third, to avoid tertiary harm, preserve the anonymity of ethnographic informants by avoiding source attribution. In theory, this will prevent identification of individuals by either combatant. With enough triangulation however, as every journalist knows, even vague descriptions can lead to a positive identification of sources (whether IRA volunteers or British Army soldiers.) The point here is that protecting sources by preserving anonymity is an inadequate strategy for harm avoidance, if that is one's goal. In any case, understanding the adversary's levels of political support, political tactics, and symbol systems will often have more value in the aggregate than an individual name with home address.

theory

Readers may find it curious and perhaps heretical that I describe my theoretical approach as 'neo-functionalism' in this dissertation. In trying to understand war as complex sociocultural system using existing anthropological theory, I discovered that the choices available to me were inadequate to the task. In the 1990s, the choice for understanding how *systems* worked was either cultural ecology or some variety of structural functionalism. Marxism might reveal something about the Britain's economic interests in Northern Ireland, but it would not explain the power of symbols to shape perception. Conversely, symbolic anthropology might illuminate the meaning of paramilitary funeral rites or the Easter lily, but it would not explain how the historical conditions led to current political configurations. Cultural ecology, while useful for explaining the duration of cultural patterns over time, might explain how individuals adapted to the physical and biological aspects of the war system, but it was inadequate as a theory to explain the war system *itself*. The writers who came closest to the approach I wanted to take were the phenomenological post-structuralist military theorists: De Landa, Deleuze and Virilio. But none of them could clearly articulate the relationship of the war machine to the society and the individual. Plus, they were way too abstract for mere mortals to comprehend.

Then I remembered the classic ethnographies we critiqued in graduate school with such naïve brutality, reviling them as ahistorical, polluted with colonialism, and inherently ethnocentric. We scapegoated the ancestors – Malinowski, Middleton, Radcliffe-Brown,

Richards, Evans-Pritchard, Gluckman, and Leach –
imbued them with our disciplinary sins and cast them
out into the wilderness. But what if we were all just
completely wrong? What if functionalism actually had
some theoretical value as a heuristic device for
explaining complex social systems? Other scientific
disciplines build cumulatively on the bones of the
ancestors in order to create knowledge; anthropologists
repudiate the ancestors, proclaim their guilt and burn
them in an intellectual *auto-da-fé*. Yet if I ignored all those
sophomoric graduate school critiques lurking at the edge
of my mind, it was clear that all of these writers had
actually tried to explain the same thing I was witnessing:
self-regulating complex systems (whether economic,
religious, political or military) intermeshed and
interacting within a society (and often with an external,
intervening political authority) over a long period of
time, which both constrained and enabled the lives of
individuals. Thus, 1994 I wrote that

> The neo-functionalist view adopted herein
> leads to the perception that systemic
> equilibrium of military cultures is not
> disturbed, *but maintained* through the death
> of soldiers during combat, as long as they
> do not exceed the limits of the system but
> remain within acceptable levels. Anything
> that maintains war at an acceptable level
> sustains conflict. Coded limitations on
> violence — strategic, cultural and
> ideological — prevent conclusion. Politics
> prevents complete release, to paraphrase
> Clausewitz.

I was embarking into dangerous territory, questioning the conventional wisdom of the high priests of anthropology, speaking the taboo words: functionalism, equilibrium, and adaptation. Does questioning disciplinary assumptions make me a heretic? Perhaps; but I could find no other way to explain how a stable death rate – an 'acceptable level of violence' in the words of British Home Secretary Reginald Maulding – maintained the war system. Military organizations, in fact, are the only human organizations designed to continue functioning after sustaining casualties; their resilient properties promote, even ensure, systemic equilibrium.

the means of production

Looking at the document now twenty years later, my recurrent thought is: 'this is what happens when you read too much Foucault and don't write an outline.' The dissertation feels disorganized and haphazard, overly theoretical, and too abstract. Did I really use silly words like 'deproblematize'? Despite its imperfections, the only changes I made to the 1994 document were minor — correcting typos and replacing British spelling with US equivalents. When I wrote this dissertation, I was ignorant about the military (as most academics are) and I had to assimilate an abundance of heterogeneous information in short order. No doubt there are errors in the document, particularly pertaining to armaments, vehicles and other technology (about which I knew very little at the time).

If this manuscript leaves so much to be desired, why bother self-publishing it? Some of the themes in the dissertation (such as the intersection between counterinsurgency and cultural knowledge) became relevant during the recent conflicts in Iraq and Afghanistan and will likely prove relevant in future conflicts. Also, the treatment of war as a socially constructed system, while now a commonplace academic approach in war studies, nevertheless illuminates much about the power of culture in creating and sustaining violence. Northern Ireland provides an interesting example of that dynamic, which can be more fully explored now in peaceful conditions by other scholars. Finally the link between University Micro Films and Proquest means that your dissertation (just like mine) is no longer being stored somewhere in a deep vault, safe from everything except a nuclear blast. Unless you request an embargo, your dissertation – with its ridiculous conclusions, bad grammar, and outdated citations – can be downloaded as a PDF for free. Self-publication, for those of you who would like retain control over the means of production, represents a sensible alternative.

in conclusion

Writing this dissertation twenty years ago took me on a deeply personal journey, in the shamanic sense of the term. Anyone writing about a subject that moves the heart will find that their deepest fears, deepest emotion, and deepest pain emerges and recedes in waves. Research is a journey inwards as much as outwards, in which we explore the contours of our own minds as we examine the subject of inquiry. Despite my attempts at objectivity, I still wonder: did I observe something real, or just impose order on the chaos of war?

preface and statement of disclosure [1994]

Word choice is a very complex issue when writing about political violence in Northern Ireland. The Provisional Irish Republican Army refer to themselves as the IRA, the Provos, the Provisionals, the 'RA, 'auchnacloys', and sometimes affectionately as 'Chuckies.' In order to distinguish them from the Official Irish Republican Army (OIRA), British Security Forces generally refer to them as 'the PIRA'. In order to bypass this stylistic problem of unintentional ideological ascriptions, I follow the typical US designation of the Provisional IRA simply as 'PIRA', dropping the definite article in front of the acronym. When referring to this organization prior to its split into Official and Provisional factions, I use the standard abbreviation, IRA.

Because technical terminology is inconsistent between academic and military sources within the British military establishment, not to mention between the US and Britain, my use of terms complies with the 1993 International Military and Defense Encyclopedia, published by Brassey's. This seems to be the most coherent distillation of current usage. A glossary of technical terminology and acronyms also is included.

Ethical considerations and on-going friendships prevent full disclosure of sources. Although I conversed with numerous soldiers, politicians, and theorists on both sides of the ideological and military divide, I have made a conscious effort to use publicly available sources, published elsewhere, in preference to direct informant statements. Although anthropologists writing about security issues sincerely claim to have concealed the identity of informants, insider knowledge often allows accurate guessing. When informant statements are used, either in quotations or as evidence to support a point, I do not attribute them, except in two cases where permission was explicitly given.

This doctoral research was financially supported by the National Science Foundation Graduate Student Fellowship (1991-94), Yale University Fellowships (1990-94), the Andrew W. Mellon Foundation pre-dissertation Fellowship (1990-91), a Council on West European Studies travel grant (1993-4), John F. Enders Fellowship (1993-4), the William's Fund (1991, 1993-4), and the International Security Program/Smith-Richardson Foundation at Yale (1993-4).

Fieldwork was conducted in 1989, 1991, 1993, and 1994. Interviews, prison visits, participant observation, correspondence, and conversations were conducted with members of the Republican Movement in the United Kingdom, the Republic of Ireland, Holland, and North America, including but not limited to the following groups: Provisional Sinn Fein (PSF), Glor na n-Gael, Provisional Irish Republican Army (PIRA), Irish National Liberation Army (INLA), Irish Northern Aid (NORAID), the Ulster Gaelic Club (UGC), and Information on Ireland (IOI).

Interviews, participant observation, correspondence, and conversations were conducted with members of the British defence establishment in the UK, including but not limited to the following groups: serving and retired members of the British Army, the Ministry of Defence (MOD), the Royal United Services Institute for Defence Studies (RUSI), the Corporation for Operations Research and Defence Analysis (CORDA), historians at Royal Military Academy at Sandhurst, and an 'independent military sub-contractor' (mercenary).

Neither members of the British defence and security establishment, nor their Republican counterparts, were aware of my research on the other side of the military lookingglass, and this manuscript may therefore come as somewhat of a surprise to them. This research was conducted informally, without the assistance or official sanction of the British Army or the Republican Movement. No members of either organization exceeded the limits of security, or jeopardized their operations, by allowing me access to classified documents. The only secret documents to which I was privy dated from 1920. This research does not intentionally benefit any partisan organization, nor was this research financially aided or abetted by any partisan organization.

I would like to acknowledge the intellectual contributions of Professor Harold Scheffler, Professor David Apter, Professor George Andreopoulos, Professor John Middleton, and Professor John Szwed at Yale University. I would also like to thank Professor Paul Bracken, Professor Graham McFarlane at Queen's University, Belfast, Dr. John C. Dolan at University of Otago, New Zealand, Omid Mantashi, Linda Angst, Peta Katz, Alistair Renwick, Brian Baer and especially Arturo Cherbowski-Lask. Thanks to Robert Bell at the Linen Hall Library in Belfast, and to John Montgomery, Head Librarian at RUSI. This project was completed in memory of James Pearson Palmer.

list of illustrations and tables

Chapter One:

The 'Tribe in the Desert':
Anthropology and the War Machine

"War is not a life: it is a situation,
One which may neither be ignored nor accepted,
A problem to be met with ambush and stratagem,
Enveloped or scattered."

-T.S.Eliot, A Note on War Poetry, 1942

On March 16, 1978 two British soldiers crouched in a covert observation post in the Glenshane Pass in South Derry, Northern Ireland. Lance-Corporal David Anthony Jones, regular of the 3rd Battalion of the Parachute Regiment, allegedly seconded to the Special Air Service (SAS),[1] and Lance-Corporal Kevin Smyth listened to the portable radio and looked through a pair of night binoculars. Both were heavily armed: in addition to the 9mm pistols strapped to their sides, Jones was carrying a 7.62mm self-loading SLR rifle, and Smyth had a Sterling submachine gun. They were wearing camouflage jackets and Doc Marten boots "favored footwear with gunmen on all sides of the Irish conflict" (Beresford 1987:108).

Through their night-vision binoculars, the Paratroopers watched a pair of armed men in camouflage gear clambering across the hedgerows of a damp field. Even in the dark, they could see that the men had guns: one man was carrying an M14 rifle and two magazines, and a .38 Special. The other man had an Armalite and a 9mm pistol. With their guns in a semi-alert position, their berets, dark clothes, and insignia patches stitched onto their combat jackets, they resembled members of the UDR (Ulster Defence Regiment). In order to avoid friendly fire casualties, the British soldiers issued a verbal challenge to the armed men. These men were not fellow British soldiers of the UDR; they were Provisional Irish Republican Army.

Dropping into kneeling positions, they opened fire. Lance-Corporal David Jones was hit by a rifle shot in the chest, and mortally wounded. Bleeding and in pain, Jones fell onto the grass. Smyth, hit in the stomach, raised his weapon, and returned PIRA fire. One of the bullets struck bone, wounding PIRA Volunteer Francis Hughes. Hughes, bleeding profusely, stumbled and crawled through the wild, damp scrub, taking cover under a dense bush. Jones, meanwhile, lay dying at his post. Waiting for helicopter evacuation, Jones whispered to another officer, "If I don't make it, make sure Anne gets all my stuff." Although surgeons operated on him immediately, Jones died the following day from shock and infection.

Jones' dying wishes raised a very complex legal problem; did his whispers constitute an 'oral battlefield will'? Prior to military deployment, soldiers are typically advised to make a will: Jones' original will named his mother as the beneficiary in the event of his death. Due to be married shortly, Lance-Corporal Jones now intended that his fiancée, Anne, receive his property, post-mortem. Following his death Anne Mannering submitted an application to the court for Jones' oral declaration to be admitted as his last will and testament. A British court then had to determine whether this was, in actuality a soldier's will, such that no written documents or signature would be necessary for the will to be considered valid. According to Section 11 of the Wills Act 1837, oral 'battlefield wills' can be effected by soldiers on "expedition or in actual military service." For the purposes of Section 11 of the 1837 Act, was the deceased engaged in actual military service? Is actual military service the same thing as participating in a war? Are British troops deployed in Northern Ireland on actual military service? And finally, the most sensitive issue: does the armed conflict in Northern Ireland constitute a war?

Sir John Arnold, who decided the case of <u>Re: Jones (Deceased)</u>, ruled that Jones had indeed produced a valid soldier's will. The precedent cited for the case was an Australian decision [<u>In the Will of Anderson</u> (1958) 75 WN (NSW) 334)], concerning the death of a soldier during the Malayan Emergency suppressing 'armed and organized aggression and violence... designed to overthrow the Government by force.' In the case of Anderson, Justice Myers had ruled that

> In the present case there was no state of war and it is difficult to see how there could have been, for there was no nation or state with which a state of war could have been proclaimed to exist, but in all other respects *there was no difference between the situation of a member of this force and that of a member of any military force in time of war*. In my opinion the deceased was in actual military service... (5 at H. Italics are mine).

Sir John Arnold, echoing the precedent established by Justice Myers, determined that the character of opposing force was irrelevant to a determination of active military service. "The fact that the enemy was not a uniformed force engaged in regular warfare, or even an insurgent force organised on conventional military lines ...cannot in my judgment affect any of those questions..." (5 at J). Active military service depends, thus, on the nature of the activities of the deceased, and on the unit or force to which he was attached, and not on the nature of the opposing force. A War Office document from 1922 also concludes that "...it must be recognized that *the troops engaged are at war although their opponents may be only irregular partisans* indistinguishable from ordinary civilians" (Jeudwine Papers 1923:142. Italics are mine).

It has erroneously been argued that these legal judgments, by upholding battlefield wills, confer a *de facto* legal status of war on the military activities in Northern Ireland, thereby rendering PIRA a legitimate army. In fact, the two judges, and the 1922 War Office document, carefully distinguish between the nature of the opponent and the nature of one's own forces; the military nature of one need not confer any military legitimacy on the other. Furthermore, armed military engagements do not necessarily a war make. Actual military service may take place outside the conditions of war, such that "The fact that there was not a state of war...was irrelevant in deciding whether the deceased was in actual military service" (1 at F).

In fact, both Sir John Arnold and Justice Myers were debating a moot legal point: since the Kellogg-Briand Pact of 1928, strengthened by Article 2(4) of the United Nations Charter 1945, 'war' as a legal category has not existed. In the absence of a legal construct to describe armed conflict between high-contracting parties, the basis for legal decisions reverts to battlefield conditions, rather than the declarations of states, as the defining feature of armed conflicts. In other words, despite the fact that a 'state of war' can never legally be said to exist, British soldiers who conduct active military operations in Northern Ireland do so in conditions which, for legal purposes, are identical to war. A war by any other name is still a war.

Do not forget that Lance-Corporal David Jones was shot and killed during active military service by a very symbolic, charismatic figure: Volunteer Francis Hughes of the Provisional Irish Republican Army. While Jones was evacuated by helicopter to a hospital near Magherafelt, Hughes remained hidden in the damp hedgerow all night, blood seeping from his bullet wounds. A patrol of heavily armed British soldiers used dogs to track Hughes, the 'most wanted' man in the United Kingdom. They followed the trail of blood for 600 yards, and discovered Francis Hughes camouflaged by the gorse. After taking a few photographs of Hughes, pale and covered with his own blood, they moved him to a military hospital. After recovering from surgery which left him permanently crippled, Hughes was subsequently convicted, and sentenced to life in prison. Hughes did indeed die in prison, on hungerstrike for special status as a Prisoner of War in 1981. Photographs of Hughes in Long Kesh Prison show an emaciated proto-corpse, a bearded specter of a man, draped in a white sheet, starving to death.

The ostensible goal of the hungerstrike was to force the British government to acknowledge the political nature of PIRA operations. Hughes' death, almost exactly four years after he shot Lance-Corporal Jones in a field in South Derry, sought to establish that the conditions of 'active military service,' which applied to Jones, applied to him as well. Hughes and nine other men died in order to secure political status, which a British judge, however unknowingly or unintentionally, had already awarded to them based on the actual conditions of armed combat in Northern Ireland during the case of <u>Re: Jones (Deceased)</u>. As Seamus Heaney once wrote about Hughes' military operations in the fields of south Derry, "the bomb flash/ Came before the sound" (1984).

Lance-Corporal David Jones made a fatal error in mistaking Hughes and his companion for UDR. By yelling a warning before he opened fire, Jones had intended to prevent a 'friendly fire' casualty. Jones fatal warning call — fatal to Francis Hughes, as well, since it resulted in his capture and subsequent death — resulted from Jones unwillingness to risk shooting another British soldier. Friendly fire, very difficult to prevent in large-scale wars, becomes exponentially trickier in low-intensity conflicts where Special Forces and paramilitaries operate in such close proximity to one another that they establish shared combatant codes, and duplicate one another's tactics and operational procedures. But why did Jones mistake Hughes for UDR? Because Hughes had short hair, was wearing a black sweater, dark jeans, and black Doc Marten boots, and was openly carrying a weapon, he resembled a soldier.

And why was Hughes dressed in a way which would immediately distinguish him from the local population, mark him as a combatant, and therefore as a target? For inclusion under Protocol I of the 1977 Protocols of the Geneva Convention, combatants must distinguish themselves from civilians by carrying arms openly during a military engagement, and during military deployment preceding the launching of an attack. Hughes' capture corresponds exactly with PIRA's campaign to gain combatant status under the international laws of war. Thus, it's no wonder that Jones mistook him for a member of a military organization: that was the goal, international law was the playing field, and the hungerstrike was the final outcome. From a military point of view, this story serves as a good example of how fear of friendly fire slows combat, even in low-intensity situations. For soldiers on active service, restraint in the use of force can sometimes be more deadly than the use of force, itself.

The story, however, is even stranger. While Francis Hughes was in prison in Long Kesh, he met and befriended a fellow PIRA Volunteer, Raymond McCreesh. During a long, cold evening sitting in a prison cell, McCreesh recounted to Hughes, as prisoners often do, how he was captured by the British Army and finally wound up in prison. On June 25, 1976, McCreesh and three other men were on an operation in South Armagh to ambush a covert OP (Observation Post). Armed and wearing black balaclavas, they moved silently across the dark green fields, and were spotted by a group of Paratroopers on a nearby hill. Waiting for a perfect shot, Lance-Corporal David Jones opened fire on McCreesh and his company. Hearing whine of the high-velocity shots, the PIRA volunteers scattered and McCreesh and another man ran into a farm house. Lance-Corporal Jones and the Paras under his command surrounded the farm house and began firing. McCreesh, knowing that the game was up, surrendered (for a comparison of the British Army and PIRA narratives of the same event, see Morton 1989:172-176 and Iris 1981:30). Thus, McCreesh found himself in HMP Long Kesh in 1981 hungerstriking to the death, with Francis Hughes, the man who had killed Lance-Corporal David Anthony Jones eighteen months later. Very shortly, all three of them would be dead.

a war by any other name...

The interwoven narratives of Lance-Corporal David Jones, Volunteer Francis Hughes and Volunteer Raymond McCreesh, and the legal and mythological residue which their deaths left behind, illuminate some complex, recurrent themes in this project. To begin with: dying for legitimacy shows the literal and lethal human consequences of terminological disputes. "War," "terrorism," and "political violence" are imprecise terms, whose definition is debated by governments, international legal bodies, scholars, soldiers and terrorists. Inventing a typology of violence entails deciding on which side of the fence one wishes to sit. Deploying and negotiating the language of war is a political act, because the definitions of the terms themselves advance inherent political claims about the legitimacy of violence.[2] The lack of a legal basis for war means that the legitimacy of, and authorization for, violence must be derived from outside the law. The laws of war were designed to control and limit the use of force by nation-states, but due to insurgent challenges to state authority (and to state law), the law itself has become a site of struggle. Jurisprudence is no longer, if ever it was, a neutral arbiter of conflict or a means of preventing violence: the legitimacy of law is now at issue (Nietzsche 1967, vol. II: 11), a point examined in Chapter Four. Furthermore — and this is the central point — war limitations and war legitimacy derived from law, or from politics, have nothing to do with war *per se,* because war is now largely autonomous.

What are we to make of the fact that the judgment in Re: Jones (Deceased) almost exactly paraphrased the War Office's judgment on the status of British soldiers in Ireland in 1922, when "the troops engaged are at war although their opponents may be only irregular partisans indistinguishable from ordinary civilians"? Not so much has changed in this long, long war: neither the operational conditions, the identity of the participants, nor the problems of legal definition, differ substantially from the situation in the 1920's. This similarity has been noted and acknowledged within the British Army (for example, Drewry 1993). British expertise in low-intensity conflict (LIC), in fact, was founded on the strategic, logistic and tactical failure of the Anglo-Irish war. When the Anglo-Irish War ended by treaty in 1921, British veterans scattered across the Empire (where, as Republicans say, the sun never sets and the blood never dries), and found employment as policemen, colonial governors, mercenaries and frontier irregulars.

Following World War II, however, the sun was setting very fast indeed, as colonies such as Malaya asserted their prerogative of self-governance (see Appendix Three). The Anglo-Irish War provided the British Army with an excellent example of how not to conduct a counter-insurgency operation, and the lessons were not lost on Malaya. Malaya, in turn, became the textbook case for counter-insurgency properly conducted, inside and outside of Britain. High-ranking veterans of Malayan operations were seconded as advisors to the US Army in Vietnam, and also to Kenya, Cyprus, Aden and elsewhere. Lance-Corporal David Jones, a professional British soldier, had seen active service in Belize and Cyprus and was not unfamiliar with the operating principles of counterinsurgency.

The counter-insurgency loop was brought to penultimate closure with a prodigal return to Ireland in 1969. Ironically, although perhaps predictably, <u>Re: Jones (Deceased)</u> depended on a Malayan case as a precedent. In Malaya, as Justice Meyers ruled "there was no state of war and it is difficult to see how there could have been," in the absence of any state against which a declaration of war could be made. Avoiding declarations which inadvertently confer legitimacy in favor of emergency legislation was a strategy derived from the negative experience of Ireland in 1921, used successfully in Malaya in the 1940's, and redeployed via the counter-insurgency loop to Ireland in 1969. This self-recapitulating counter-insurgency loop is the subject of Chapter Two.

This strange story of Hughes, McCreesh and Jones also sheds some light on the synchronicity and durability of war as a system. Hughes, from Tamlaghtduff, County Derry and McCreesh, from Camlough, County Armagh, were both sons of staunchly Republican families and had been raised in conditions of war, experiencing terror as a banal fact of everyday life (Taussig 1987). Hughes and McCreesh grew up seeing blood on the pavement; they had been arrested, interrogated, interned, detained and physically assaulted by security forces. Before the hungerstrike, McCreesh was 'on-the-blanket' in Long Kesh. Because he refused to wear prison clothes, McCreesh had no visitors for over four years. The grim, fanatical determination and intractability of men like McCreesh and Hughes is as typical of PIRA Volunteers as it is of SAS members. They had more in common than their black Doc Marten boots: like Lance-Corporal David Jones, they were soldiers, and they believed it was their duty to serve their country. Hughes, especially, was part of the military force tradition in the Republican movement, and had little tolerance for political initiatives. Despite any ideological convictions they may have had as individuals, all of these men were produced by cultural systems which expected, required, and rewarded their participation *as soldiers*.

Political violence in Northern Ireland is widely acknowledged to have cultural permanence. According to the Diplock Report, in the Republican strongholds of Ardoyne and Lenadoon, children are "growing up in an environment of violence and destruction. To them a battlefield is always at their door" (1972:35). The cultural permanence of a long, long war has implications for security policy. In order to sustain conflict as a cultural norm, violence must be made manageable. The chaos of war must be deflected and rationalized by a countervailing system of order to promote longevity. The secret of efficient war is not expenditure of energy, but conservation. How is it possible that the war system in Northern Ireland is self-sustaining, and capable of self-reproduction? Cultural longevity depends on successful adaptation to conditions and on the sensible utilization of local resources. In Northern Ireland, death is plentiful and predictable. Republican culture has very sensibly made a virtue of a necessity: death, especially sacrificial deaths like those of McCreesh and Hughes, has become a means of cultural regeneration.

The 1981 hungerstrike is an excellent example of the long-term cultural effects of violence, and the inter-systemic functionality of death: at the beginning of the hungerstrike about 80% of the prisoners in Long Kesh and Armagh were under 25 years old. Some prisoners came from second and third generation Republican families. Because martyrdom contributes to the maintenance of cultural vitality, the conditions of re-production are military targets. Lord Wolseley once said "In planning a war against an uncivilized nation...your first object should be the capture of whatever they prize most" (Gwynn 1934:40). Clausewitz called it a "center of gravity target". If death is the mechanism of cultural continuity, mortality too must be contained. Counter-insurgency, as we will explore in Chapter Six, seeks to abolish the capability of the war system to reproduce itself by interfering with cultural regeneration achieved through martyrdom.

Hughes, McCreesh and Jones had much in common; they all died in a forgotten war over a political issue which few people know or care about, aside from those whose lives are directly affected. A recent study has concluded that political violence in Northern Ireland has almost no impact on the daily lives of the majority of residents (O'Malley in NY Times, 29 Nov., 1993). As the saying goes, the only people who care about criminal law are criminals and lawyers.[3] But criminals and lawyers, like soldiers and 'terrorists,' know a great deal about one another because *it is their business to know*: "The terrorist is very well identified. We are on sixty per cent first-name terms with the main terrorists in this area and they are, likewise, with us" (Sergeant, UDR, cited in Arthur 1988:223).

Not only do soldiers and terrorists share in the creation and re-production of violence, they often draw the same conclusions about the impossibility of purely military solutions. Compare, for example, a statement of Bobby Sands: "It is purely political and only a political solution will solve it" (1981: Day 10) with a Royal Marine officer: "it would never ever be a military victory...It's going to be a political victory" (cited in Allen 1988:278).

As early as 1937, it was known that "it is not the business of the armed forces to find the cure..." to the 'problem' in Ireland (Simpson 1937:35). Members of the British armed forces and the Republican movement arrive at the same conclusions about the nature of the conflict, and about possible solutions, over a period of more than sixty years because they are constituent elements of the same war system, and because the same war system is still slithering along.

The fundamental principle of military strategy is to know your enemy. But knowledge of other people, especially intimate knowledge, is also the basis of compassion and empathy, emotions from which soldiers, despite their training, are not exempt. According to one border officer,

> Francis Hughes was a man for whom the police and the Army had the highest regard. He was a really *professional terrorist*... when they came over the border at us they were really very, very effective (cited in Hamill 1985:268-69. Italics are mine).

Despite the rhetoric, these soldiers developed relationships with one another as counterparts in the counter-insurgency game.

> I don't know why they make those jokes up about the Irish. I think they're about the brainiest guys I've met yet. Really clued up. These guys know that we know them so there's no way you'll find anything on them. We have all the information about

them, whether they've been in jail, or if they
are INLA, PIRA, Official IRA (Corporal, 1
King's Own Scottish Borderers, cited in
Arthur 1988:1988).

Until policy dictated otherwise, PIRA and the
British Army dealt with one another as soldiers:
documentary photographs show British soldiers at PIRA
funerals as late as 1973 paying their respects by *saluting*
the corpse of fallen Volunteer Charlie Hughes. This is the
terrible paradox at the heart of military life: successful
killing depends on intimate knowledge, but the greater
the knowledge, the greater the inhibition to killing
(Booth 1979). One could conclude that ethnocentrism —
bad anthropology — interferes with the conduct of war.
But does good anthropology contribute to better killing?
I return to this question in Chapter Three.

Although soldiers and terrorists may develop a
grudging respect for one another, cultural prejudice and
political animosity do not necessarily evaporate. War,
especially of the proto-colonial type, often has a racial
component. Counter-insurgency and internal security
(IS) in Ireland exasperated soldiers because the enemy
was superficially the same as them: not distinguishable
by skin tone.

We'd been in Aden and Borneo, and as far
as we were concerned that was easy
because they were dealing with a person of
a different colour, which at that time,
during the decline of the colonial era, was
significant (Captain, 1 King's Own Scottish
Borderers, cited in Arthur 1988:20).

The Irish, racially indistinguishable, were sometimes marked off culturally as different. According to one soldier:

> I felt the Catholics weren't British at all. Not because they were on a different island to us, but because they shared none of our natural attributes. To me they were more Mediterranean in their temperament: terribly volatile, a woman-dominated society (Captain, 2 Coldstream Guards, cited in Arthur 1988:149).

Counter-insurgency, in particular, depends on accurate cultural knowledge. COIN utilizes local knowledge and 'on-the-ground' operations — terms which have crept into the jargon of anthropology (see Geertz 1983)— to infiltrate the 'hearts and minds' of the opponent. Local knowledge is often derived from tribal people: Sir Robert Thompson, grand wizard of counter-insurgency, advocated the use of special forces to make contact with "aboriginal tribes, living in the jungles and mountains" and "to recruit and organize them into units for their own defense on the side of the government" (1966:108). In Northern Ireland, this pattern persists in the form of both the useful knowledge of local security forces, and in the recruitment and use of informers in the intelligence war.

> The advantage of the UDR is, of course, that an Irishman can pick out a goodie or a baddie. They know what to look for because there are certain indicators that they've grown up with and they can see

them intuitively. Now, we go over looking for our own indicators of the sort of people they are, and they're different in Ireland, which is why you end up with bank managers being thought of as terrorists... (Major, 1 King's Own Scottish Borderers, cited in Arthur 1988:224).

The successful conduct of counter-insurgency operations depends, to a great degree, on accurate cultural knowledge of local informants. Intuitive use of cultural indicators by local ethnic groups has, not surprisingly, been studied by an anthropologist (Burton 1979).

Despite the fact that Lance-Corporal Jones was a highly-trained member of one of the world's most technically proficient and militarily sophisticated armies, he was killed by a high powered rifle. Although Britain now has non-terrestrial satellite defence capabilities, the cause of death for soldiers on the ground is often an old-fashioned lead bullet. Ironically, the simultaneous dearth of large-scale wars makes the capability of computerized defence networks to deploy nuclear war-heads anywhere, anytime, rather redundant. The 'no contest' battles of the Falklands and the Gulf no longer pose much of a technical challenge for western military systems.

Meanwhile, armies are occupied with numerous low-intensity operations short of war in compact geo-political theaters, such as Northern Ireland. Cyber-armaments and predatory machines designed for the electronic battlefield are, for the most part, useless against hit-and-run banditry. Rudimentary technology still predominates in LIC. Military intelligence (MI) in low-intensity operations is thus often exasperatingly and atavistically dependent on human sources. Hooking the big PIRA fish swimming in the Republican waters of South Armagh depends on knowing which species inhabit the pond. To this end, military intelligence requires the penetration and observation of contra-state groups, and the detailed recording of informant information about 'on-the-ground' practices; information requirements which sound suspiciously like anthropology. The death of Lance-Corporal David Jones typifies the historical continuity of combat mortality resulting from absurdly low-tech lead bullets, despite the existence of far more glamorous modes of killing. This theme surfaces again in Chapter Five.

bloody theory

The systematic nature of war, involving large population groups over extended periods, makes war an appropriate object for anthropology. On the other hand, anthropologists have tended to become more directly involved with their subject, which is perhaps the nature of the discipline. Many exquisite ethnographies have been produced by anthropologists whose first contact with their subjects was, so to speak, on the other side of the rifle sight. A few anthropologists whose participant observation was geared more towards participating than observing have been executed either as informers by rebel groups, or as rebels by state forces. Numerous anthropologists have followed the footsteps of Victorian proto-ethnographer Sir Richard Burton, and used their fieldwork as a cover for covert intelligence gathering. This activity was wide-spread enough during the Indo-China/Vietnam era, to induce the Ethics Committee of the American Anthropological Association to pass a resolution against using anthropological fieldwork as a front for intelligence operations (see Wakin 1992).[4] The status of anthropology as the 'handmaiden of colonialism' has also been subjected to relentless scrutiny (e.g., Asad 1975). As documents are slowly declassified, more recent political and inter-organizational affiliations between academic anthropology and the war industry are being exposed, greeted by the bored yawns of cynics who expected the culpability of anthropology from the get-go.

Although anthropologists have certainly participated in the war system, war and violence remain underrated and problematic academic subjects. War is clearly a product of culture, and the conduct of war is culturally determined. Boundaries between cultures are often established by martial means (Ferguson and Whitehead 1992), culture itself may even be a consequence of war. Despite this, numerous anthropologists have noted that their discipline is "strangely inarticulate on the subject of war" (Freid, et al. 1968: xxiii; see also Falk and Kim 1980; Ferguson 1989; Foster and Rubinstein 1986). Anthropologists are not quite wholly silent about 'primitive' war, and about the colonial wars which are the bedrock of the discipline. Although the foundation of the Royal Anthropological Society of Great Britain and Northern Ireland was concurrent with the heyday of British colonialism in Africa, most de-historicized ethnographies from the period neglect to mention the political context of research. Fortes and Evans-Pritchard introduce <u>African Political Systems</u> by noting that changes occurring "as a result of European conquest and rule" will not be examined, since the authors are "more interested in anthropological than administrative problems"(1940:1).[5]

Anthropology, as a discipline, is bloody from war. Major John Wesley Powell, veteran of the Indian wars, founded the Bureau of Ethnology to investigate American Indian cultures. Thus, "the earliest accounts of Native American political organization were written under conditions of war on the frontier in the 1860's" (Vincent 1990:40). Culture contact decimated tribal populations by introducing chicken pox and guns; anthropologists frantically took notes on vanishing tribal cultures, while simultaneously cooperating in systematic extermination, as for example in the Solomon Islands (Bodley 1993).

'Enemy' cultures been subject to analysis during wartime; anthropology has thus contributed directly to war. During World War II, American anthropologists began producing ethnographies on the Axis powers which allowed "prediction based on national character" (Mead and Metraux 1965:121), such as Benedict's (1946) study of Japanese national character, The Chrysanthemum and the Sword and Ladislas Farago's German Psychological Warfare (1942) (see also Bateson 1942; Metraux and Hoyt 1953). 'Culture and personality' methodology formulated "a set of hypotheses about those aspects of national character that would be relevant to the way we dealt with them as enemies.... we concentrated on those aspects of national character that would be relevant to wartime problems — attitudes toward victory and defeat, relative strength and weakness, ...expectations of death or survival in battle, etc." (Mead 1979: 149). The Cross Cultural Index, which later became the Human Relation Area Files (HRAF) at Yale, was used during World War II to provide information on Japanese-occupied former German territories of Micronesia (Mead 1979:152).

The contribution of anthropologists in the machinery of war is often unabashedly direct, and even appropriate: Geoffrey Gorer, known for his work on Japanese culture (1943), moved from the Office of War Information to British Political Warfare (Mead 1979:152). During World War II, social scientists worked for Naval Intelligence, the Office of Strategic Services, the Office of War Information, and so on with follow-up money provided for wartime studies provided by the Office of Naval Research, RAND and subcontracted to MIT (Mead and Metraux 1965:123). Project Camelot, developed by the Department of State, was taken over by the Department of Defence in the mid-1960s. As an "insurgency prophylaxis," Project Camelot recruited social scientists to study the causes of insurgencies (Ferguson 1989:156). Lucian Pye, in "The Roots of Insurgency and the Commencement of Rebellions" sought, for example, to understand how human behavior relates to the causes of insurgency in order to develop policies to cope with the threat of internal war (1964:179).

The importance of anthropology became especially evident during the era of counterinsurgency, where tribal people often seemed to be at the center of the war effort. Robert Tilman, in a seminal article in Military Review in 1966, pointed out quite clearly that British counterinsurgency in Malaya succeeded because it took account of tribal and ethnic distinctions, while similar US efforts in Vietnam were bound to fail because they lacked anthropological finesse. Lucian Pye, who was permitted to visit Malaya while a graduate student, later produced Guerrilla Communism in Malaya (1956), a very sophisticated analysis of tribal factionalism and political ideology. Salemink's (1991) recent work on the invention and appropriation of Vietnam's Montagnards by the French and eventually the CIA confirms Tilman's point. Wakin's (1993) recent indictment of anthropological involvement in counterinsurgency in Thailand points out the continued involvement of anthropology in the war system. Testifying before the U.S. Congress in 1965, Dr. R.L. Sproul, director of the Defense Department's Advanced Research Projects Agency (DARPA) said:

> it is [our] primary thesis that remote area warfare is controlled in a major way by the environment in which the warfare occurs by the sociological and anthropological characteristics of the people involved in the war, and by the nature of the conflict itself (cited in Wakin 1992:85).

Anthropology, however, has not only been useful to government forces, but to resistance movements as well. Malinowski was quite correct when he wrote "[t]he truth is that science begins with application" (1945:4-5). Little did Malinowski know that his former student from the London School of Economics (LSE), Jomo Kenyatta (whose ethnography of the Kikuyu, <u>Facing Mt. Kenya</u> (1938) Malinowski had greatly praised in his "Introduction" as an example of indigenous ethnography) would later allegedly lead the Mau Mau resistance and became the Kenyan head of state.

Why in light of so many connections between anthropology and war, have anthropologists had such a hard time developing a conceptual, analytical framework for exploring war? Quite possibly, "[t]he real problem is the relationship between anthropology as an ongoing enterprise and, on the other hand, empire as an ongoing concern" (Said 1989:217). Theoretical paradigms (Marx's "ruling ideas" 1956:78) certainly "helped to justify colonial domination at particular moments" (Asad 1975). The more overt connection between academics and the war system is no secret: the Ministry of Defence funds academic posts in Britain to over £150,000 a year and encourages and facilitates research work on defence and security issues (Aldred 1993).

Anthropology may be the epistemological product of a colonial war apparatus, but it is equally true that the philosophy of war which animates and informs military decision making is connected to and derived from notions of 'civilization', 'primitives', and 'savagery' generated by anthropological discourse. Anthropology does not merely justify the civilizing mission of imperialism, nor is it simply the corrupt intellectual product of empire. Rather, the underlying ideas of war and anthropology cross-pollinate and illuminate one another, and in some instances are indistinguishable.

War, in anthropology, is conceptually and analytically linked to conflicts within the socio-cultural realm.[6] Distinctions between types of violence and aggression (interpersonal violence, riots, feuds, terrorism, and warfare) are drawn, not according to the scale of violence, the number of participants, duration or severity, but according to organizing structural principles which govern the violent interaction. Whereas feuds occur where the contesting parties are part of a unified political group, wars occur between distinct and unrelated groups. Nadel's (1947) definition is similar to Wallace's (1968) "sanctioned use of lethal weapons by members of one society against members of another society".

Ferguson (1990) defines war as "organized, purposeful group action, directed against another group...involving the actual or potential application of lethal force. War...is a condition of and between societies" (26).

References to warfare were quite common in early functionalist ethnographies (Malinowski 1920, 1926; Hocart 1931). Conflict was believed to have a positive function (Simmel 1955); social systems remained, in spite of conflict and sometimes because of it, in a state of equilibrium. Amongst the Lugbara tribe, in what is now Uganda, "though a state of conflict between them is normal," the system remained in equilibrium (Middleton 1966:149). Political conflict between the Zulu and the colonial government was characterized by Gluckman as a balance of complementary forces, resulting in a stable political structure: "the balance between them is the dominant characteristic of the political system" (1940:4). Although structural-functionalist theory has long been unfashionable within anthropology, the oscillating equilibrium model (with the addition of population dynamics and catastrophe theory) is now being used in military operational research for predicting the oscillating force-structure patterns in counterinsurgency (Dockery and Woodcock 1994). Strangely enough, tactical combat patterns in low-intensity conflict seem to resemble the tribal warfare systems described by anthropologists long ago.

Structural-functional anthropology rarely focused on anti-colonial warfare; monographs emphasized 'culture contact' and assimilation between tribal and colonial groups (Kuper 1937). Indigenous resistance to assimilation was scarcely mentioned. Inter- and intra-tribal conflict, however, was of interest (Evans-Pritchard 1940; Gluckman 1940). Evans-Pritchard wrote "Between tribes there can only be war, and through war, the memory of war, and the potentiality of war the relations between tribes are defined and expressed" (1940:161). Culture was seen a consequence of warfare, and society itself as a means to war (see also Nietzsche 1968:53). "[I]f it were not for warfare," said W. Lloyd Warner, "Murngin society as it is now constituted would not exist" (1931:457).

Characterizing conflict as "the readiest means of defining political structure" (Radcliffe-Brown 1940: xx), entails assumptions about the 'natural' state of man-kind. The 'noble, peaceful savage' was opposed to the 'war-like, fierce savage,' engaged in a war of 'all against all.' Amongst the pre-industrial savages inhabiting the Hobbesian universe, warfare was an imbedded part of social structure and armed inter group conflict was normal amongst hunters and gatherers (Fortes and Evans-Prichard 1940). A chronic state of war was said to have caused the development of rules of war amongst the Nuer (Evans-Pritchard 1940) and New Guinea tribes (Strathern 1979), and the development of a warrior class among the Masai. In either the peaceful-savage-better-off or fierce-savage-not-better-off construct, historical change appeared as a transition from one abstract pole to another (Roseberry and O'Brien 1991; Wolf 1982:5).

Biological, monocausal explanations of violence also normalized conflict. Man, according to Lorenz (1966) and Ardrey (1961, 1966), has an innate, genetically programmed, instinct for violence: "Man is a predator whose natural instinct is to kill with a weapon" (Ardrey 1966:316). That biological rationalization of violence was common anthropological currency during Vietnam War (Washburn and Lancaster 1968) provokes certain suspicions about the degree to which theories absolve moral responsibility for social behavior (Montagu 1968). The gene-driven, "man-the-hunter/killer" model is flawed in a number of obvious ways: extrapolations about human behavior are generated from ethological studies, the innate aggression, 'territorial imperative' hypothesis discounts the instrumental purposes of violence, such as hunting for food out of hunger (Montagu 1968), and denies cooperation as a factor in evolutionary history (see Slocum 1975, for a feminist critique). Although biological theories tend to discount learned, transmitted behavior (Barnett 1968; Leach 1968) — what anthropologists usually call 'culture'— biological explanations of human behavior often posit 'civilized society' as a check on killer instincts, barely restrained in primitive society (Gorer 1968).

Monocausal, biological explanations of warfare have been replaced by sociobiological explanations. Sociobiology, "the systematic study of the biological basis of all social behavior" (Wilson 1975:595), has been called by its founder, the "antidiscipline" of anthropology (Wilson 1976:2). In a sociobiological model, cultural evolution would be seen as a result of biological selection; a 'cultural selection' theory of war would see it as an adaptive behavior selected to maximize fitness (McCaulay 1990). Conflict over material resources, such as food, land, goods and reproductive opportunities would appear adaptive (Chagnon 1974; Ferguson 1990). Stretching the theory to the breaking point, patriotism could be seen as an evolutionary response (Shaw and Wong 1989). Materialist, biological models of warfare can be historicized by incorporating the idea of 'pre-adaptation' (McCauley 1990), and the concept of "causal primacy of infrastructural factors" (Ferguson 1990).

War has also been correlated with social development (Ottenberg 1978), and appears to be a natural part of 'primitive' social life. Worsley (1986:299), for example, argues that warfare does not emerge with the state, and that notions of 'peaceful' societies should be "dismissed as pastoral myth." Otterbein's (1970) massive cross-cultural study of war in traditional societies examined the relationship of warfare and political organization, concluding that centralized authority within traditional societies correlated with military professionalization. On the other hand, groups such as the I Kung, the Toda of southern India, and the Eskimo, who do not wage war, nevertheless have extremely high levels of internal violence (Ottenberg 1978; Wolf 1987). Thus, for example, the 470 Kung San in the Dobe region of the Kalahari Desert, who have been called the 'harmless people', have twenty-two documented cases of homicide from 1920 to 1955 (Lee 1984).

Whether war increases or decreases with industrialization is also subject to debate in anthropological circles. St. Simon and Comte believed that industrialization increased the opportunity for freedom and curtailed violence. Modern anthropologists like Cohen (1986:265) agree that "as societies become more industrialized, their proneness to warfare decreases." Conversely, World War I era evolutionary anthropologists such as Holsti (1914) and Hobhouse, Wheeler and Ginsberg in their The Material Culture and Social Institutions of the Simpler Peoples (1915), rejected warfare as a normal condition of 'primitive' societies, and saw it as a result of industrialization.

Marxist anthropologists saw war as an import of capitalist cultures, seeking to appropriate resources and means of production (slavery) and markets for industrial goods. War as confrontation, as collision of modern and traditional societies, is seen as a process of 'articulation' of capitalism with peasant culture (Wolf 1969; Bradby 1975; Dupre and Rey 1968; Geschiere 1985).[7] War in this model is a product of industrialized societies, which have a surplus production of materials and services to support conflict. Subsistence-bound economies do not have the agricultural base with corresponding urbanization, or the complex economic system, to make war. The level of cultural development determines whether societies experience war. In certain complex societies, war between classes becomes a permanent social condition.

To what degree warfare, conflict and violence represents an instrument of order, or a crisis of disorder, is also an anthropological issue. The functionalist 'peace in the feud' (Gluckman 1960) approach assumes that conflict sustains order: "war is order" (Greenhouse 1986). War can also appear to be a form of Durkheimian social pathology, occurring when other methods of conflict resolution have failed, appearing as a surrogate for the exchange of goods, services and women (Mauss 1954; Malinowski 1926; Clastres 1977). War as a result of the absence of normative, legal order, has been postulated in ethnological jurisprudence and in comparative legal theory (Barkun 1968; Worsley 1986).

Symbolic anthropology views culture as a check on violence, turned to useful purpose through the symbolic processes of religion and ritual (Leach 1976; Douglas 1966; Turner 1969; Lienhardt 1961; Balandier 1986; Corbin 1986). Girard (1977) saw sacrifice as the central instrument for the transformation and domestication of violence (Hamerton-Kelly 1987). Social order was established by an original act of violence, expelled through ritual scapegoating. "In the beginning was violence, and all history can be seen as an unending effort to control it" (Balandier 1986:499). Conflict, the expression of which would otherwise be too dangerous for the social order, "is given dramatic and symbolic form" during the enactment of rituals (Turner 1967:25). Ritual makes conflicts between norms and desires, between social subsystems and the total system, apparent. Symbolic enactment resolves conflict: "the raw energies of conflict are domesticated into the service of social order" (Turner 1967:39). Because this seemed to be an overly Appolonian view of conflict, Turner later (1969) conceded that structural break-down, social inversion and the collapse of institutionalized norms do occur, and proposed to call this 'communitas' or 'liminality' (see also Douglas 1966).

To what degree 'communitas', or anti-structure, exists outside of structure is unclear. To what degree it *can* exist outside of structure is also unclear. Models of stable, normative socio-cultural systems inevitably exclude endemic social disorganization and this is the real epistemological problem that anthropology as a discipline has with war. Social theory cannot cope with chaos. Anti-categories, such as 'anti-structure' merely reinstates chaos as a form of temporary (dis)order. The need for discrete analytic categories is an epistemological problem infecting all scientific inquiry[8]: "the whole aim of theoretical science is to carry to the highest possible and conscious degree the perceptual reduction of chaos..." (G.G. Simpson, cited in Levi-Strauss 1966: 9). Anthropological war paradigms do not admit, to paraphrase Leach, 'the limits of their anthropological competence'. War, fundamentally chaotic, is a most difficult subject.

If anthropology is an epistemology of political power, then in a nihilistic sense, all oppositions between synchrony/diachrony, fierce savage/peaceful savage are about the relation of anthropology to itself, a form of intellectual onanism. With the end of the cold war and a dearth of colonial subjects, anthropologists find themselves on an empty stage, apologizing for their embarrassed silence. So sensitized to their own relations to power, anthropologists have discovered violence as a 'spectacle', no longer about resources, development or equilibrium, but about itself. Geertz's (1980) 'war as theater' and Bourdieu's (1977) 'symbolic violence' approach both focus more on the performance of violence than on the concrete political economic realities which lead to its performance. At the other end of the spectrum, anthropologists have resisted the intellectual tautology of violence as a spectacle by become involved in 'hands on' advocacy for armed insurgent groups (Swedenburg 1991), and as "critics of contemporary counterinsurgency theory and practice" (Sluka 1992:287).

making war on culture

If anthropologists have had a difficult time coping theoretically and methodologically with the disorder and chaos of war, military historians and theorists have had a much worse time. That chaos — "fog and friction" — is inherent in war has become almost a truism. The search for ordering principles of war has thus been abandoned in favor of understanding its inherent chaos (Newell 1991:7). Characterizing war as chaos does not imply a lack of constraint in the conduct of war (1991:8), nor does it negate the development of explanatory theoretical models. On the contrary, the introduction of chaos theory to the study and practice of warfare has enriched the complexity of theoretical models, especially at an operational level (Simpkin 1985; Newell 1991; DeLanda 1992; Dockery and Woodcock 1994). With the acknowledgment of chaos as a principle of war, the conduct of war and the smooth functioning of command and control systems have probably improved. Devices such as *'Auftragstaktik' or* de-centralized mission-oriented control schemes are ways not of eliminating chaos from the conduct of war, but of harnessing it. The ability to utilize and cope with chaos during military operations, interestingly, appears to be a cultural predisposition. Anthropology and war could be said to be intersecting somewhere on the culture/chaos grid.

Culture theory, especially anthropology, can serve as a theoretical tool for understanding war. Cultural knowledge as a means to improve military prowess has been sought after since Herodotus concerned himself with the opponents' conduct during the Persian Wars. T.E. Lawrence's Seven Pillars of Wisdom, (1926/1949) for example, is essentially an ethnographic text, concerned with the customs and conventions of desert dwellers. Early attempts by military thinkers to rationalize the cultural forces of colonial campaigns, such as Callwell's Small Wars (1906) and the recruitment of 'martial races', relied heavily on evolutionary Victorian concepts of culture derived from anthropological sources. Occasionally, the British military establishment became concerned about the influence of exotic customs on military prowess, as for example a 1923 paper in the Army Quarterly, "The Veiled Wives of Indian Soldiers: A Problem" (Bell 1923). Knowledge of the locals and adapting local knowledge have always been classic tricks of the military trade.

Military use of culture categories coalesces in "ethnic state security maps" (Enloe 1980). According to Cynthia Enloe, in her excellent and underappreciated study Ethnic Soldiers, state elites employ operational concepts of ethnicity to promote, recruit and/or exempt certain groups from military service. Ethnic criteria are thus instrumental in military policy decision making: "Military policies operationalize state security maps. Official choices in recruiting, promotions, assignments and field deployments all articulate what state elites are usually restrained from spelling out programmatically" (1980:16). Enloe argues that military institutions not only reflect ethnic conditions but shape them, what she calls the "Ghurka syndrome" (1980:26). 'Martial races,' such as the Ghurkas, Scots, Sikhs, Berbers, Maoris, Mongols, Kurds and even the Irish, have been developed, propagated, and exploited by the state for the sake of achieving military goals: "state-mobilized cannon fodder" (13). Enloe demonstrates how cultures can be constructed or invented as a by-product of military recruiting (as in the case of the 'Ghurkas', actually a mish-mash of tribal people of Nepal, 39-42), how internal regimental cultures utilized ethnic ascriptions to mobilize soldiers (for example, the wearing of kilts in the Scottish regiments of the British Army as a mark of distinction, 35-37) , and finally how 'local knowledge' can be incorporated into the state's war machine by recruiting and promoting the bearers of that cultural information (for example, the Montagnards in Vietnam, armed and trained by US Special Forces, 207). The concept of 'martial race' itself demonstrates an awareness that war-making capacity and warfighting style are aspects of culture.

Since the Vietnam War military theorists have been confronted with an increasingly complex set of problems relating to culture. An emergent trend in the war system is death as a way of life, or in other words, the cultural permanence of war on a global scale. Northern Ireland, like Chad, the Sudan, South Africa, Ethiopia, Israel, Pakistan, India, Burma (Myanmar), Cambodia, the Philippines, Guatemala and El Salvador, has experienced more than twenty years of war (Kidron and Smith 1991:23). The longevity of these conflicts requires a cultural explanation: "the most durable disputes establish a culture or tradition of conflict which then structures future action" (Hill and Rothchild 1992:189). Conceptual models are needed to explain the culturally determined features of war. (Ironically, low-intensity conflict which shows the most direct cultural, as opposed to technological or political, features is also the most resistant to feasible operational modeling.) The statistical frequency of low-intensity conflict means that cultural factors have become an important variable in political risk analysis (Overholt 1982; Calverley 1985).

In this era of low-intensity insurgencies, civil wars, proxy wars and wars of de-colonization, cultural knowledge, especially local knowledge, has become a pre-requisite for successful counter-insurgency. As Beckett and Pimlott succinctly point out, a "significant feature of much post-1945 COIN theory was the considerable space given to understanding the nature of an insurgency as a preliminary to its eradication" (1985:5). What does local knowledge, a term probably familiar to most anthropologists (see Geertz), mean in a military sense? The concept is clearly explicated by Sun Tzu: "We are not fit to lead an army on the march unless we are familiar with the face of the country — its mountains and forests, its pitfalls and precipices, its marshes and swamps. We shall be unable to turn natural advantages to account unless we make use of local guides" (1983: 32). Exploiting local knowledge means, for example, that at the beginning of the Malayan Emergency, men from the Iban tribe in Sarawak were brought to Malaya to impart their deep-jungle tracking methods to the SAS (Seymour 1985:276; Geraghty 1980:55-56). Local ethnic groups, such as the Senoi, an aboriginal people from the interior of the Malaysian peninsula, were also used by the British during the 1956 Emergency against the Communist insurgents. Similarly, the Tho in North Vietnam have been recruited for military service since 1940 by the Viet Minh for their skills as jungle trackers and scouts (Enloe 1980:207). US forces in Vietnam, inheriting the French colonial policy of appropriating local military labor, recruited numerous tribal people referred to as 'Montagnards' (Salemink 1991).

That war is an extension of culture, as well as politics, is generally accepted by military theorists. With the recognition of culture as a variable in war, anthropology has begun to have an osmotic influence on military historiography (Shy 1993). Military historians sometimes have an odd, pre-conceived notion of anthropology as a way to explain 'irrational' wartime behavior (e.g., Harris 1975): seventeenth century Iroquois attempts to minimize casualties (Richter 1983) and the Dani of Irian Jaya (Western New Guinea) who cease hostilities after a limited number of casualties (Gardner and Heider 1968) are well known cases. Anomalous western military behavior is now being subjected to the same variety of scrutiny, especially how social, ideological and cultural factors contribute to the development of programmatic, erroneous military responses, for example the 'cult of the offensive' during World War I (Snyder 1984), US counter-insurgency policy in Vietnam (Shafer 1988), and the analogical derivation of post-WWII US security policy in Korea (Khong 1992).

Sir Michael Howard derides such 'cult of the offensive' explanations for war (1986), preferring the view that war is the result of rational, instrumental decision making at a policy level, rather than the result of ideological prescriptions or 'natural' war-like impulses. Poo-pooing the innate aggression hypothesis of violence, however, often leads to an assertion that war expresses supreme instrumental rationality. Bellamy, for example, in The Evolution of Modern Land Warfare "agrees wholeheartedly with anthropologists" that man has no innate aggressive impulse (1990:8) and deduces that because man is not naturally violent, war must represent an expression of human rationality: a "reasoned, cold, calculated application of organized military force" (1990:8). Although a concept of 'cultural evolution' may have disappeared from the vocabulary of main stream social anthropology, evolution predominates as a metaphor in military discourse. Warfare, which emerged "primarily a product of ...nomadic herding, societies" (240), expressed the logic of capital accumulation and appropriation. According to Bellamy, as war between states becomes increasingly rare and recedes to less-developed areas of the globe, the baroque arsenals of large-scale conventional land warfare ought to be consigned to the scrap heap of history. Nomadic warfare, as I argue, although it transpires at the margins, is not anachronistic legacy of history, but an atavistic military form which states are now compelled to adopt

(Deleuze and Guattari 1992:360; Huntington 1993), a problem returned to in Chapter Five.

Military sociology, especially the sub-branch of civil-military relations, has used 'culture' to explain the professionalization of armies as an evolutionary social process (Janowitz 1960; Huntington 1957), such that professional soldiers represent culture in both senses of the term (see Hackett 1983). Regiments, which could be characterized as soldier's tribes, were the obvious repository of culture, and were studied as such. This paradigmatic 'fife and drum' regimental history approach in military historiography, has, under the influence of Keegan's classic Face of Battle (1976), given way to a experientially oriented, 'war and society' approach (Beckett 1994, communication) focused on 'the actualities of war' (Holmes 1986).

In a wider sense, cultures of conflict, and of détente (Whitfield 1991), have also been opened up to study, including how national identity may be predicated on popular representations of soldiers (MacKenzie 1992), how social conditions have affected the conduct of European land warfare (Bond 1975; Strachan 1983) and how war may become paramount in the construction of nationhood (Aulich 1990). The mechanics of indigenous military response to colonialism has been subjected to operational level analyses (Crowder 1978), as well as the contribution of social and ideological factors in the development of mechanical weapons, such as the machine gun (Ellis 1975).

The influence of cultural factors on strategy formulation, what has been termed 'strategic culture', illuminates the intersection of policy and highest level military praxis. Derived partially from the political culture literature (Pye 1965; Welch 1993), and the British 'way of warfare' approach (Liddell Hart 1935; French 1990; Howard 1983), 'strategic culture' analysis examines how a states' national culture determines particular 'styles' in the use of force and seems (superficially at least) to resemble Mead's and Benedict's World War II era predictive studies of the national character of Axis power belligerents. The 'strategic culture' school's intellectual alliance with anthropological methodology receives its finest expression in Ken Booth's neglected Strategy and Ethnocentrism (1979). Following the lead provided by Dixon (1976), Booth examined the contribution of ethnocentrism to errors in strategic thinking.

Newell, approaching the problem of military ethnocentrism from an operational level, suggests that ethnocentrism may interfere with joint military operations, such as NATO or UN peace-keeping. "Each service's culture produces different patterns of professional knowledge, belief, and behavior which in turn influences the way officers learn and transmit information amongst themselves" (1991:80). Offerdal and Jacobsen (1993) have also produced a recent empirical study on the correlation of cultural attitudes and class dynamics with performance during decentralized combat operations. Maneuver warfare capability can be correlated with certain cultural predispositions; successful decentralization of command and directive control depends on cultural attributes of initiative and innovation (Behagg 1993:125-6).

The cultural features of war have become so glaring that Huntington (1993), very influential in military-security circles (and frequently cited by them, as for example in Major-General Willcock's 1994 speech), has recently asserted that in the next century the dominant global form of conflict will no longer be political or ideological cleavages, but cultural fault lines between civilizations (Huntington 1993:22). Cold war political boundaries having proved less immutable than cultural characteristics, the question 'which side are you on?' has been replaced by 'what are you?' "And as we know, from Bosnia to the Caucus to the Sudan, the wrong answer to that question can mean a bullet in the head" (1993:27). The emergent Confucian-Islamic military connection (in McCurdy's words, "'a renegades' mutual support pact") to promote the acquisition by members of weapons technologies (47) is an example of inevitable de-Westernization and strategic re-alignment. Despite Huntington's' anxiety-prompting agenda about Western socio-military security, his concept of 'culture-war 'is consistent with other recent attempts to explain the occurrence of violence at cultural boundaries and the internationalization and diffusion of domestic, communal and ethnic violence (Midlarsky 1992). This pushing of violence to the margins has spawned a military "neoprimitivism" of terrorists, pirates, mercenaries and multi-nationals (Deleuze and Guattari 1992:360). As the old world order of nation-states deconstructs under the pressure of ecological catastrophe, engineered famines, and tribal war (Virilio 1986; Kaplan 1994), the role of the armed force in remote "pockets of anarchy" (e.g., West Africa) is under intense scrutiny by the US and the British military.[9]

Possibly the theoretically most refined model of cultural influences on actual military operations Major RAD Applegate and JM Moore's, "Warfare — an Option of Difficulties: An Examination of Forms of War and the Impact of Military Culture" (1990), in which they develop a model of military culture based on a typology of war-fighting forms. Their concept of 'military culture' might be seen as a fragment of a 'state security map' in Enloe's terms. Simply, military culture determines military style:

> The ability of an army to choose the most appropriate form of war is not only a product of the prevailing external conditions and circumstances, but more importantly the army's perception of these conditions and circumstances which is a product of its military culture (16).

The inter-relationships of regimental culture, military organizational structures, the deterministic influence of general culture, persistence of prescriptive theoretical models (policy), battlefield 'style,' and war-fighting doctrine all combine in a military culture, which can be defined as:

> the collection of ideas, beliefs, prejudices and perceptions which constitute and determine the relationship between the constituent parts. As a result, the military culture determines internal conditions such as selection and promotion criteria, training, education, the allocation of resources and the vocabulary of military debate; these in turn combine to give a distinct character to a military organism and determine the nature of operations they can carry out and hence the form of war they adopt. Thus the internal conditions of an army (i.e. its culture) can have a greater influence on the form of war adopted than the external conditions and

circumstances — this can be the path to defeat. At times the cultural values are codified in a clearly stated doctrine, but more often they are enshrined in an informal set of conventions and practices formed over time which are difficult for the uninitiated to comprehend (1990:16. Compare this with Shafer 1988 and Khong 1992).

A military culture, according to Applegate and Moore, can adopt four distinct forms of war-fighting: positional war, maneuver war, long range penetration (LRP) and guerrilla war. However, "the differences in culture determine the forms of war which an army can use successfully and the manner in which the forms are implemented" (1990:18), and furthermore, "the response to technological change is... governed by culture" (1990:19). Positional war seeks to maintain systemic cohesion by exploiting ground and fortifications to increase fighting power, while limiting opportunities open to the enemy (1990:14). Positional war is characterized by physical destruction of enemy tactical units and the exhaustion of resources (e.g., Dien Bien Phu, Normandy). Maneuver war typically exploits movement relative to the enemy, and aims to increase firepower relative to the enemy's cohesion (e.g., Prussian, French Napoleonic armies). Attacking from unexpected directions is believed to lead to decisive success in battle, thus the traditional German battle of annihilation following encirclement (Sedan, Bryansk) or envelopment (Konnigratz, the Schlieffen Plan).

Long-range penetration (LRP) seeks operational and strategic victory through the internal collapse of the enemy's military structure. "Thus a subtle but important distinction emerges between manoeuvre war and LRP: the latter seeks to undermine the enemy's fighting power not, as in manoeuvre war, to increase its own" (1990:14). LRP strives to disrupt the enemy from within, and is characterized by close operations support and exploitation of deep operations (e.g., blitzkrieg, Special Forces). Guerrilla operations seek to destroy the cohesion of an enemy through protracted psychological stress; military action is secondary to survival and maintaining a threat. The object is to force the enemy to 'fix' himself into a protective posture: "guerrilla war induces in many armies a tendency to positional war, when in fact the only answer is to conduct either a counter-guerrilla war, or to change the circumstances and conditions underpinning the guerrilla action" (1990:16).

Mobile cultures, practicing either guerrilla warfare or LRP, go for the enemy's jugular, attacking the physical and psychological weak points to destroy the cohesion of the enemy's "military organism." The offensive is viewed as the only decisive form of war fighting; security is a product of superior tempo and certainty is impossible, because no plan will survive contact with the enemy. Mobile cultures devolve command and decision-making to level of contact, promote flexibility and responsibility in commanders and forego consolidation to exploit advantages. Attritional cultures, on the contrary, believe that victory comes from attacking the enemy's strength, which is achieved by amassing a superior killing potential. Defence is perceived as the strongest form of war-fighting: thus an intrinsic importance is attached to occupation, consolidation, field defences, emphasis on 'the plan,' mathematical approaches, compiling quantitative measurements (Vietnam body counts are an excellent example of this type of military pathology. Secretary McNamara's 1962 statement that "every quantitative measurement we have shows that we're winning the war" summarizes the viewpoint. Cited in Applegate and Moore 1990: 17).

Applegate and Moore's conceptual development of attritional and mobile cultures anticipates certain post-structural approaches to nomadic and sedentary war-machines (Deleuze and Guattari 1992; DeLanda 1991), future threat analysis of netwar and cyberwar (Arquilla and Ronfeld 1993), and antidiplomacy and purewar (Der Derian 1992; Virilio 1986), and is more fully examined in Chapter Five. It is sufficient for the moment to note that their model attends to the relationship between armies, societies, and warfighting style and sees culture as central to the development and use of the military art and science.

Cultural aspects of low-intensity conflict are increasingly relevant, and yet distressingly slippery, for military analysts and practitioners. For the British Army, high-intensity warfare means the use of heavy armored forces and mechanized infantry. Low-intensity "refers to the usage rate of resources, rather than limitations on the employment of weapon systems and platforms" (MacMahon 1992:26). Thus, although light weapons used in Lebanon resulted in 151,000 dead, the conflict is officially categorized as LIC. According to Major-General M. Willcocks, intensity categories give a false impression of step changes and do not accurately describe conditions. When considering intervention, the warfighting capabilities of the belligerents should be examined, as well as the basis of the conflict. Interest based conflicts, over trade or resources are amenable to the application of military force. Value based conflicts, because they are cultural, will tend to be intractable and should be left well alone (speech, 1994; see also Huntington 1993).

Furthermore, small wars do not go over well with big organizations. "NATO in general and the United Kingdom in particular," according to Major M. McMahon, writing in the <u>British Army Review</u>, "has an operational warfighting doctrine designed for a conflict that may never be fought and a doctrinal gap for the small wars in which it is more likely to be involved" (1992:25). The persistent conflagration of 'brush fire wars' has forced the British Army to reevaluate the theoretical concepts underlying the art of war, as well as operational and tactical doctrines and procedures used to fight war. The current series of Army Field Manuals (AFM) and British Military Doctrine (BMD) do not offer a comprehensive definition of low-intensity conflict, although the AFM divides the conflict spectrum into Civil Disorder, Revolutionary War, Limited War and General War. The US Army's <u>Field Manual 100 — 20 (Low Intensity Conflict)</u>, 1989, defines LIC as:

> a politico-military confrontation between contending states or groups below conventional war and above the routine, peaceful competition among states. It frequently involves protracted struggles of competing ideologies. Low Intensity Conflict ranges from subversion to the use of armed force. It is waged by a combination of means, employing political, economic, informational and military instruments. Low intensity conflicts are often localised, generally in the third world,

but contain regional and global security implications (cited in MacMahon 1989:26).

The Joint Chiefs of Staff further specify that "Low-intensity conflict is generally characterized by constraints on the geographic area, weaponry, tactics, and level of violence" (US Department of Defence Memoranda 1985, cited in Vetschera 1993:1578). Low intensity conflict, which can occur at the tactical/operational level as well as the strategic/political level (Vetschera 1993:1582), can be broken down into a number of categories, including stabilizing operations, combating terrorism, support to insurgencies, counterinsurgency operations and peacetime contingency operations (MacMahon 1989). It should be noted that the terminology "operations short of war" has recently replaced "low-intensity conflict" in official British Army usage.

Counterinsurgency, as a type of LIC, includes "the political, economic, social, military and paramilitary measures that indigenous governments and associates use to forestall or defeat revolutionary war" (Yarborough 1993:1576). The British Army defines COIN as "those military, paramilitary, political economic, psychological and civic actions taken to defeat insurgency and thereby sustain an existing legitimate government" (ADP 1993:7-2). Dr. John Pimlott and Dr. Ian Beckett, instructors at RMA Sandhurst have been instrumental in developing widely adopted principles of COIN, derived from historical and contemporary sources. The "six principles" (as they are known) include: recognizing the problem as essentially political, imposing a civilian-dominated coordinating machinery, establishing an intelligence organization for gathering and collation, isolation of activists from supporters through PSYOP and physical barriers, elimination of the isolated activist through appropriate military action, and implementing long-term political reform to prevent resurgence (see Beckett and Pimlott 1985).

Cultural and political rather than military aspects of COIN are often considered most relevant since "the main characteristic which distinguishes campaigns of insurgency from other forms of war is that they are primarily concerned with the struggle for men's minds..." (Kitson 1977: 290). COIN, nevertheless, does have a military aspect. "Tactical operations by conventional military forces against guerrillas call for: 1. mobility comparable to or greater than that of the guerrillas, 2. small-unit leadership of exceptionally high quality, 3. intelligence not always available through purely military collection systems, and the 4. ability to live in the field with minimal resupply for protracted periods" (Yarborough 1993:1577). These tactical procedures, what in British Army parlance would be termed 'operational art', have far more in common with the style of warfare typically associated with guerrillas, than with large-scale attrition battles fought by conventional armies.

Military operations based on psychological indirectness, tempo, surprise, exploitation of weakness and mobility, are referred to as 'maneuver warfare'. Maneuver warfare doctrine, often seen as the opposite of attrition, is intended to apply to chaotic high-intensity battles of massed armored formations. Does maneuver warfare doctrine work in low-intensity conditions? The operational level of command and operational art in general, is traditionally focused on high-intensity combat, to the exclusion of more common, lower intensity operations short of war. Operations short of war, often neglected in warfighting doctrine, do not give commanders or soldiers much latitude to demonstrate military prowess. Maneuver warfare doctrine thus appears irrelevant to low-intensity scenarios, although, in fact: "A warfighting doctrine should be applicable to all levels of the conflict spectrum including Low Intensity Conflict" (MacMahon 1993:30). British soldiers often express frustration that, as a result of police primacy and legal constraints in Northern Ireland, operational art suffers: "the campaign in NI in terms of coordination and direction appears to require a more ruthless and incisive approach" (1993, unattributed). A ruthless and incisive approach at the operational level, though delightful to soldiers, would be a political and legal disaster in the absence of a long-term political goal. On the other hand, maneuver warfare doctrine may provide the most appropriate model for operations in Northern Ireland. "Success in Low Intensity Operations is predicated upon a sensible interpretation of the tenets of Manoeuvre Warfare" (MacMahon 1992:31). Maneuver warfare may be most successful in low-intensity conditions for an explicitly cultural reason: it provides the most similar operational structure to that of PIRA.

Although the military aspects of COIN seem to be quite distinct from the soft "hearts and minds" cultural aspects, military culture determines, to a large degree, the types of doctrine which an army will be capable of accommodating and utilizing. Whether or not a large, cumbersome state army will prove capable of adopting maneuver warfare to operations short of war, will depend to a large degree on their own military culture (Applegate and Moore 1990), and how they perceive the culture of their co-belligerents.

Finally, cultural interpretations of violence have penetrated popular discourse on Northern Ireland. In addition to popular accounts of the contemporary British Army (Parker 1985; Barker 1981), and organizational biopsies such as Stanhope's The Soldiers: An Anatomy of the British Army (1979), first-person narrative accounts of military service in Northern Ireland are a booming industry for publishers. Although some of this material is quite sympathetic to British involvement (Allen 1990; Lake 1990), some is more ambivalent (Clarke 1983) and some vaguely hostile (Asher 1991), all of these accounts seek to explain the causes, nature and effect of violence on soldiers. Such accounts consistently cite racial and cultural factors as central in soldier's experiences of the conflict:

> To me, it's a sickening conflict, because you can't identify the enemy....Over there you're in a corner of what's supposed to be Britain and you can't tell one from another. It goes against the grain to be thrown into an environment where people of your own race are trying to kill you. We've adapted and geared ourselves to counter-revolutionary operations now, and we're very sophisticated at it, but that doesn't alter the fact that we're fighting an enemy within, an enemy we can't identify. ...People say Northern Ireland's a good training ground, but to me it's sad when conflict within the boundaries of you own country provides training. I really can't understand all the bravado... (Sergeant, 45 Commando, cited in Arthur 1988:245).

These first-person narrative accounts, despite their (mis-)understandings, ascriptions of 'savagery', and ethnocentrism are often illuminating as pseudo-ethnographies, and contribute, albeit informally, to an anthropology of war in Northern Ireland produced by the military in its own fashion.

the anthropology of political violence in Northern Ireland

Although the chaotic nature of war may create methodological anxiety for anthropologists who require orderly social systems as objects of study, armies that fight wars do not suffer from a lack of order. On the contrary, armies often approximate perfect systems, sublime in their complexity. Why do anthropologists eschew armies as analytical objects? Military structure, and the relations between military and civilian social orders (Janowitz 1960; Huntington 1985) are normally considered as the province of military sociology (Berk 1993),[10] and this may reflect a general tendency in anthropology to avoid complex social organizations within complex societies.

While ethnographies of colonialism may occasionally touch on military intervention as one aspect of an overall condition of social violence, often they focus on the civil society within colonial regimes. Protracted anti-colonial guerrilla wars are often approached through small scale group studies of rebel insurgents (Lan 1985). This has been particularly true in the case of Ireland, where village studies which assess the level of violence, or lack thereof, have been the norm. Clearly, ethnographies that demarcate small scale social organization as their province of study do not necessarily need or benefit from discussing the interaction of the state with local histories. By default, the spotlight often comes to rest on rebel groups rather than states or state armies as the objects of study.

Ethnographic research in the 'post-colonial' era has typically concentrated on the 'colonized,' especially resistance to imperialism. Fewer anthropologists have focused on the 'colonizers' themselves as subjects worthy of study: "comparable attention has not been paid... to the consciousness and intentionality of those identified as 'agents' of domination" (Comaroff 1991:9). While exceptional studies have examined medical practices as 'discourse' in Africa (Vaughn 1990), the missionary 'colonization of consciousness' in South Africa (Comaroff 1991), the relationship of gender and colonial power structures in Indonesia (Stoler 1991) and seen colonizers themselves as 'victims' of the colonial power structure in South Africa (Crapanzano 1985), few ethnographic studies have been directly concerned political violence, or with the complicity of anthropologists in such violence.

Arensburg and Kimball (1940) study of Clare failed entirely to mention the recent civil war, thereby ignoring violence as a constitutive aspect of rural Irish culture. Family and Community in Ireland, begun as a Harvard University Field Project in 1931, and published in 1940, exemplifies the ethnographic non-intermeshing of big structures and little processes. Ireland appeared as an object of ethnographic inquiry much earlier than other European cultures, probably because the Irish occupied the same structural category of 'savages', later reserved for African and Asian peoples (Szwed 1975:20). Methodologically grounded in Durkheimian comparative sociology, Arensberg and Kimball saw Clare as a bounded territorial unit (1968: xxvii), representative of a traditional European 'primitive' society, and without a significant history of its own (Wolf 1982:13; see also Roseberry and O'Brien 1991:3).

Arensberg and Kimball introduced a number of influential tropes into Irish ethnographies, such as synchronic village studies and necrographic fetishism of 'dying' cultures. Scheper-Hughes' <u>Saints, Scholars and Schizophrenics</u> (1979) recapitulated certain theoretical assumptions of Irish society as a stable, integrated, kin-based society, characterized by demographic patterns of dispersal, emigration, late marriage, and stem family kinship structure.[11] The persistence of the structural equilibrium model, in ethnographic vogue in 1930, results partially from a cultural nostalgia about Ireland. Peace writes that "the touchstone for estimations of what is culture is always some mythical, holistic, past form--by which standards any contemporary social activity is necessarily found wanting"(101). Change, conflict and adaptation are seen as abnormal, and interpreted as a sign of culture-death. Cultural decrepitude is assigned to the cultural present, while health is assigned to the past. Levi-Strauss called this type of anthropology, which seeks to record and preserve cultures before they 'die' necrography. Scheper-Hughes declared: "I share with other recent ethnographers...the belief that rural Ireland is dying and its people are consequently infused with a spirit of anomie and despair" (1979:4).

The 'death of the Celtic fringe' trope is elemental to the mythology of the Irish (Thompson 1967).[12] The death of cultures then becomes a phenomenon of nature rather than a phenomenon of history,[13] and legitimates a particular kind of domination. The efficacy of the myth lies in its ability to deproblematize the subject; Ireland seems to be dying of its own accord, so no ethnographic autopsy is needed. The transmission of the myth into anthropological models discounts war as an endemic problem, and prevents war from serving as a forensic explanation of cultural-death. Peace writes that "the essential fault lies in discoursing about culture anthropomorphically, in giving culture a life of its own and thus also the ability to die: a culture (and a language) can only be said to have died when the population which possesses it has itself been exterminated" (105). We return to this ghoulish ethnographic feast in Chapter Six.

The tropes in Irish ethnographies effect the way violence is explained: actual political and military conditions were often ignored in favor of Durkheimian community studies, within which conflict appeared as abnormal and destructive. Anti-historicism, generally combined with anti-militarism, severely truncated any real possibility of explaining the nature of the conflict. A bizarre romantic myth of Ireland as the land of leprechauns and green clover, combined with the simplistic, unthinking, pacifist notion that 'war is bad' makes British military intervention seem illogical and bellicose, and obscures Ireland's position as a cog in the war-machine.

Nevertheless, the amount of material published on political violence in Northern Ireland is immense, and quite diverse. The Social Science Bibliography of Northern Ireland (Rolston, et al. 1983) listed more than 5000 texts and Whyte estimates that if it were updated it would contain closer to 7000 (1991:viii). Social anthropological studies of ethnicity and national allegiance have been conducted by attitude survey (Rose 1971; Moxon-Browne 1983) and through participant-observation (Harris 1972), the mode of research most favored by anthropologists. Social psychological interpretations of attitudes and interaction (Fraser 1973; Doob and Foltz 1973), travel writing (Buckley 1986; Belfrage 1987; Conroy 1988), oral history (Griffith and O'Grady 1982; Parker 1993), and personal narrative (Adams 1982; Farrell 1988; McCann 1980) have also contributed to a public accounting of the conflict.

Typically, the research question posed is 'why is there continual conflict in Northern Ireland?' Whyte suggests, however, that a better question is 'why isn't the conflict worse?' (1991:15), which Leyton (1974) and Darby (1986) have done, thus contributing to the exploration of systemic limitations on conflict. Anthropological community studies of conflict in Northern Ireland have stressed the normality of life in the region, while political scientists and historians have stressed dysfunction and violence (Darby 1986:5-7). Ethnographies have focused on the Provisional Irish Republican Army's position within the Republican community (Sluka 1989; Burton 1978), the relationship between Catholic and Protestant paramilitary groups (Feldman 1991), or the absence of violence in particular communities (Bufwack 1982).

Whyte in Interpreting Northern Ireland (1991) develops a conceptual scheme for sorting out the exponential number of possible positions that can be adopted regarding Northern Ireland. He writes that "the traditional nationalist view of Northern Ireland can be summed up in two propositions: (1) the people of Ireland form one nation; and (2) the fault for keeping Ireland divided lies with Britain" (1991:117). Although this viewpoint has undergone reassessment and revision in accord with political circumstances, the model has influenced the way Irish history and nationalist biographies are written, and the way partition is explained. The traditionalist-nationalist view, according to Whyte, is that Britain is actively preventing Ireland from exercising its right to self-determination, colluding with Unionists in the maintenance of a sectarian state and practicing neo-colonialism (for example, Gallagher 1957; Adams 1988). The majority of scholarly writers have rejected the traditional nationalist view that the conflict is primarily between Britain and Ireland, substituting instead a view of the Protestant community, rather than Britain, as the obstacle to future settlement (1991:141).

The traditional Unionist view on the other hand is that "(1) there are two distinct peoples in Ireland, unionist and nationalist (or Protestant and Catholic); and (2) the core of the problem is the refusal of nationalists to recognize this fact..." In this model, Britain — perfidious Albion — assumes the guise of an unreliable ally prone to capitulate to nationalist demands (1991:146). Whyte sees the traditional nationalist and unionist views as functioning within the same paradigm, but essentially disagreeing about the natural unit of self-determination (1991:172).

Marx and Engels (1971) extensive writing on Ireland contributed to the development of a paradigmatic interpretation of the conflict (distilled into the Irish Republican ideology via figures such as Connelly) as an embedded colonial situation (de Paor 1970). The colonial explanation, also known as the 'settler/native syndrome' model, sees the conflict as the result of cultural rather than religious affinities (Boserup 1972; MacDonald 1986), brought into being as a result of the establishment of a settler state in non-metropolitan parts of state territory (Schaeffer 1990; Weitzer 1992).

"The fact that Northern Ireland is legally not a colony, but part of the United Kingdom, does not destroy the analogy" (Whyte 1991:178).

According to Whyte, the 'zone of ethnic conflict' model is an alternative to the colonial model; although the colonial situation could be resolved by departure of imperial power, ethnic conflict would probably be worsened by that departure. This colonialist model is probably more appropriate to analyses of Ireland as a whole prior to 1921 (Hechter 1975). The Marxist focus on economic determinism has been criticized by authors like Rose (1971) who believed the conflict was especially intractable because it was not economic, and Burton (1978) who felt that Marxist explanations were simplistic.

Whyte also identifies the emergence and recent predominance of what he calls the 'internal-conflict approach', also known as the 'two-nations hypothesis', which stresses the endogenous relationship between Ulster Protestants and Irish Catholics (O'Brien 1974) over exogenous factors. The source of the problem is seen to lie within Northern Ireland (Barritt and Carter 1962). The Cameron Commission Report (1969), in its conclusions regarding the rioting and civil disturbances of 1969, did not cite any causes for unrest external to Northern Ireland (although it did cite Catholic grievances about housing, electoral gerrymandering, sectarian policing, and so on, which seem to be the long-term result of external factors), so this viewpoint could be said to have been absorbed into 'official discourse' (Burton and Carlen 1979). Intractability of internal conflict is sometimes seen as a result of social, economic and geographical divisions (Boal and Douglas 1982; Darby 1976) so that violence appears as a psychologically embedded (Fraser 1973; Heskin 1980), self-sustaining aspect of culture (Burton

1978; Feldman 1991). Relations between the Republic of Ireland and the United Kingdom (O'Malley 1983), 'internationalization' of the conflict (Guelke 1988), and nature of Ireland as a besieged 'frontier region' (Wright 1987) can be seen as external, contributory factors.

Violence is self-sustaining, and has to a great degree become an institutionalized part of culture. Conflict persists, and has not been resolved, because *violence is not intolerable.* How is this possible? Personal interactions and relations at the local community level make violence tolerable. Countervailing tendencies within the Republican community may make the reproduction of a culture of violence sustainable (Burton 1978). As the Khmer Rouge showed, even cultural genocide is survivable, and may become a normalized part of sociopolitical identity (Fields 1977). Certain characteristic outlooks of ethnic groups, such as prejudice, may help to sustain conflict (Heskin 1980; MacDonald 1986). Life, as they say, goes on. Continuing to live on the 'narrow ground' of a war zone has produced effective yet mundane control mechanisms, such as intimidation (Darby 1986). Republican culture, just like the British Army, seeks to countervail the chaos of war and make violence manageable.

Frank Burton and Pat Carlen's Official Discourse (1979), very difficult to classify, is possibly the most interesting of all anthropological treatments of Northern Ireland and the state. Burton and Carlen examine the textual products of impartial government inquiries into breakdowns of law and order, and miscarriages of justice. These Commissions are established to ascertain the facts, allocate a quasi-judicial form of culpability and recommend institutional reforms (13). Official, judicial discourse seeks to maintain the credibility of the criminal justice system: "judicial discourse in its pristine form is concerned with providing the jurisprudential justifications for the coercive and administrative practices of the state" (70). The central discursive task is to repair the state's fractured image of administrative rationality (51): "this form of intellectual collusion is a strategy of discursive incorporation through which legitimacy crises are repaired and the reforms they engender are publically presented" (8). Burton and Carlen analyze the interior discursive logic of the Diplock and Cameron Reports, and conclude that the official paradigms for explaining miscarriages of justice are tautological and incoherent products of a pragmatic legal instrumentality.[14]

the war machine

Violence in Northern Ireland appears to be self-sustaining discursively and socially, and in certain ways, remains abstracted and removed from politics. The conceptual device that allows violence to be discussed as a separate, autonomous system is the 'war machine', explicated by Deleuze and Guattari in A Thousand Plateaus (1980, also 1987, 1992). The 'war machine' concept, which presupposes that war is external to the state, derives from the anthropology of Pierre Clastres (1977).[15] Clastres, in Society Against the State (1977), argued that if one accepts the Weberian definition of the state as that entity capable of monopolizing legitimate violence, then primitive societies are stateless because violence is not concentrated in the hands of any "state", but diffused throughout the society. Clastres concurred with Mauss and Levi-Strauss that in primitive society women, words and material goods are signs to be exchanged. When and if this economy is disrupted, violence becomes the mechanism of "exchange," after a fashion (1977:19-37).[16]

Deleuze and Guattari absorbed the idea that primitive society entails certain preventive mechanisms for warding off the development of states (1992:357-360). War inhibits the instillation of stable power structures and is, therefore, against the state: "either against potential States whose formation it wards off in advance, or against actual States whose destruction it purposes [sic]" (Deleuze and Guattari 1992:359). The war machine "seems to be irreducible to the State apparatus, to be outside its sovereignty and prior to its law: it comes from elsewhere" (1992:352). The chaos of war is antithetical to the order of the state; states must seek to tame and control the war machine.

> The State has no war machine of its own; it can only appropriate one in the form of a military institution, one that will continually cause it problems. This explains the mistrust States have towards their military institutions, in that the military institution inherits an extrinsic war machine. Karl von Clausewitz has a general sense of this situation when he treats the flow of absolute war as an Idea that States partially appropriate according to their political needs... (1992:355).

In the Clausewitzian tradition, war serves *as a tool* for the conduct of politics. For Clausewitz, armies had political aim of their own, but had to be made to serve the political aim of the state. In the language of Deleuze and

Guattari, political aim is therefore said to be external to the war-machine. Because the war machine escapes the political control, or political use, of the state, the war-machine is anti-Clausewitzian.

Actually, the concept of a war-machine probably derives far more from Clausewitz' arch-nemesis, General Ludendorff, whose <u>Der Totale Kreig</u> (1935) (translated in 1936 as <u>The Nation at War</u>) not only advanced the thesis that politics was now subservient to war, but that war ought to be conducted for its own sake. Ludendorff postulated that the rise of nationalism, and of nationalist armies, precluded the use of force as merely an instrument of policy. Nations, as total cultural entities, conducted warfare not in order to satisfy policy goals, but tooth-and-claw as a survival mechanism.

> The days of Cabinet wars and of wars with limited political aims belong to a bygone age. Such wars were marauding and predatory expeditions, rather than morally justified combats....Clausewitz imagined in his time the relations of politics and the conduct of wars. He had, however, in mind only foreign politics, which regulate the relations between States, declare war, and conclude peace. Of another kind of 'politics' Clausewitz did not think at all. He placed the importance of foreign policy high above that of war... (Ludendorff 1936:17).

Blockades and propaganda subjected populations to the operations of war. Indiscriminate targeting erupted the distinction between civilian and military. The conduct of totalitarian war, therefore, involved not only armies, but whole cultures.

> The nature of war has changed, the character of politics has changed, and now the relations existing between politics and the conduct of war must also change. All the theories of Clausewitz should be thrown overboard. Both warfare and politics are meant to serve the preservation of the people, but warfare is the highest expression of the national 'will to live', and politics must, therefore, be subservient to the conduct of war (Ludendorff 1936:24).

The war-machine is anterior to the state, belonging conceptually and anthropologically to the world of the nomad (see also Applegate and Moore 1990). The war machine always retains an element of the nomadic, as Pick (1993) says "something of 'the tribe in the desert': 'And each time there is an operation against the State — subordination, rioting, guerrilla warfare or revolution as act — it can be said that a war machine has revived, that a new nomadic potential has appeared'" (Pick 1993:259, citing Deleuze and Guattari 1992:386). PIRA are the neo-nomads in this scenario; tribal membership is acquired by virtue of nativity and genealogy. Special Forces in low-intensity conflict, specifically counter-insurgency in Northern Ireland, appropriate the war-machine of PIRA, their own tribe in the desert.

The Clausewitzian tradition holds that war should always be subordinate to political aims (see also Huntington 1957; Janowitz 1960),[17] the tradition of

Deleuze, et al. (and, of course, Ludendorff) recognizes war as autonomous. If war is conceived as autonomous sphere, no longer intersecting with politics, COIN emerges as the finest expression of an autonomous, nomadic war-machine appropriated by the state. Special Forces, the mainstay of counter-insurgency, are especially prone to exceed the limits of the state's harness on the military. Special Forces in counter-insurgency show most clearly the tendency for the 'tribe in the desert' to revert back to the basic, chaotic, primordial war-machine form. Political aim is the prime directive of COIN no matter when, or who, or why the policy is being formulated.[18] The very stress on political aim demonstrates that counter-insurgency threatens order and provokes disorder inside of the war-machine. War tends to exceed limits, and is restricted. COIN tends to exceed limits, and sometimes no restrictions are imposed, except the limits of force itself.[19]

Postulating an autonomous war machine does not imply a neo-fascist aesthetic *kreig an sich*. COIN praxis demonstrates clearly that the political control of war by states is nothing more than an ideal. To believe in the invulnerability of state control of the military would be profoundly naive. Armies may serve policy ends, but first and foremost, they are military apparatuses and they obey their own logic. To this end, I agree with Huntington that the military virtue of an army, *any army*, is not found in the nature, or the cause, or state, for which it fights. "The ends for which the military body is employed are outside its competence to judge" (1957: 57). Separating politics from military praxis, and maintaining this conceptual distinction, prevents me from taking any easy ideological position regarding PIRA or the British Army. In any case, as some of the material shows, these institutions relate to each other as military practitioners rather than political agents. A pragmatic, neo-functionalist recognition of the inevitability of bellecism does not exclude compassion or preclude nihilism, and may contribute to the exploration of the causes, consequence and ideas which inform the practice of military violence.

Chapter Two:

Counterinsurgency and Lethal Exchange

"The only logical course was to condemn the war, but to take the people who fought it on their merits."
—Kitson 1977:37

"Violence is neither right nor wrong, it's merely an aspect of the situation"
—Malcolm X

Prior to the twentieth century, war and religious conversion were the primary forces for sustaining large-scale cultural interactions between divergent groups (Appadurai 1990:1; Tilly 1990). Since 1945, conventional large-scale wars fought between Western nation states have become the exception; numerous, small-scale, "informal" (Reynolds 1989:204) wars,[20] fought in (ex-)colonial territories (Brown and Snyder 1985) have become the norm.[21] Since 1945, seventy-five percent of wars world-wide have been waged by political entities that were not states, without the use of regular armies. In 1992, *none* of the more than twenty wars world-wide was being waged between state-owned, regular armies with heavy, modern weapons (at least not on both sides) (Van Creveld 1992:62). Political violence, according to Prime Minister Heath, speaking to UN General Assembly in 1970, is being "preached and practiced not so much between states as within them" (cited in Bunyan 1976:68fn.).[22]

Despite the superior fire-power, technology and military capability of state forces, guerrillas and insurgents have prevailed in these low-intensity conflicts almost without exception. The French were defeated in Indo-China in 1954, and in Algeria in 1962; the US Army was humiliated in Vietnam and the Vietnamese, in turn, were forced to withdraw from Cambodia; Castro destroyed the Batista Regime in 1960 and since 1975 the Cubans have failed in Angola. The Soviets also lost in Afghanistan. Low-intensity conflicts in developing regions were often proxy wars of East-West bloc (Sereseres 1985:161); while the actual victims tended to be members of economically poor and politically weak states (see Kidron and Smith 1983, 1991).

The emergence of low-intensity conflict as a dominant warfighting form, and the consistent pattern of insurgent victories, deserves an explanation. Low-intensity conflict may have resulted in part from the permanent state of détente among nations with nuclear capability. The international military order of Superpowers, NATO signatories, and client states has created a world-wide state of "ambient insecurity"; political relations between wealthy, Western nation-states are regulated by the threat of war rather than by war itself (Kidron and Smith 1983). According to one school of thought, if proxy wars, wars of decolonization, third-world revolutions and civil wars are discounted, the era since 1945 has apparently enjoyed the longest 'peace' among European states since the Roman empire (Mueller 1991:23; Mueller 1989). Armed détente, of course, may easily be mistaken for peace.

Is the absence of major war a result of political stabilization? Has political order made war obsolete? In the classical tradition, war is controlled by politics; the absence of war signifies political accord. According to Clausewitz: "[p]olicy...is interwoven with the whole action of war, and must exercise a continuous influence upon it..." (1987:119). Political concerns restrain the deployment of military force: "politics prevent complete release" (Virilio and Lotringer 1983:48, see Clausewitz 1987:118-119). Social stability, in this model, derives from the subordination of the military to the political, so that military reins are held by civil hands. According to Sir Michael Howard, for example, the concentration of legitimate social violence into military structures and "the subordination of military force to the political government, and of the control of a government in possession of such force by legal restraint and the popular will" (1957:12) is the *basis* of civilization. This connection between war and order, soldiers and civilization will crop up again.

Alternatively, the stable military relations between Western states may be explained as a by-product of self-regulating military system. Not politics, but the conservative nature of warfare, itself, restrains war. The Clausewitzian paradigm of rational, instrumental war in pursuit of political goals is, for military practitioners anyway, a trite cliché: "we are not all Clausewitzians in the sense that even where Clausewitz is thought relevant, Clausewitzian ideals may not be attained" (Booth 1979:74). Politics has ceased to exert a restraining influence on war because war between states is no longer anchored in political life (Reynolds 1989). Warfare is a separate, self-contained, self-regulating realm: a *kreig an sich*.

According to Virilio in <u>Pure War</u>, an autonomous "war economy" (1983:10) is emerging, which "has more to do with the logistical machinery of other nations than with its own civilian society" (1983:14). This "a-national military model" (1983:14) of social organization is fundamentally anti-Clausewitzian; the war-economy concept presupposes that war has exceeded the boundaries of politics. Politics ceases to be the arbiter of conflicts and the bellecism of the military-industrial complex becomes an autonomous, self-generating social order. Because war is everywhere an element of social life, the civilian and the military realms become conflated.

Technological imperatives are the diesel that keeps the a-national military-industrial machine chugging along. The end of political restraints on war signifies the era of "the absolute offensive," which is "the fruit of technology and nothing else" (Virilio and Lotringer 1983:70). The greater the annihilation potential of total war "the more...political rationale becomes subordinate to the level of violence" (Reynolds 1989:5). Because the technology for war now exceeds the capacity to control it once unleashed, the circulation of war between states has ended. Listen, for a moment, to Brigadier Richard Simpkin: "The advanced world, too vulnerable to survive a war of attrition or mass destruction, must learn to conduct its affairs by the rapier —by the threat or use of small specialized forces exploiting high tempo and strategic surprise" (1985:180).

An absolute offensive can be deployed only locally and internally under highly controlled conditions: "what corresponds to the end of classical war is a kind of exacerbation of local conflicts" (Virilio and Lotringer 1983:47). War must now be limited to local conflicts, which provide, according to military analysts, a limited space for war (Newell 1991:149). The military view corresponds with Baudrillard's assertion that "If military power once again finds a theater for war, a restricted space...for war...it will again be possible to exchange war" (1990:15). Potential 'war space' is expanding as the range and lethality of weapons increases (Bellamy 1990:46-48), while simultaneously, the actual cost of occupation of global 'war space' is decreasing (Kidron and Smith 1991:54).[23]

Economy, exchange, and war all involve circulating streams of materials, knowledge, and technologies. The rise of the modern nation-state as an economic entity is so inextricably bound to the conduct of warfare (Tilly 1990) that one may reasonably presume that "[t]he distinction between the economy and warfare is, in fact, a modern distinction" (Kaldor 1991:179). Exchange is a prevalent metaphor for war: Simpkin in <u>Race to the Swift</u> writes that "...fighting is essentially an exchange of energy. Firepower is the ability to transfer energy to the enemy [sic], survivability the ability to avoid or absorb this energy" (1985:82). Dixon points out, "wars are only possible because the recipients of this energy are ill prepared to receive it and convert it into a useful form for their own economy" (1976:26). Survivability of weapon systems as a concept should be emphasized, because it lies at the heart of military exchange. Military pragmatism dictates that a tank destroyed on the first-strike would be a ridiculous waste of money and manpower. Delivering firepower is not enough; in order to function efficiently in war, a weapon system needs to be able to give and receive blows. War exchange would be impossible without survivability.

In another sense, military art and science as an abstract pedagogy contribute to the survivability of the war system because they produce and promote orderly exchange of violence according to set principles with which one's antagonists certainly should be familiar. Doctrines and protocols of land warfare instruct how to absorb the energy of the antagonist. Principles of war provide a typology of the forces of war: concentration of force, unity of command, economy of effort, security, and endurance. Military art provides a language for the action of war: envelopment, encirclement, desant, interdiction (see Fuller 1925). Military art also exemplifies how a professional class develops doctrines and protocols as a means of managing and ordering the application of violence. These protocols have a long history, jealously guarded by initiates and handed down in Staff Colleges, Royal Military Academies, and Naval War Colleges. Officers object to terrorists because *they have not been properly instructed* in the alchemy of violence. The belief that terrorists are outside the code is phrased as a "lack of honour."

Terrorism, which defies the deep structural logic of military art, "is a logical response to systemic closure...a refusal to fight by the 'rules of engagement', to accept the fixed set of entities and actions of a systems discourse" (Edwards 1989:156). Terrorism violates a fundamental exchange relationship. Because states are usually obligated to uphold the pacifist pretenses of liberal democracy, and are formally prohibited by international law from engaging in acts of counter-terrorism (which are known as reprisals), terrorism essentially prevents equivalent use of counter-force. "The art of deterrence prohibiting political war favors the upsurge, not of conflicts, but of *acts of war without war*" (Virilio and Lotringer 1983:27. Italics in text). Terrorism interferes with the codes for political war between states by promoting and conducting acts of war without the status of war.

Suspending for the moment a fuller discussion of what constitutes terrorism suffice to say that terrorists are members of (para-) military organizations which do not have air forces.[24]

Though terrorism is hardly absent as a *technique* in Northern Ireland, guerrilla warfare better describes the warfighting style and organizational structure of PIRA.

Guerrilla warfare does not violate the codes and protocols of military art, but establishes an independent exchange system, putting a new spin on the concept of survivability (a problem returned to in Chapter Five).[25] Insurgency and counter-insurgency are described in the literature as an "economical technique" of warfare by proxy (Paget 1967:18). The conflict in Ireland, in particular, "showed the rest of the world an economical way to fight wars, the only sane way they can be fought in the age of the nuclear bomb" (Foot 1973:69).

In anthropological theory, war is sometimes said to occur when exchange relationships break down; war thus replaces exchange. Mauss (1967), for example, characterized warfare as a consequence of a failure in a system of exchange:

> In the systems of the past we do not find simple exchange of goods, wealth and produce through markets established among individuals. For it is groups, and not individuals, which carry on exchange, make contracts, and are bound by obligations ...further, what they exchange is not exclusively goods and wealth, real and personal property, and things of economic value. They exchange rather courtesies, entertainments, ritual, military assistance, women, children, dances, and feasts; ...Finally, although the prestations and counter-prestations take place under a voluntary guise they are in essence strictly obligatory, and their sanction is private or open warfare (1967:3).

Political hierarchies are established and maintained through the exchange of gifts, words, or women (1977:14) and when exchange fails, war ensues. Perhaps exchange is already a form of warfare? Despite the fact that the gift appears voluntary, it conceals a political challenge: "Every exchange contains a more or less dissimulated challenge, and the logic of challenge and riposte is but the limit toward which every act of communication tends" (Bourdieu 1977:14). A "spirit of rivalry and antagonism" dominates exchange between tribes, which may result in the "purely sumptuous destruction of accumulated wealth in order to eclipse a rival chief" (Mauss 1967:4). Total war, or war which has transcended its own internal limits, should be considered as a form of Potlatch.

The military view that "fighting is essentially an exchange" (Simpkin 1985:81), the anthropological view that fighting is a consequence of failed exchange (Mauss 1967), and the post-structuralist view that war can no longer be exchanged (e.g., Baudrillard 1990), can be reconciled if we recognize that war *as the primary mode of exchange* has been replaced by war as one element *within a system of exchange*. War is no longer the central, focal mechanism of global cultural interaction. If major wars appear to be absent from the repertoire of international relations, this is because military technique and technology has been dissolved, subsumed and re-incorporated into every level of social life. Ethnoscapes, technoscapes, mediascapes, finanscapes, and ideoscapes are now the basis of the trans-national exchange of material resources and information (Appadurai 1990:11). War has permeated all of these once separate, civil realms. All history is now military history.

Despite the fact that war, and images of war, seem to be everywhere, the praxis of war is obviously still very concrete. Military hardware and military technique are still being bought, sold, bartered and stolen by and between states. But the arms bazaar has recently been crashed by some uninvited guests. Revolutionaries, guerrillas and insurgents of all sizes and shapes have scammed invitations to the soiree and disrupted the state monopoly on military equipment and resources.

> The world-wide spread of the AK-47 and the Uzi, in films, in corporate and state security, in terror, and in police and military activity, is a reminder that apparently simple technical uniformities often conceal an increasingly complex set of loops... (Appadurai 1990: 15).

Insurgents and counter-insurgents have established a distinct 'war system' with its own complex logic of conflict. Tactics, strategies and technologies of violence, which once circulated mostly between states, have been hijacked by contra-state groups — the terrorist network (Sterling 1981) — and are being exchanged on the military black market of counter-tactics and counter-ideologies.[26] The 'kula dance' of war is now circling in both directions.

Finally, the 'absence' of war is a deceptive chimera: in the conditions of military post-modernity, peace and war are very much the same. In 1978, in the British Army Review, Major GR Durrant very insightfully identified permanent, covert war an emergent norm. The military inertia produced by high technology, Durrant argued, has resulted in the sublimation of military force into covert combat (Durrant 1978). In this Hobbesian world, the unanimous 'warlike disposition' of nation-states has settled into a 'metastable condition', which is neither peace nor war. "The states one knew as peace and war are progressively merging into a metastable condition threatened alike by shifting and by collapse" (Simkin 1985:272). The primary threat to the system is not external, but 'shifting and collapse' of the system, itself. "In a social configuration whose precarious equilibrium is threatened by any ill-considered initiative, security can henceforth be likened to the absence of movement" (Virilio 1986:125). The hyperreal threat of movement which induces systems collapse is usually identified as the terrorists' prerogative (Der Derian 1992:81; Simkin 1985:272, 320-1).

Despite the deceptive placitude of a neo-Kantian 'perpetual peace', military force is still being used, but not in the service of "genuine and limited politico-economic war aims..." (Simpkin 1985: 274). Rather, military force has become a tool for political re-stabilization, with the adjunct effect of producing new geo-strategic equilibrium.

> The proper aim of conventional armed forces may therefore be not to defeat the enemy but to restabilise the situation at some different level, thus allowing some form of negotiation or mediation to resume. In sum, *diplomacy becomes a continuation of war by other means* (Simpkin 1985:276. Italics are mine).

In <u>Antidiplomacy</u>, James Der Derian — despite his post-structuralist ancestry — is very much in agreement with Simpkin's anti-Clausewitzian model. According to Der Derian, traditional notions of diplomacy have imploded under the influence of chronopolitical and technostrategic forms of surveillance and terror, such that Western diplomacy "normalizes relations by continuing *both* war and peace by other, technical means" (Der Derian 1992:32. Italics are his).

As was suggested earlier, low-intensity conflict — perpetual, invisible and permanent — has emerged as a dominant warfighting mode. Despite the plethora of taxonomies generated to account for LIC, certain characteristic features are consistently identified. Although extensive diplomatic, psychological or economic activity may occur in a low-intensity conflict, the primary feature is covert, and sometimes hesitantly *overt*, use of 'armed force short of war', a description consistent with current terminology in the laws of war (International Military and Defense Encyclopedia 1993). Non-combat missions which involve the military in LIC's include peacekeeping, training and advisory assistance. At the other end of the spectrum, 'wet' operations, such as counter-insurgencies and limited contingencies, tend to rely on special forces as "the decisive instrument of offensive action at operational and strategic levels" (Simpkin 1985:289). Special Forces, as opposed to regular armies, are the central figures in the counter-threat scenario of low-intensity conflict. Current strategic threat assessment identifies LIC as the primary task facing modern armies: the MOD defence 'White Papers' demonstrate the influence which threat assessments exert over fiscal budgets. For the UK, although continuing NATO duties are the military priority, internal defence and LIC are the runners-up.

the counter-insurgency loop

The protracted American military engagement in Vietnam is sometimes mistaken as the first politically motivated low-intensity conflict directly challenging the legitimacy of an imperial state. The pop-culture atrocity exhibit of Vietnam-the-myth has guaranteed the permanence of the war in popular memory, but deflects attention from Vietnam-the-war as military history. In other words, the <u>Deer Hunter</u> and <u>Apocalypse Now</u> can only tell us so much about a history which is stranger than most fiction. While the US Army, accustomed to deploying massive squadrons of tanks, submarines and heavy artillery, napalmed the jungle and slaughtered villagers, the British Army had already quietly perfected a different sort of LIC war-technique. Having recently fought a war in Malaya, the British Army was esteemed for its precise, delicate and sophisticated jungle warfare counter-measures. The US Army, always looking for ways to build a better mouse-trap, decided to try the 'Malay Model' in the forlorn hope of actually catching the Vietcong. British military advisors were forthwith invited to the newly established jungle-warfare training schools in Johore. In 1962, Major General RB Mans, then a colonel, lectured the American military advisors in Vietnam on Britain's long-haul, limited force approach in Malaya. Sir Robert Thompson, formerly of the colonial Malayan civil service, and co-author of the 'Briggs plan,' advised the South Vietnamese government in Saigon, later serving on the British Advisory Mission in Vietnam, and as a consultant to the National Security Committee. The Briggs' Plan "strategic hamlet" approach was a direct adaptation from Malaya, as were many other operational techniques.

Mockaitis points out in <u>British Counter-insurgency</u> that no one "has adequately explained why the British alone among the nations confronted with insurgencies were so adaptable" (1990:10). Despite force reductions following WW II, which substantially weakened British low-intensity conflict capabilities, "the British Army had to respond to many insurgencies occurring abroad, and its response was surprisingly generally strong and effective" (Hoffman and Taw 1991: v). While structural pre-dispositions within the Army, itself (Mockaitis 1990; Pimlott 1988), or continual experience of low-intensity engagements (Charters 1989) may have been contributing factors,

> The answer is that the British had been conducting internal-security operations very similar to counterinsurgency for at least 30 years prior to the Malayan emergency....*Britain's first exposure to insurgency came not in Malaya but in Ireland 30 years earlier* (Mockaitis 1990: 10. Italics are mine).

Counter-insurgency methodology originated in British imperial history (Pimlott 1985:16; Beckett and Pimlott 1985:13), specifically the British experience in Ireland. The Irish have been described as the "model internal enemy" (Faligot 1983:2) and "the struggle to govern Ireland may fairly be regarded as Britain's longest counter-insurgency campaign" (Townshend 1986:45). The Green Howards, raised in 1688 to defend the Boyne, were back in [London]Derry three hundred years later, doing pretty much the same thing (see Soldier Magazine 13 June 1988). The 'Standing Army to Deal with the Problem of Ireland,' has remained garrisoned there, in one form or another, since 1689 (Muenger 1991).[27] Ireland's use as a training ground for the British Army has been the subject of left-wing paranoia since Marx was writing by candlelight in London (Marx 1972; Ackroyd, et al. 1980; Manwaring-White 1983; Bunyan 1976, etc.).[28] Calling for a resolution on Irish amnesty in 1870, Marx wrote: "Ireland is the only pretext the English government has for retaining a big standing army, which, if need be...can be used against the English workers after having done its military training in Ireland" (1972:163).

Since Northern Ireland was established by partition in 1921, armed force has been used to maintain a precarious peace. Although most European nation-states have undergone 'internal pacification' (Giddens 1987), "the withdrawal of the military from direct participation in the internal affairs of the state," political order in Northern Ireland (according to many observers) depends on the continued presence of the British Army. This viewpoint holds that the withdrawal of the Army would result in a chaotic blood bath, a civil war between Protestants and Catholics. On the other hand, it is sometimes asserted that the presence of the Army and British security policy in general, actually sustains the conflict (O'Duffy 1993).[29]

As was suggested in the previous section, the military stalemate among Western nation-states has resulted in military attention recoiling inwards towards internal enemies (Virilio and Lotringer 1983:95; Der Derian 1992:115). According to Virilio and Lotringer, "colonial extensiveness is replaced by endo-colonial intensiveness" (1983:157; see also Hechter 1975; Gonzales-Casanova 1965).[30] Whether Northern Ireland is 'internal' to the United Kingdom depends on the political status of the territory, sometimes 'inside,' and sometimes 'outside' the boundaries of the nation-state.[31] Whether Britain has a colonial relationship, either of an 'endo-' or 'exo-colonial' type, with Northern Ireland is another, highly debatable, matter altogether.

Internal pacification does not involve an actual decline of the occurrence of war,[32] but "a concentration of military power 'pointing outward' towards other states in the nation-state system" (Giddens 1987:192). Internal use military force is typically made redundant by the consolidation of administrative power:

> the expansion of surveillance capabilities and internal pacification, radically lessens the dependence of the state apparatus upon the wielding of military force as the means of its rule. The distinction between the military and civilian police is symbol and material expression of this phenomena (Giddens 1987:192).

Despite the expansion and consolidation of bureaucratic surveillance networks, no nation-state has successfully established the hegemonic, ideological legitimacy of its rule in Northern Ireland (cf. Gramsci 1971:80). Consequently, the distinction in function between the military and civilian police is often weak.[33] The taxonomic and functional permeability of the police and the military resulted, at one point, in an actual turf war between the RUC and the British Army in Northern Ireland, discussed in Chapter Five.

Furthermore, the conflation of the martial and the civil policing realms has also made the definition and categorization of political violence extremely flexible. In

Ireland, political violence is sometimes treated as a 'war' fought by armies, empowered to act by sovereign political units. At other times, political violence is treated as a criminal activity ('terrorism') to be investigated by the police and punished by civil courts. Rather than recommend the legitimacy of either position, which has already been the subject of too many ideological position papers, I mean to suggest that bizarre taxonomic schemes and snarling bureaucratic in-fighting are structurally related phenomena, interesting in their own right, which can be investigated as a coherent system.

What is certain is that the qualitative flexibility and interchangeability of the civilian police and the military, a by-product unsuccessful internal pacification, has been adapted to British counter-insurgency praxis. The development of British COIN has been rigorously flogged in numerous historical studies (Townshend 1986; Pimlott 1988; Charters 1989; Mockaitis 1990; Jones 1992), and in in-house British Army publications (Perkins 1981), where it has been generally agreed that colonial (imperial) policing contributed directly to the development of a doctrine of counter-insurgency (Pimlott 1988:17-18). Mockaitis argues that internal security operations in Ireland were the basis of COIN praxis and doctrine (1990), exactly recapitulating Jefferies' contention in The Colonial Police (1952) that 'colonial policing' doctrine derived from the experience of quelling Irish rebellion in 1916-1921. In one way or another, Ireland seems to be an important node in the counterinsurgency loop.

Despite the long time lags between the Anglo-Irish war, the Malayan Emergency, the Vietnam war and the current deployment of British troops in Ireland, the cognitive content of security policy has remained relatively stable (Shafer 1988), as have the legal conditions of British 'intervention' in Ireland, not to mention the actual status of the British Army as a 'Cinderella service' understaffed, underfunded and under enthusiastic (Drewry 1993). Inevitably, any analysis of the persistence, circulation and exchange of COIN ideas will be full of gaps, especially since many of the military documents remain classified for security reasons. Nevertheless, it may be possible to suggest, rather than to prove, that the "lessons" of the British in Ireland were transferred and disseminated to Malaya, to the US in Vietnam, and then re-circulated back through the counter-insurgency loop to Northern Ireland, decades later. The penultimate source of 'lessons' of counter-insurgency was the British experience in Ireland.

muzzling dogs/suppressing rebellion

During the Victorian era, domestic peace-keeping during non-emergencies became the province of the police (Palmer 1988). While the Army was busy maintaining 'public order' and suppressing riots, the London Metropolitan Police was concentrating on 'crime'. The 1829 Police Act, proposed by Home Secretary Robert Peel and influenced by Jeremy Bentham, established the London Metropolitan Police, with similar police forces developed in the boroughs in 1835. As Sir Charles Napier observed, "In wartime and in peacetime the uniformed police are the first line force of internal security" (cited in Bunyan 1976:71) and the Army the last. The creation of a quasi-military police force in London by legislative fiat transformed the police into a sort of Praetorian Guard for the central state (Vogler 1991:96-97; Melville Lee 1901:395). Although the police were institutionally and functionally separate from the Army, the Commissioners of the LMP during its first hundred years of existence were from military or colonial backgrounds. Successful LMP commissioners, such as Sir Nevil Macready, who broke police strike in 1919, were rewarded with posts in Ireland.

The London Metropolitan Police, despite the fact that they increased the institutional presence of the central state in the civil arena, represent the best example of consent-based ('bobby on the beat') policing. Unarmed, locally recruited and 'serving' a peaceful community by protecting property and public morals, the police officer "represents not the central government but the local community, discharging duties and exercising powers which by common law belong to all citizens" (Sir John Moylan, cited in Jefferies 1952:24). The police were not meant to control disorder after the fact, but to pre-empt it. Sir Richard Mayne, the first Commissioner of the Metropolitan Police wrote in the General Orders that "the primary object of an efficient Police is the prevention of crime; the next that of detection and punishment of offenders if crime is committed" (cited in Jefferies 1952:24). Crime, which had been earmarked as an individual rather than a class problem, necessitated a new type of policing based on the apprehension of individuals. The new police were "to prevent rather than to investigate crime. Whilst investigation meant interference and the invasion of privacy, prevention...demanded no more than a public pressure on the streets" (Baldwin 1982:9).

Like Bentham's Panopticon, 'pre-emptive policing' emphasized surveillance rather than force, and exerted 'disciplinary' power in order to re-form and prevent deviant behavior through continuous observation

(Foucault 1979). The 'civilizing influence' of the Metropolitan Police on the dangerous classes of industrial Britain (Bunyan 1976) translated to colonial situations. Pre-emptive policing was one aspect of "the colonizing nature of disciplinary power" (Mitchell 1990:35). Police acted as the agents of disciplinary colonial power: "Police, therefore, occupy a position of vital importance in the Commonwealth; ...the restraining influence exerted by a good police system is necessary...to the welfare of society" (Melville Lee 1901, cited in Jefferies 1952:211). European disciplinary institutions, too harsh to be used at home, were sometimes deployed experimentally in the extremities of empire. Bentham's two contributions to the art of surveillance, the Panopticon and pre-emptive policing (Jefferies 1952:22), were first used in the colonies (Mitchell 1992).

Prior to the establishment of the LMP, Ireland was clearly a 'testing ground' for pre-emptive policing. Following the Union of 1801, when Ireland was incorporated formally into the United Kingdom, British police were introduced into the province. Intended to alleviate the sectarianism of the Protestant police, whose predominant role heretofore had been to maintain the Protestant ascendancy (Hawkins 1991:25), police centralization transferred control to the British. The police were thus detached from the divisions of Irish society and allowed to "exhibit the law as acting for the protection of every class of the community" (Hall cited in Hawkins 1991:25). The Irish Peace Preservation Act of 1814, instigated by Irish Secretary Sir Robert Peel,[34] authorized the recruitment of an organized constabulary force in any disturbed area in order to deter crime. 'Disturbance' being such a common condition, numerous independent constabularies mushroomed. 'Self-policing' in communities with sectarian divisions often became a *de facto* mandate for murder (Farrell 1983). In 1822 and 1836, further Constabulary Acts encouraged centralization by providing for permanent uniformed and armed constabularies headed by a provincial inspector-general, who was a direct subordinate of the central government (Hawkins 1991). Centralization, which was intended to remove sectarian bias in the police force, increased the efficiency of the Royal Irish Constabulary (RIC), forerunner of the Royal Ulster Constabulary (RUC), as a paramilitary organization.[35]

The RIC certainly resembled an army: they were armed, housed in barracks, recruited from areas outside of the communities they patrolled, and organized as a territorial rather than a local force. Their uniforms were modeled on the Rifle Brigade's, their ranking system followed that of the Army, their furniture, clothing and weapons were supplied through the War Office, and their arsenal of standard issue carbines (with sword bayonet), revolvers, shotguns, repeater rifles, grenades and automatic weapons for special duties were military issue. The Secretary of State was advised in 1938 that, "in some Colonies the police uniform was so like that of a soldier that the public did not realise the fundamental difference" (Jefferies 1952:49).

Although Irish-style policing was too military to be appropriate in metropolitan England, it was ideal for export.[36] Charles Jeffries in The Colonial Police (1952) advanced the thesis that nineteenth-century colonial policing was based not on the London Metropolitan Police, but on the "Irish model." LMP style policing was also "...exported, but the Irish model dominated the colonies" (Palmer 1988:3). As Jeffries writes:

The fact is that the really effective influence upon the development of the colonial police forces during the nineteenth century was not that of the police in Great Britain, but that of the Royal Irish Constabulary... from the point of view of the colonies there was much attraction in an arrangement which provided what we should now call a 'paramilitary' organisation or gendarmerie armed and trained to operate as an agent of the ...government in a country where the population was predominantly rural, communications were poor, social conditions were largely primitive, and the recourse to violence by members of the public who were 'against the government' was not infrequent. It was natural that such a force, rather than one organised on the lines of the purely civilian and localised forces of Great Britain, should have been taken as a suitable model for adaptation to colonial conditions (1952:30-1).

Although the importance of the 'Irish model' in colonial police forces is well documented (Jefferies 1952; Broeker 1970; Palmer 1988),[37] man-power costs made duplication of the patrolling strength of the RIC outside of Ireland unlikely (Hawkins 1991:26-27).[38] Nevertheless, contemporary informants often claimed that their police were based on the RIC.[39] Sir John MacDonald not only modeled the Canadian North West Mounted Police (who patrolled the Yukon during Gold Rush), but "many of the officers were graduates of the Royal Military College...some seeing action in the Fenian raids of the

1860's" (Morrison 1991:95).[40] Commissioner Cahill of the Queensland Force in 1899 and Sir Charles Napier, commanding in Sind in 1843, restructured their respective police forces "on the model of the RIC" (Hawkins 1991:21). Likewise, the Egyptian gendarmerie (D. Johnson 1991:164) and the Canterbury force in New Zealand (Hill 1991:60) were modeled after the RIC. In the Gold Coast not only did "the RIC offer... a structural model" for the local police (Killingray 1991:112) but each constable was supplied with handcuffs "to be worn on the right side like the Irish Constabulary" (Killingray 1991:107).

The police were often the sole representative of the colonial state (Anderson 1991:198) and, in India and Africa, carried out a prominent internal security role (Anderson and Killingray 1991:11). Colonial police forces, following the Irish model, were designed to be self-sufficient in hostile, foreign societies (Jeffries 1952)[41] and therefore, were inevitably responsible for a wide variety of tasks. According to one officer in 1881: "Everything in Ireland, from the muzzling of a dog to the suppression of a rebellion, is done by the Irish constabulary" (H.A. Blake cited in Hawkins 1991:24). A by-product of the political autonomy of the police from the local community was that, as one RIC officer pointed out in 1881, "Not one in five hundred of the community...looks upon the commission of crime as a matter affecting anybody but the government" (cited in Hawkins 1991:26).

English constables were recruited for work in Ireland, while other colonial forces engaged 'reliable' members of tribal or ethnic groups to patrol rival groups, such as Indians in Mauritius and Fiji, Chinese and Indians in Malaya and Sikhs in Hong Kong (Anderson and Killingray 1991:7). Former RIC constables were recruited to work colonies like Trinidad (Johnson 1991:83) and in Kenya officers received RIC training (Anderson 1991:184). Whether or not the Irish model is totally coherent, "it is clear that individuals and ideas passed along the line on a significant scale" (Anderson and Killingray 1992:8).

In Ireland, the constabulary originated as a semi-military force designed as the primary defense against insurrection (Townshend 1992:26). Because "it was essential that the civil power should not have to rely on military assistance" (Hawkins 1991:29), the RIC had to be capable of carrying out the duties of an army in everything but name. Jefferies points out, "these semi-military forces were not only suitable for peace-time police work under the conditions then existing, but were capable of being mobilized as defence units in time of war" (1952:32). The legal standing of the RIC was so ambiguous, that during WW I, the War Office assured nervous RIC officers that in the event of deployment they would be entitled to combatant rights under the Hague Conventions (Hawkins 1991:29, fn. 34). The RIC was militarized to the extent that in the 1860's, their efficiency as police was questioned. "They are bad soldiers, because they have never been trained to act in bodies; they are bad constables, because they have never acted alone (Francis Trench, 1858, cited in Hawkins 1991:28). In a report of 1866 Committee of Inquiry, Richard Mayne, the former head of London Metropolitan Police recommended that the constabulary 'should as far as possible be divested of a military character" (cited in Hawkins 1991:29).

Although the RIC were "an army of occupation on which is imposed the performance of certain civil duties" (C.D. Clifford Lloyd, 1892, cited in Hawkins 1991:28), they could never be legally defined as, or allowed to function as, an army. The British Army, conversely, resembled a police force more than a conventional land army. "For most Western armies large-scale conventional warfare has been the *modus operandi*...The British Army was in many respects an imperial police force for which conventional war was the interruption" (Mockaitis 1990:173, 147).[42] Suppressing internal rebellion and imperial policing were the norm for the British Army in the twentieth century (Bidwell and Graham 1982).

The ambidextrous capability of British Army to conduct both internal security operations, and fight big, conventional wars, results from a combination of their imperial history, political geography and certain peculiarities of the common law. Because Britain is an island nation, naval warfare capacity has been a crucial element in geo-political survival strategy (Howard 1975), while the Army has been assigned the role of "defending the outposts of Empire" (Charters 1989:176) and suppressing rioting and civil unrest in Britain (Babington 1990).[43]

A great deal of mistrust surrounds the use of troops to combat political disorder within the United Kingdom, evident in the vagaries of the common law's approach to the use of armed force.[44] Lacking any device comparable to a concept of 'siege' (Townshend 1986), martial law tends to be declared by statute or proclamation. Until 1919, domestic deployment of the Army was guided by the Riot Act, only repealed in 1967. The Riot Act was "'a law to abolish law', a kind of modified martial law against rioters" (Vogler 1991:2). The paradoxical and vague legal climate vis-a-vis domestic political disorder was exacerbated by incessant dispute about who would actually be in charge of running the public disorder show. The Tonypandy Riots (1910) initiated a shift toward unified military control of police and troops during emergencies (Evelegh 1978:16; Vogler 1991:79). Throughout the 'Macready Era' (1910-26) "police structures were being coupled and reorganized to permit a temporary flow of authority from the central state via the army" (Vogler 1991:82).[45]

Regiments on imperial duty were primarily engaged in combating irregulars,[46] rather than 'conventional' adversaries (Mockaitis 1990:146). Victorian era expeditions and campaigns were undertaken by disciplined soldiers "against savage and semi-civilised races" to add "the territory of barbarous races" to the British crown (Callwell 1906:21). Despite the lax application of a minimum force doctrine and opponents who generally preferred to run away, small wars were often characterized by military faux-pas. The shabby condition of the Army and the foot soldiers lack of initiative, many of whom were Irish, contributed to failure (Judd 1973). The Duke of Wellington said regarding the soldiers under his command, "I don't know what they do to the enemy, but by God they frighten me" (cited in Judd 1973: xvi). The regimental roles of the Army during the mid-1800s resembled Irish parish records (Featherstone 1973:16). The Passenger Acts of 1842 and 1847 by the US Congress restricted emigration (Woodham-Smith 1980:215, 239) during the Famine (1845-48), and joining the British Army was seen as preferable to death by starvation. The Coercion Act of 1846[47] had introduced British troops into Ireland to control possible insurrection by the starving laborers (Woodham-Smith 1980:69), insuring that many of the new recruits already had first-hand familiarity with the Army.

The Cardwell reforms of 1873,[48] intended to improve the Army's colonial era reputation as "an unprofessional coalition of arms and services" (Bidwell and Graham 1982:2), officially defined the Army's primary role as imperial policing, increased political control by elevating the Secretary of State above the Army Commander-in-Chief, and introduced the 'linked-battalion' system.[49] The linked-battalion system, by retaining one battalion in Britain while the other was engaged in imperial duty, assured that troops with irregular combat experience would always be available (Charters 1989:177) for overseas and domestic duties, should Peel's 'new police' prove inadequate.

Following World War I, the British Army resumed imperial policing as its main task (Bidwell and Graham 1982:150; see also Cole and Priestly 1936; Jones 1991), but was nevertheless unprepared for the unconventional nature of the Irish rebellion. Although guerrilla tactics had been sporadically employed in otherwise conventional land battles, the Anglo-Irish War of 1919-1921 was the first successful example of a prolonged guerrilla war.[50] And despite superior fire-power, equipment and training, the British lost the battle and the colony. Failure has been explained as a result of forgetting how to conduct low-intensity operations during WWI (Mockaitis 1990), ignoring TE Lawrence's accounts (1920, 1927) of guerrilla warfare (Jones 1991), and misapprehending the conflict as a conventional engagement (Charters 1989:189; Mockaitis 1990). Although all of these factors certainly contributed, failure was practically guaranteed by the severe restraints placed on the application of force by political considerations. In 1921, keeping Ireland as a colony while conserving British political legitimacy, and suppressing the rebellion as quickly and as painlessly as possible, entailed very different approaches. Whether to use the Army in 'aid to the civil power,' or to proceed as if it were full-scale military engagement, became a perplexing conundrum which was to have dire consequences.

walk softly: constraints on the use of force

From 1870-1900, colonial campaigns were conducted as 'external wars' rather than 'internal disturbances.' Small wars, by virtue of being exempt from European honor codes of war, rendered punitive action legitimate (Mackenzie 1992). An implicit imperialist ideology of force (Jones 1991), supported by Victorian racial attitudes towards non-Europeans, contributed to the toleration of colonial brutality. Callwell (1906) wrote,

> to filch the property of irregulars when they are absent is not the true spirit of waging war...; the proper way to deal with them is to kill them or wound them, or at least to hunt them from their homes and then to destroy...their belongings (146).

Contradictory, patriarchal attitudes about colonial insurgents sometimes offered the only protection against brutality. Insurgents, naughty schoolboys needing to be disciplined, were insignificant and no political significance was attached to civil unrest. William Hay Macnaughton, Secretary in Kabul (1840-41) said regarding the Afghans: "These people are perfect children and should be treated as such. If we put one naughty boy in the corner, the rest will be terrified." Macnaughton was eventually murdered by Afghans; his head stuck on pike and carried through the streets of Kabul (cited in Mockaitis 1990:65).

After 1900, colonial unrest, now treated as an internal problem, came under a whole new legal ambit. Common law requires all citizens to 'aid the civil power' in the event of civil unrest, using only the minimum force necessary to restore order. The King's Regulations state that the task of aiding the civil power is "not the annihilation of an enemy, but merely the suppression of a temporary disorder, and therefore the degree of force to be employed must be directed to that which is essential to restore order, and must never exceed it" (Duties in Aid of the Civil Power, 1923:3).[51] Duties in Aid of the Civil Power does not distinguish between civilians and soldiers (although soldiers were called upon before private citizens). Soldiers are, therefore, bound to use minimum force.

The trial of General Dyer served as a powerful object-lesson for the advisability of minimum force.[52] At Amritsar, India in 1919, General Dyer ordered a company of fifty riflemen and forty Ghurkas to fire into a crowded protest meeting of five thousand people and killed more than two hundred fleeing Indians. In his defence, Dyer claimed that he believed the Punjab to be in state of insurrection, and that military law was in effect. Since English common law does not recognize a 'state of siege', when power would be automatically transferred to the military, "martial law was the absense of all law" (Mockaitis 1990:22). In a rather odd logical inversion, if no law was in effect, Dyer could claim to be acting legally. The Hunter Committee's report concluded that Dyer had violated the requirement minimum force for quelling internal disturbances by not giving advance warning and continuing to fire after disbursement: "the employment of excessive measures is as likely as not to produce the opposite result to that desired" (Hunter Committee Report 1920:1034-5).[53] Amritsar demonstrated the politics behind the use of force: "In the inter-war years the Cabinet oversaw all internal security operations, to prevent any politically damaging use of force" (Jones 1991:44).[54]

While the government was considering what sort of force should be used in Ireland, General Dyer was being court-martialed in London. Lord Balfour said in regard to Ireland: "The Dyer debate has not helped us to govern by soldiers" (cited in Jeffery 1984:86). The government, fearing that a misapplication of military force would result in a situation like Amritsar, was perhaps overly conservative in the initial application of force to civil disturbances in Ireland. Initially, the uprising in Ireland was treated as just another internal disturbance, easily put down by the police. The RIC, under the Chief Secretary's remit and directed by the Irish Office, had become so de-militarized that it was derisively sneered at by military commanders.[55] The RIC's failure to quell the unrest led to a debate between proponents of military solutions and other factions disposed towards milder methods.[56] In May 1919, Sir Henry Wilson reported to the Cabinet that "Sinn Fein...controls some 100,000 well-organized, though indifferently armed, men", which made it necessary "to contemplate no mere police measures, but active military operations of a serious nature" (Jeffery 1984:78). Macready pointed "the futility of relying on the civil power to quell a rebellion" (Jeffery 1984:83) and suggested the introduction of eight battalions.[57]

The Cabinet found such 'military' measures too military and suggested enlisting ex-servicemen into a newly formed Auxiliary Division of the RIC, the 'Black and Tans'. Clothed partly in the black uniform of the RIC and partly in British khaki, the Black and Tans bridged the gap between the military and police: the first 'third-force'. Since November 1919, Major General Sir Hugh Tudor, chief of RIC during Anglo-Irish war of 1920, had been recruiting non-Irish demobilized soldiers into the RIC (Townshend 1975:46, 110). The RIC was reorganized into divisions under "disbanded military officers who carried the life and death attitude of the fighting man of the 1914-18 war into the day-to-day relations of the police with the civil population" (Duggan 1968:93, cited in Townshend 1992:34).

In January 1920, when British troops intervened, the war-time Defence of the Realm Acts were still considered sufficient. At this point, the insurgency was still being treated as a conventional rebellion, and the Army was using a conventional approach.[58] Only after Sinn Fein actually usurped politico-legal authority by establishing the Irish Republican Police Force, and installing Republican courts, was the more stringent Restoration of Order Act passed. By late 1920, despite the Restoration of Order Act and the reinforced, semi-militarized RIC, martial law was considered as an option. Lloyd George believed there was: "...good deal to be said for declaring a state of siege or promulgating martial law in that [Western] corner of Ireland" (cited in Jeffery 1984:87). The legal status of the Army in Ireland and the chain of command regarding mixed forces was confused

(May 22, 1920 and July 4, 1920 PRO WO 35/90). According to one military commander:

> As I understand it we soldiers in Ireland are there in aid of the Civil Power, and we are therefore in no sense responsible from a military point of view. If the Cabinet ever decide to have martial law the Civil Power, i.e. the Police, will come under us. But as it is we are under the Civil Power. It is therefore quite unfair to abuse the Army for anything that may go wrong in Ireland. I keep pointing this out on every available occasion, but of course it does not prevent anybody thinking that the soldiers are mugs (Jeudwine Papers 72/82/2, correspondence from H.Q. November, 23, 1920).

Military commanders favored martial law as a means to end this confusion: "It seems to me that the outstanding advantage of Martial Law is the unity of command under one head — civil and military" (Jeudwine Papers 72/82/2, correspondence from December 4, 1920, marked SECRET). In December 1920, martial law was declared in Cork, Tipperary, Kerry, and Limerick, and then in Clare, Kilkenny, Waterford, and Wexford. Political considerations prevented the imposition of martial law over the entire island. In the areas where it had been declared, martial law created administrative confusion and caused divided operational control of police and troops, exacerbated by the lack of a permanent system of liaison.

Despite the introduction of martial law, the Army was hemmed in by politics. The Army "continued to insist that a state of war existed in Ireland and that the troops should be allowed to deal with the IRA accordingly" (Mockaitis 1990:149). "From the War Office down to the common soldier, the army believed that it was fighting a war, not quelling disturbances" (Mockaitis 1990:20). While capable of a military victory, the Army's natural tendency towards military solutions was a very delicate matter. Sir Henry Wilson explained to the Cabinet that:

> The reduced garrison will, under such circumstances be practically impotent as an instrument for the maintenance of law and order, unless it can be employed for operations of a more military nature, which would only be possible if the Sinn Fein movement were proclaimed as a rebellion (cited in Jeffery 1984:85).

Military authorities were eager to recognize the belligerent status of the IRA under the 1907 Hague Agreements in martial law areas (Jeudwine Papers 72/82/2, Record of the Rebellion in Ireland, vol. I: 56-7),[59] because at least then they would have the opportunity for a proper scrap.

Hancock correctly argues that the British government was unwilling to declare a state of war since this would mean recognizing the Dail Eireann as a belligerent authority, and thus Irish claims to national sovereignty (Hancock 1937).[60] According the Record of the Rebellion in Ireland in 1920-21, "recognition [by the military authorities] of the IRA as belligerents, might *ipso facto* be said to involve the Imperial Government in the recognition of an Irish Republic" (Jeudwine Papers 72/82/2). Calling the conflict a 'war' would have transmitted legitimacy to Sinn Fein and threatened the political legitimacy of the British government and of the Union, itself. Lloyd George said in April of 1920, "you do not declare war against rebels" (cited in Jeffery 1984:86). The use of force to combat insurgency is constrained by the definitions of force, itself.

counterinsurgency cul-de-sac

Jones (1991) points out that there is considerable debate about the influence of the Irish rebellion on the subsequent development of counter-insurgency doctrine. Beckett argues the "Ireland provided lessons that should have been noted" (1985:4). Jeffery, likewise, writes that "scarcely any [counterinsurgency] lessons...were drawn from Ireland" (1987:121). Mockaitis, however, assets that "...the methods developed to combat the IRA contributed to the eventual development of British counterinsurgency," (1990:12), although the transmission of this knowledge was highly informal (see also Shafer 1988). Jones, himself, contends that Mockaitis overestimates the importance of Ireland in the construction of counter-insurgency theory. Nevertheless, "defeat promotes analysis" (Pimlott 1993, private conversation).

The political aspects of the insurgency were well-understood at the time. A document called <u>The Record of the Rebellion in Ireland in 1920-21 and The Part Played by the Army in Dealing with It</u> was produced for future doctrinal reference. The Foreword to the <u>Record</u> apologizes for emphasizing political activities in a military history, but "political and military activities were so closely interwoven that it is impossible to disentangle them" (Record 1922:A3). Furthermore "[i]t was not suggested now, or at any other time that any military action could finally pacify the country or solve the Irish problem" (1922:23), because "the solution of the Irish problem...was not, and never could be, the task of soldiers" (1922:53).

Immediately following the war, it was also quite clear that public representations of the conflict were equally, if not more, important than how it was fought. Sinn Fein had capitalized on the extremist tactics of the Army; collective punishments, such as house-burnings, had helped swing international public opinion against Britain. "[S]ince the British had decided that the gunmen were mere criminals, they assumed that their criminality would be obvious to law-abiding people everywhere" (Mockaitis 1990:68). In fact, creating a mask for the enemy to wear is essential for psychological warfare. This error would be rectified in future COIN operations.

The Anglo-Irish demonstrated forcefully that political restrictions on military deployment (especially on using the Army in conjunction with martial law to form a totally militarized zone) weakened British effectiveness in controlling the rebellion. Because pure military means create a legitimacy risk, it is necessary to have a force trained like an army and capable of defeating a rebellion, which at the same time is not an army. Not only must such a force be capable of hiding from the law, but it must be able to turn the law to its own favor, an issue which will crop up again fifty years later.

The last 'lesson' which should be touched on at this point is military intelligence (MI). Until the end of 1920, the British Army, believing itself to be acting in a limited capacity in aid of civil power, had considered intelligence as a police matter. The IRA, meanwhile, targeted the already ineffectual British intelligence structure: Michael Collins, IRA Director of Intelligence, "eliminated virtually the entire undercover branch of the Dublin Metropolitan police in a single day" (Mockaitis 1990:74): the original 'Bloody Sunday'.[61] Collins' brilliant tactical move presaged a fundamental transformation in the nature of war. By targeting the information and communications system of the Army, Collins' (and the IRA) pinpointed information as the most valuable asset of the war-machine. Their aim was to disrupt, damage, and modify the intelligence network, through deception, assassination and interference in the information system of their opponent, in this case the British Army. While the IRA was functioning as a loose, coalition-based network, the British Army, traditional, hierarchical and cumbersome, was constrained by its institutional nature, and by the legal apparatus which surrounded it. Arquilla and Ronfeldt (1993) propose calling this type of information-based guerrilla combat 'netwar', and suggest that military victory depends upon mastery of this form of coalition war. The adaptation of counter-insurgency to 'netwar' organizational structure is analyzed in Chapter Three and Chapter Five.

Ideas generated during the Anglo-Irish War were transmitted through journals and academy curricula (Jones 1991). Lowe's 1922 "Some Reflections of a junior commander upon the campaign in Ireland, 1920 and 1921," offered practical advice regarding low-intensity operations. Major BC Denning's highly circulated

"Modern Problems of Guerrilla Warfare," (1927) compared the Anglo-Irish war and the Boer war concluding that

> with modern methods of communication and publicity, combined with the progress which has been recorded in the civilization of most of the great powers, it is inconceivable that the forces of the Great Power will be able to display that ferocity in their conduct of the struggle, whatever the guerrillas may do, which has been such a potent ally in the past in the task of putting down insurrection (1927:349).

According to Denning, "guerrillas have...a new weapon, political propaganda, which draws blood upon the home front of the Great Power" (1927:349). The susceptibility of military atrocities to public scrutiny imposed new limitations on force. In the Army Quarterly 'prize essay' of 1933, Captain D. Wimberley criticized the Army's organizational structure, which prevented systematic incorporation of military 'lessons': "exactly the same mistakes made in [Ireland in] 1920 were made again [on the subcontinent] in 1930. In Ireland the British Army learned rapidly by experience without much guide from official manuals" (1933:212). The North-West Frontier had, by this time, superseded Ireland as the most important arena for small wars training (Jones 1991:62).

Major-General Sir Charles Gwynn in Imperial Policing advocated the minimum use of force as a means to avoiding an "aftermath of bitterness" (1936:14), and recommended the use of martial law which accounted for the "continuity of necessity" (1936:16-17), providing for a margin of expediency in the use of force.[62] Gwynn was also an advocate of the short, sharp shock of the machine-gun (1936:31) for suppressing armed rebellion, since pre-emptive, aggressive measures nipped rebellion in the bud (17-20). According to Notes on Imperial Policing, 1934, a field manual almost certainly written by Gwynn, unrest could "no longer be dealt with by a series of isolated actions in aid of the civil power, and their suppression demands a concerted military plan of operations" (1934:5). Gwynn's militarism eschewed political context and political resolution (Charters 1989:190; Mockaitis 1990:182), the necessity for a minimum force doctrine,[63] and overlooked the emergent hybrid civilian-military nature of insurgents (see Jeffery 1988 on Gwynn and the War Office doctrine of colonial wars).[64]

HJ Simpson's British Rule, and Rebellion (1937) invoked the lessons of Ireland to map out a plan for countering Jewish insurgency in Palestine. Conceptualizing insurgency as a form of political warfare, Simpson wrote,

> modern rebellion has assumed a form which makes its prompt suppression essential. It has become a dangerous malignant growth which attacks the whole framework of government. It is subversive, stealthy, and secret, and depends for success on intimidation...Rebellion is too good a name for the thing. It is proposed to call it sub-war (1937:34).

Simpson clearly recognized the political-psychological nature of insurgency, and the woefully inadequate intelligence set-up in Palestine and in Ireland, but was apparently neglected by the Army (Charters 1989).

Although all of these early articulations of 'sub-war,' contributed to the development of LIC doctrine, quintessential counter-insurgency theory was distilled at an obscure department in the War Office. This department directly links British experiences of WWI desert frontier wars, the Irish rebellion, the Malaya Emergency and special operations during World War II. In 1938, JCF Holland, a major in the Royal Engineers, was posted to General Staff (Research) [GS(R)], a tiny research branch at the War Office. Having won a DFC serving as a pilot with Lawrence's irregulars in Arabia, followed by a posting in Ireland (Jones 1991:74), Holland had a strong inclination to study irregular warfare at the War Office.

In 1939, after Hitler's' occupation of Prague, the War Office bureaucracy restructured and transformed GS(R) into Military Intelligence (Research) [MI(R)], and assigned to it the study of guerrilla warfare. Holland chose Sir CM Gubbins, with whom he had studied at Woolwich, as his assistant (Foot 1986:31-32). Colin Gubbins had served as a gunner on the Western Front, as a major in Ireland in 1921-22, and on the North-West Frontier in the 1930's. Holland and Gubbins were joined by Gerald Templer who had served in Palestine in 1936-7, later became the Governor of the British Zone in Occupied Germany 1945-6, and acted as the Director of Operations at the Malaya High Command in 1952-4. They applied their combined knowledge of counterinsurgency to what became the Special Operations Executive (SOE) (Cookridge 1966).

SOE thought was the primary influence on counterinsurgency doctrine in the post-1945 period (Jones 1991:47; Morris 1986:73-5), and laid the foundation of British Special Forces (Seymour 1985:6-7). Weirdly, SOE training was directly based on studies of the IRA (Beckett and Pimlott 1985:17). Both Holland and Gubbins, having served in Ireland, "saw the advantages, in the economy of life and effectiveness of effort, of the Irish guerrilla...both determined that next time, guerrilla [tactics] should be used by the British instead of against them" (Foot 1973:68).

> [W]hat [Michael] Collins did in Dublin had a noticeable impact, in the end, on British secret service method....The Irish can thus claim that their resistance provided an originating impulse for resistance to tyrannies...Irish resistance...showed the rest of the world an economical way to fight wars... (Foot 1973: 69).

In 1921, after the Anglo-Irish War was ended by treaty, the country partitioned and the Irish civil war began, Collins was assassinated by anti-treaty pro-de Valera forces. Ironically, because the IRA had effectually won, Collins and his colleagues joined the Free State (Republic) government, ending up on the same side of the fence as their former enemies: Gubbins, Holland and other British Army officer. Gubbins wrote in a private letter that he was not disappointed at the end of the war by treaty; however, "what I did not like was having to provide a gun carriage and six black horses for the funeral of Michael Collins..." (cited in Wilkinson and Bright Astley 1993: 27). By war's end, Gubbins had the only matched set of black carriage horses in Dublin.

During World War II, the SOE supported European anti-Nazi resistance with arms and training (Foot 1973:68). As early as 1940, concepts such as the "commando" squad had been invented and actively advocated by Holland at SOE. Many post-war counterinsurgencies were conducted by men trained by the SOE; Orde Wingate, who conducted counter-insurgency operations in Palestine and General Templer were influenced by SOE training (Foot 1984:206-7). SOE's unorthodox and experimental approach established a trajectory which carried straight into Malaya, and in turn, to Vietnam.

By the late 1960's, British methods, of which the SOE was the ur-source, had failed to produce any positive results in Vietnam. Failure may have resulted from institutional resistance within the US Army, accustomed to conventional wars based on fire-power, to fighting a low-intensity, low-technology counterinsurgency (Shafer 1988:23; see also Krepinevich, Jr. 1986). Perhaps they were not even trying: "it has been argued that the US Army never seriously attempted counterinsurgency in Vietnam" (Beckett and Pimlott 1985:7, cf. Cincinnatus 1981:60). Or perhaps the US Army was replicating of a false model of Maoist insurgency in Malaya (Beckett 1982:207, see also Tilman 1966), despite the fact that the Malayan model may have been fundamentally inapplicable to Vietnam (Fall 1966). According to Sir Robert Thompson, however, in No Exit from Vietnam (1969), "I myself and my colleagues on the British Advisory Mission to Vietnam, who had spent all twelve years of the Emergency in Malaya and subsequently nearly four years in South Vietnam, all remarked that we had found nothing new in Vietnam except in scale or intensity" (1969:133). According to persnickety Thompson, Vietnam was exactly like Malaya and if the US had just followed his plan, cogently explicated in <u>Defeating Communist Insurgency</u> (1966), the war would have been won.

Thompson, drawing on an explicitly Maoist tri-phase model of contra-state war (dictating the behavior of the 'communist' opponent), delineated appropriate counter-insurgency responses[65] to a range of conflict situations from subversion to guerrilla warfare (1966:29-34). Thompson identified insurgency not simply as a rebellion against current political conditions, but as a threat to the very structures of state power, motivated ideologically by "anti-colonialism" (1966:21). Recognizing insurgency as politically motivated, Thompson argued that the insurgents political aim of controlling population and military aim of neutralizing the government demanded a joint political-military organization (1966:30), headed by a senior military commander (1966:82-3), and capable of developing an overall plan. The political nature of insurgency required a reassessment of the numbers logic of conventional warfare: the task of "defeating political subversion, not the guerrillas" (1966:55) meant that victory would not be tallied in mega-deaths, but by discrediting the insurgents.[66]

Some COIN methods, despite failure in Vietnam, were transferred back to Northern Ireland in the early 1970's. The assignment of Brigadier Frank Kitson to 39 Belfast Brigade provoked a sea-change in the orientation of the war. Kitson had developed counter-gangs in Kenya for use against the Kikuyu Land Freedom Army and introduced the 'pseudo-gang' approach into the "Mau Mau jungle" (Burton 1978:132) of Northern Ireland in the form of 'mobile reconnaissance forces' as early as

1971 (Geraghty 1980:186). Kitsonian counter-insurgency emphasized transforming low-grade 'background' information into 'contact' information, the bureaucratization of the war through systematic filing of intelligence reports, and using the law to reinforce security policy (for a British Army traditionalist criticism of Kitsonian legal modifications, see Evelegh 1978:91-92, for a Marxist conspiracy theory of Kitson see Faligot 1983). Kitson de-emphasized large-scale deployment of the Army in favor of Special Forces; he allegedly proposed the use of the SAS to the War Office (1977:198).

City redevelopment schemes in Belfast (the "destruction of militant ghettos and deportation of their population to the outer circle of the city" in the words of one critic (Faligot 1983:120), recapitulated the "strategic hamlet" approach of Malaya and Vietnam. The strategic hamlet concept required that villagers be moved into walled enclosures to isolate the guerrillas from their support base. From 1969-1976, around 60,000 people, or 12% of the population were forced to move out of certain Belfast neighborhoods because of the rioting (Boal and Murray 1977; for an account see DeBaroid 1990). 'Peacelines' have since been established at the interfaces of Catholic and Protestant communities; the peaceline in the Clonard district in the Lower Falls was "initially made of corrugated iron, over the years it has been extended in length and height until now it consists of a brick wall up to four metres high surmounted by corrugated sheeting and barbed wire" (Jarman 1993:112). Barricades and barriers segment and divide space, within which ethnically segregated populations conduct and exchange a limited war (for a general account of Belfast geography see Boal and Douglas 1982).

The concentration of surveillance in crime 'zones,' and the physical isolation of these zones by road-block or cordon is often a feature of policing during civil disorder (Vogler 1991:108), and are designed to heighten the security forces' control of physical, urban space (Mallory and Ottar 1973). On newer estates in Belfast, "peacelines have been charmingly replaced by cul-de-sac urbanism..." (Lelievre cited in Faligot 1983: 120).[67] "The destruction of [Victorian grid-style road layouts] and the creation of circuitous routes and dead ends aimed both to reduce attacks and ease the work of the security forces not only in pursuing terrorists, but also in sealing off areas for blanket house searches..." (Jarman 1993: 113). Road access within estates can be blocked by using steel fences, pedestrian access gates, or horizontal crossing barriers. "Controlling vehicle access is the primary aim of these structures...The creation of cul-de-sacs and no-through roads is intended to reduce the use of cars in terrorist attacks" (Jarman 1993:112). Cul-de-sacs truncate space and cut the paramilitary off from easy access to a run-back, which consists of a "network of alleyways, double-entry building, streets systems, and highways that permit the evasion of police/army patrols, checkpoints, and antagonistic paramilitary units" (Feldman 1991:42). Spatial control means that violence is channeled "*into specific formats, times, and spaces*" (Feldman 1991:29. Italics are his).

Postponing a more detailed discussion of technique until the next chapter, the only other aspect of the COIN loop which deserves mention is interrogation technique. British interrogation methods, refined by the Joint Services Intelligence School at Maresfield, were derived from the experiences of British internees during the Korean War. The one-time CLF in Northern Ireland,

Major-General Farrar-Hockley, had been subjected to interrogation and torture by the Chinese and North Koreans as a young soldier during the Korean War. "As a result it had been decided to train soldiers how to resist them, particularly the Special Air Service Regiment" (Hamill 1985:66). The 'five techniques,' including sensory deprivation, 'group dynamic' and stress manipulation, were used in Aden and in other colonial contexts (see Charters 1989; Cunningham 1972). The techniques were imported to Northern Ireland by the CIGS, General Sir Geoffrey Baker, after it became apparent that the bumbling RUC had no concept of proper interrogation method. Army Intelligence trained the RUC, but the interrogations were conducted by the Special Branch. According to one officer:

> So internment and this very, very small scale interrogation was set in train and both were eventually talked out--because the world has become a more talkative place than it was when we used these techniques in colonial situations (cited in Hamill 1985:67).

Interrogation and torture can be seen as elements in the 'economy of violence': torture, a mode of extracting information, is a logical state response to the political-economy of information scarcity (Taussig 1987:58). In a conflict where power is information, and information is strictly controlled, "surveillance reduces human values to a single exchange-value: information as the alienated commodity of the intelligence market-place" (Der Derian 1992:59).

the dogs of war

Ken Booth writes in <u>Strategy and Ethnocentrism</u>, "the dogs of war, when unleashed, may sometimes exhibit Pavlovian behaviour, but they also have a pedigree" (1979:23). After countering insurgencies in Malaya, Kenya and Cyprus, the British Army returned to Ireland in 1969, lugging with them the methodological baggage of fifty similar campaigns. Armies almost always replicate past experience (Shafer 1988; see also Beckett 1988), but Northern Ireland recapitulated earlier experience on the same battleground in very similar politico-legal conditions. The kind of war on which the Army was embarking had changed remarkably little over time. In 1921, Lloyd George stated, "we will not negotiate with rebels." In 1981, Margaret Thatcher declared, "we will not negotiate with terrorists." Sixty years later, the fear of conferring a *de facto* legitimacy on rebel factions still persisted. The British counter-insurgency model, albeit with minor alterations, was to be redeployed to its point of origin in order to cope with a problem which showed no sign of vanishing. Like Oroborus, the snake which eats its own tail, the repatriation of counter-resistance methods to their point of origin closed the loop.[68]

The longevity of the conflict gives the appearance that the COIN system is somehow self-sustaining. All systems, including military ones, implicitly strive towards self-reproduction. War machines sustain themselves in a number of ways. The NATO doctrine of *sustainability* refers, at a strategic level, to the ability of a nation or other political unit to continue a war over a protracted period of time. Following the extremely high rates of expenditure of equipment and munitions during the 1973 Arab-Israeli war, member nations of NATO investigated ways of improving the viability of military forces and combat effectiveness, through improvements in logistics, equipment reliability, and resource usage. Sustainability at a strategic level may also have a subjective, cultural component, such as morale. In a tactical sense, sustainability describes the ability of a military force to continue combat operations despite inevitable losses. Tactical sustainability can thus be compared with weapon system survivability (see Murphy and Blandy 1993:2636-2640).

War machines can improve the sustainability of combat by fielding chaos effectively. Armed forces often appear to structure and organize social violence, so that the emergence of professional, disciplined armies "has had the effect of transforming violence into a mode of order and making its victims appear to be destructive threats" (Feld 1977:16). Legal, professional violence is, therefore, associated with the production of order (Dandaker 1990; Reynolds 1989). In this model, military professionalism reveals itself as the proper management of inevitable violence (Hackett 1983). War, the most complex form of large-scale legitimate violence, thus appears to be a supremely rational act:

> conflicts between states...have normally arisen not from any irrational and emotive drives, but from almost a superabundance of analytic rationality. Men have fought during the past two hundred years neither because they are aggressive nor because they are acquisitive animals, but because they are reasoning ones (Howard 1984:14-15, 22).

Despite the multiplicity of theoretical ordering systems generated by military scientists, the actual practice of war is chaos (Newell 1991). Management of military violence entails coping with chaos on the battlefield. Since chaos can neither be predicted nor contained, stress produced by chaos within the system must be dissipated. Effective war-fighting depends, at the most basic level, on the ability to cope effectively with disorder.

The Tao of war encourages economy of effort (in other words, the conservation of strength) in conjunction with concentration of force, principles which form the basis of 'sustainability'. In British military doctrine,

> The corollary of concentration of force is economy of effort. It is impossible to be strong everywhere and if decisive strength is to be concentrated at the critical time and place, there must be no wasteful expenditure of effort where it can not significantly affect the issue (Design for Military Operations: The British Military Doctrine, Army Code 71451, 1989).

Both the US and the UK have identified 'economy of force'/ 'minimum force' as an essential component of internal security and counter-insurgency operations (Rawlins 1993:55).[69] "Limited war," according to William V. O'Brien, "emphasizes the principle of economy of force. During limited war, the open-ended objective of doing all possible injury to the enemy is ruled out. Each application of military power must be tailored to a specific military objective based, in turn, on specific political objectives" (1979:64-72).

The subtle nature of LIC pre-disposes it to conserve fire-power, except in highly controlled bursts. Limitation on violence embodied in the minimum force doctrine results in a *very high degree* of systemic *sustainability* of conflict. Because the actual level of violence is very low, and very precisely directed, force is concentrated and effort is conserved. War-chaos is thus imminently manageable, and to a certain degree tolerable.

> Both sides are locked into the conflict. The British Army, we can go on forever. It's built into the system, has been for years now: people are programmed to go there, and the casualties we suffer per annum are no more than soldiers killed each year in road accidents (Major, 1 Para cited in Arthur 1987:254).

The British defence industry is perfectly organized to guarantee economic sustainability of an LIC netwar. A brief over-view suffices to demonstrate the strength of the system. The United Kingdom is one of the few states with capacity to develop a major weapons project, including a nuclear one, from conception to deployment, and Britain also owns a high percentage of the conventional weapons distributed worldwide. From 1985-1989, Britain exported 7.7 billion dollars of armaments (Kidron and Smith 1991:67). In 1978, British firms sold defence communications equipment to Saudi Arabia, valued at over £200 million. In May, 1977, the Ministry of Defence sold equipment and services to the Shah of Iran, also valued at £200 million. These systems, with internal security, anti-terrorist and counter-terrorist applications, probably absorbed certain refinements developed 'on the ground' in Northern Ireland.

Britain has a very high expenditure on military R&D as a percentage of the total defence budget (1991:85) and at the end of the 1980 the military was absorbing 12.5% of the total central government budget (87). Despite a world-wide recession in heavy manufacturing industries, Britain has maintained a strong industrial base for electronics, shipbuilding and aerospace, including armaments production and development. Britain is a leader in surveillance technology, riot control equipment and computer storage of intelligence input, industries which accelerated during the 1970's. By the mid-1980's, Britain spent a higher proportion of its Gross Domestic Product (GDP) on defence than any member of the Western European Union (WEU) with the MOD as the primary customer (Taylor and Hayward 1989) to the tune of $28.66 billion in 1981 (Wyllie 1984). Yet, despite this militarized economy, Britain had less than 5000 deaths in 1989-90 directly attributable to war. "militarized nations...can do without armies" (Virilio 1986:128) and they can have a war-economy with hardly any body count.

The function of Northern Ireland within the war economy is the subject of endless speculation and controversy. Although, the conflict in Northern Ireland has a beneficent effect on the defence industry, the economic cost of continued involvement is high: the local economy is supported by the British government at a cost of over £1.5 billion per annum, the policing of Ulster requiring a further £500 million and more than £500 million has been paid out in compensation to civilians. On the other hand, terrorism has created a boom in law enforcement and rebuilding.

There are so many people making money out of the troubles, so many people. A large percentage of the population is employed in security work and if it stopped tomorrow what would these people do? (Private, UDR, cited in Arthur 1987:251).

Economically, the conflict seems to be sustainable, if not self-sustaining (see Rowthorn and Wayne 1988 for a political-economy approach to Northern Ireland). War-tech has definitely improved through LIC in Northern Ireland. The conspiracy theory of Northern Ireland as a 'training ground' has already been mentioned. The stunned, incredulous dismay of those who advance this theory demonstrates a naivety about what amounts to unexceptional military pragmatism: of course Northern Ireland is a 'training ground' for counter-insurgency and internal security operations.

It's a cheap war, let's face it; we're getting practical experience, which can only be good for any armed force; the logistics side is easy to handle; plus, if you look on the bad side, it doesn't even cost much to fly bodies back home (Corporal, 40 Commando cited in Arthur 1987:244).

Northern Ireland, however, is an exceptionally bad training ground for high-intensity scenarios. Counter-insurgency training, which emphasizes company rather than platoon commanders and restraint, does not transfer very well to NATO-style defence duties and led to problems during Falklands conflict (Charters 1989:234). According to Col. Michael Dewar, "incorrect procedures can be learned in the Emergency conditions and much time is spent 'unlearning' many of the techniques of urban IS operations on return to BAOR" (1985:178). COIN does not into transfer to conventional warfare, and vice versa (Mockaitis 1990:163).[70] Big wars, in fact, interfered with the smooth evolution of counterinsurgency methodology. COIN, although certain components may be useful in high-intensity situations, is a different animal, albeit a predictable one, having regular feeding-times and consistent habits: COIN theory, as Shafer points out in Deadly Paradigms (1988), is relatively stable and autonomous. Paradigm persistence, despite continual, incontrovertible failure, signifies the prescriptive, explanatory and ideological utility of the model (1988:279; Kuper 1988).

Sustainability of conflict may cause social conditions to resemble a culture of violence. Death as a way of life — in other words, the cultural permanence of low-level armed conflict — is an emergent trend in the war system. Northern Ireland, like Chad, Ethiopia, the Sudan, South Africa, Ethiopia, Israel, Pakistan, India, Burma (Myanmar), Cambodia, the Philippines, Guatemala and El Salvador, has experienced more than twenty years of war (Kidron and Smith 1991:23). All of these LICs are confined to limited theaters; netwars will disappeared from public view, except where they are consolidated into "zones of attrition" (Virilio 1986).[71] Such a 'zone' may appear shocking to new arrivals, unfamiliar with the conditions of the sub-war battlefield:

> I think the most surprising thing to me was that the people there had grown up with violence. It was nothing new to them, but I'd come from a fairly sheltered middle-class

background, and the shock was enormous. I'd be on the streets, and someone would be trying to literally take my life, for no apparent reason. It was a weird feeling, very, very strange... (Corporal, 40 Commando cited in Arthur 1987:86).

A culture of violence, which resembles a magnified and expanded weapon system, ensures the duration, continuity and sustainability of conflict. This raises a curious issue: whether the 'object' of war is destruction of the opponent through battle, or whether the object of war is self-reproduction with limited expenditure of energy. According to Deleuze and Guattari, "war in the strict sense...does seem to have the battle as its object, whereas guerrilla warfare explicitly aims for the nonbattle" (1992:416). States often seek the annihilation and destruction of the enemy — the war of attrition. Guerrillas and paramilitaries prefer the limited skirmish to maximize their crafty and fluid strength. State armies thus must devote intellectual and military effort to devising ways of bringing 'natives' to battle. Only when 'natives' are engaged in combat can they be defeated by the state's superior firepower: "In remote regions peopled by half-civilized races or wholly savage tribes, such campaigns are most difficult to bring to a satisfactory conclusion" (Callwell 1906:26). The lower the intensity, the more intractable the war.

In Northern Ireland, certain members of the British Army are still eager to have belligerent status declared for insurgents (see Simpkin 1985:319-321) since it promises a fair military fight. At one time, it is very possible that PIRA would have enjoyed exactly the same thing: a real war fought with proper equipment, uniforms, and training. According to Provisional Chief of Staff Sean MacStiofain, writing to a gunrunning contact in NORAID in 1971, "There is one hell of a difference between what you are buying and the heavy gear...the position is this, we must have the heavy stuff to win. We are not going to be beaten but at the same time just a few of those would make all the difference" (cited in Clarke 1987:5). At the present time, both PIRA and the British Army seem to prefer military pirouettes to grand jetes. When states behave like guerrillas, conducting raids or skirmishes, they can be said to have absorbed an aspect of the nomad war machine — the tactics of maneuver. Sun Tzu's advice that 'the acme of skill is to subdue your enemy without fighting' has been incorporated into British Army doctrine as 'economy of effort' and dominates on-the-ground application of force in Northern Ireland.

The ideology of 'acceptable losses' defines the war machine. The philosophy of war often seems to intersect with institutional structures. Whether one is a by-product of the other is indeterminate; however, philosophy and institutions both provide the COIN system with a stable 'food' supply, contributing to systemic, autophagic sustainability (a subject which is touched on again in Chapter Six). The capability of war-systems to feed themselves has been identified as a historical trend: "With the great geo-strategic revolution of the nineteenth century...the phenomenon of war begins to feed itself by creating the sources of its own conflicts and multiplying them; they are still dying for Suez or Panama" (Virilio 1986:51). In this sense, no war ever terminates, because war *as a system* is eternal. To what degree "life is a consequence of war, society itself a means to war" remains to be explored (Nietzsche 1968:53).

Chapter Three:

Force and Counter-Force

*In the world of visible things, the principle of
opposition makes possible the differentiation by
categories through which order is brought into
the world.*
— Hexagram 'Tun', I Ching

War requires a competitive intelligence, otherwise
it is murder — or at best, hunting. Military organizations
may aim to kill each other, but even the application of
lethal force is a type of interaction. The capacity to
interact, even in a hostile manner, is predicated on prior
knowledge of the enemy. Repeated interaction between
military units leads to the development of codes and
norms for the use of force. Knowledge of the enemy
leads to a refinement in knowledge of how best to kill the
enemy. Knowledge also leads to imitation; force and
counterforce often come to resemble each other. It is the
nature of states to "absorb... the techniques of the nomad
war-machine" (Virilio and Lotringer 1983:89).

Although force and counter-force seek to
annihilate their enemy, in the process of killing (and
developing codes for killing) they also construct each
other. Foucault calls this an "irreducible
opposition" (1990:96). Virilio and Lotringer call "a fatal
coupling" (1983:120).[72] According to a former SAS officer
who served in Northern Ireland: "The enemy is what
makes you yourself. Without the enemy you are
nothing" (Asher 1991:91). No matter what secondary
targets they may shoot at, the primary aim of the
Provisionals is to end the British occupation of Ireland.[73]
This is their reason for existing, and always has been. The
British Army, in relation to PIRA, are "not only its inert or
consenting target; they are also the elements of its
articulation" (Foucault 1980:212-213).[74] This chapter

explores how the British Army constructs the Provisionals as its primary enemy, and in so doing, constructs itself.

A few contradictory impulses mark the self/enemy relationship: on one hand, conducting war appears to depend on social-psychological construction of an enemy through propaganda, de-humanization or outright invention (Volkan 1988; Keen 1986). According to Brigadier Richard Simpkin of the Royal Engineers, "Just as it seems psychologically necessary for the people of a nation to have a traditional enemy, so the doublethink... can only be maintained in [the] face of a palpable threat. If a threat did not exist it would be necessary to invent one" (1982:101). On the other hand, accurate knowledge of the enemy is fundamental to war. Clausewitz wrote that "the first, the supreme, the most far-reaching act of judgment that the statesman and commander have to make is to establish...the kind of war on which they are embarking..." (1976:88). Or, as Sun Tzu puts it: "If you know the enemy and know yourself, you need not fear the result of a hundred battles" (1983:18).

"To know the enemy has always been a cardinal tenet of strategy" (Booth 1979:16). Yet, as TE Lawrence remarked, military commanders often have a "fundamental crippling incuriousness" about their adversaries (cited in Dixon 1976:339). This blind lack of curiosity may be the result of military anti-intellectualism (Dixon 1976:288-330) and preference for brute force. Goering, for example, said, "Wenn ich das Wort 'Kulture' hore, dann ergreife ich mein Revolver." Booth, in Strategy and Ethnocentrism, identifies this attitude as a "pathology of strategy" (1979:32), where the cognitive content of the military world view is based on crude generalizations of national character (1979: 33), racial explanations of fighting skill (see Callwell 1906) and an ethnocentric conviction of national superiority.

Ethnocentrism, an inadequate and dangerous basis for the development of policy (Booth 1979:17), contributes to misperception of the 'enemy' in strategic studies and on-the-ground operation. Most soldiers who serve in Northern Ireland, for example, have little prior knowledge regarding political conditions, except reports glimpsed on television and scanned in the newspapers. Soldiers are certainly not immune from the ethnocentrism projected in popular culture: "All these [Catholic] estates are horrid, but largely because so many of the inhabitants have no standards and live in them like animals...these inhabitants could turn Eaton Square in Belgravia into a slum" (Morton 1989:68). On the other hand, most misperceptions induced by the media are quickly dissipated by actual operational conditions:

> So you wondered how they did the killing and got away with it for years. They had to be reasonably good. I don't know if it's the right word, but I respected them. They could and did pick their time. They were better than us; very clever people (Sergeant, 1 Welsh Guards, cited in Arthur 1987:164-5).

Academic social science, rife with a more urbane ethnocentrism, provided much of the intellectual content of security policy (Shafer 1988, see also Lindblom and Cohen 1979:78-9). Development and modernization theory, which emphasized political order (e.g., Huntington 1968) as a necessity for 'stable' political change (Shafer 1988:85), contributed to US counter-insurgency policy. Upsets in the political arena were countered with hardline policies that reinforced the state's power to maintain stability (Wakin 1992:16-17). While the theoretical underpinnings of US counter-insurgency were primarily derived from political science, anthropology was a secondary source. Malinowski's cultural diffusionism, and Mead and Benedict's cultural relativism meshed with the dominant developmental paradigm, allowing Britain and the United States to play the role of administrators of "modernization" (Shafer 1988:54-55).

Anthropology contributed 'useful knowledge' in great measure to the British COIN paradigm (Shafer 1988:49fn.). While the institutional linkage of anthropology to British colonial administration in the 1920s and 30s is well documented (Asad 1975; Wakin 1992), anthropological involvement was often more direct. Sir Robert Thompson suggested that anthropologists recruit aboriginal tribes as partisans: "Insofar as they were contacted and befriended during the period of colonial rule, the greatest success was most often achieved by the eccentric European with a vocation for making the study of such people his life-long work and interest" (1966:150). According to Lucien Pye:

> It is indeed not impossible that this area [guerrilla warfare and counter-subversion] may prove to be a more fruitful one for social scientists than many other aspects of military strategy. This is because the problems posed by such forms of warfare and violence are intimately related to questions about the social structure, culture, and behavior patterns of the

populations involved in such conflicts. Without question social science research is in a strong position to contribute useful knowledge in designing and developing internal security forces (1963:155).

As Bernard Brodie pointed out regarding the defeat of the French Army in 1914, "This was neither the first nor the last time that bad anthropology contributed to bad strategy" (1959:52). Misunderstanding the nature of the enemy apparently results in poor returns on the battlefield, yet "outmoded scholarship" continues to have an impact on the development of strategy (Shafer 1988:10). Why? Policy makers misunderstood the nature of past insurgencies, prescribed and implemented inappropriate solutions, which continually failed (Shafer 1988:279-280). The persistence of autonomous ideas about 'enemies,' no matter how inaccurate, is partially sustained by the need to find solutions generated by illusory logic problems internal to the system (Kuper 1988), partially the result of the need for ideologically comforting parsimonious explanations of social complexity (Shafer 1988), partially the result of the anti-intellectualism of military establishments (Dixon 1976), and partially the result of ignoring discrepant information in favor of the non-ambiguous (Khong 1992). The most convincing explanation of persistent "analogies of war," however, is that "ethnocentrism contributes to the 'vilification of the human' which makes killing easier" (Booth 1979:98).

The fundamental contradiction between 'knowing' your enemy in order to develop effective strategy, and de-humanizing him in order to kill efficiently, is a theme to which we will return. Suffice to say, that the dogs of war do have a pedigree, which is often "anthropological" and that counter-insurgency strategy depends, not just on practical experience on the battle-field, but on historically derived analogical models of prior conflict. Paraphrasing Levi-Strauss, enemies are not only good to kill, enemies are good to think.

The fatal coupling of PIRA and the British Army depends on how they 'think' each other. Mutual military strategies or "analogies of war" replicate prior experiences, consume 'anthropological' concepts, and lock the combatants into a "terrible paradox" (Booth 1979:100) of both needing to know and to revile the opponent. In order to fight them, sedentary state armies incorporate the nomad paradigm of war (DeLanda 1991:11-13, see also Gellner 1991), and this adaptation is fraught with anxiety and cognitive strain. Interaction between combatants occurs at different levels. New 'terrorist' methods force continual tactical and equipment improvements (Dewar 1985:122; Barzilay 1973:75). Anti-riot paraphernalia, bomb disposal devices and protective clothing have been developed to cope with PIRA (Styles 1975:160). Military technology in "Ulster where rocks were being hurled with more and more accuracy" (Barzilay 1973:71) is developed to counter the increasingly sophisticated weaponry of PIRA.

> Ulster...has turned into an excellent proving ground for pieces of Army equipment and technology of all kinds, and numerous advances have been made to keep pace with the situation (Barzilay 1973:75)

Effective counter-tactics depend on adaptation.[75] The technological hardware of counter-terrorism, developed 'on the ground' in Northern Ireland is absorbed, refined, and exported by the war machine.

A "mutual imitation" of regime and rebels (Townshend 1986:32) may also occur during military operations. Observing the organization, methods and tactics of the IRA "flying columns," and imitating them, allowed the British to counter them (see Mockaitis 1990:151). Imitating either the enemy's equipment of technique in order to fight efficiently (a derivation of 'same element theory,' see Thompson 1966 and Simpkin 1985) requires refinements in technical devices, and psychological adaptation to new weapons systems and battlefield conditions. The 'Winthrop Method,' developed by a Royal Regiment of Fusiliers officer in early 1970's, used mimicry as a weapon:

> Winthrop, a subaltern in Northern Ireland, led us to a great breakthrough in arms discoveries. One day, a bit like Archimedes, he sat down and thought: "How is it that people picking up arms caches can recognise where to look from some veiled speech? Caches must, therefore, always be near a prominent feature, viz.: 'Go to a field, look to a break in a wall, arms cache fifty metres to the right.' So he employed what we now call the Winthrop theory for searching out dangerous kit....You think of these things and try to blend with the background as much as possible... (Captain, 1 Para, cited in Arthur 1987:180).

Soldiers were trained to perceive the landscape as a 'terrorist' would; lone trees should not be thought of as cover from sniper's bullets, but potential hiding places for weapons caches (Dewar 1985:223). Internal Security (IS) training involves the replication of the battle environment of West Belfast:

> There is a little village consisting of about two acres of old quarters and new houses...where the population is mostly under twenty-five and ninety per cent male. There a ten bombs a day exploding, a riot every three days, lots of stone-throwing, petrol bombs, murders galore, kneecappings, you name it. It is the most violent village west of Beirut. Before going to Northern Ireland each battalion spends three days in this English village, sometimes called Tin City. They live in this mock Army base which is all set up like the Northern Ireland situation....Everything is filmed because every square inch of the village is covered by cameras and the NITAT (Northern Ireland Training and Advisory Team) monitor how everyone reacts... (Captain, 1 Coldstream Guards, cited in Arthur 1987:159).

Understanding the possible intentions of the enemy entails being able to think like the enemy; in other words, successful pre-emptive counter moves depend on simulating the strategy of the opponent:[76]

> We were quite ruthless: we didn't want to get hurt, or our friends hurt, and we also knew who was naughty. So we imitated

known bad boys, no doubt about it (Private
2 Para, cited in Arthur 1987:116).

The imitation and incorporation of 'rebel' thought is most obvious in the otherwise marginal units of regular armies: "The term 'special operation' and 'special forces' has come to describe the most complete adaptation of style to insurgency...in which security forces literally replicate the operating methods of the insurgents" (Townshend 1986:33), such as British 'flying columns' in 1921, Wingate's counter-gangs in Palestine, and Kitson's pseudo-gangs amongst the Mau Mau in Kenya (Clayton 1976; Kitson 1977). Similarly, the formation of the post-war SAS resulted from the amateur anthropological explorations of the Malayan jungle by Brigadier Calvert, who during a six-month trek, identified "the need for small parties of specially-trained men to infiltrate into the jungle, live there for long periods of time, to win the hearts and minds of the Sakai inhabitants and to isolate them from the guerrillas" (Seymour 1985:270).[77]

State armies and paramilitaries also construct conceptual models of antagonism (in other words, they 'think' each other in cultural terms) which affects their behavior. Changes in strategic thought can be correlated with the changes in constitutional status of Northern Ireland, political policy, the attribution of political or a-political goals to the opposition, and the state's interpretation of its position vis-a-vis its opponents. How the deployment of a counter-force was conceptualized affected 'on the ground' behavior of soldiers, police and their opponents. Between 1969 and 1982, three force/ counter-force systems can be identified: peacekeeping, with the Army acting as benevolent mediator between hostile tribes; counter-insurgency, when ideology was identified as a motivating factor; and criminalization, where police (and other paramilitary forces) encountered a de-politicized opponent.

peacekeeping

On August 15, 1969 the Queen's Regiment of the British Army was deployed in Belfast, Northern Ireland. They intervened between groups of rioters from the Catholic Falls Road and the Protestant Shankill Road, establishing what became known as the 'peace line.'[78] Initially, the British Army were perceived not as neutral, but pro-Catholic, a situation which was to alter radically by 1972. A British Army Officer confirmed that:

> I think we were aware of the political dimensions. And it is only fair to point out that we as officers...were anti-Stormont, anti-orange Order and very probably anti-Protestant as well! We had a feeling there was injustice over housing, jobs, education and even justice. I think we certainly felt that we were on the side of the Catholics...there was a huge amount of sympathy for them. That lasted a long time and it was probably the ham-fistedness of the Army as much as the politicians that put paid to that (cited in Hamill 1985:81).

British troops were perceived as providing a buffer between Catholic and Protestant populations, and occasionally Catholics and local police, becoming in effect, a "pig in the middle" (Hamill 1985:21). Lord Paget, speaking in the House of Lords in 1975, remarked that Her Majesty's Government have "regarded themselves as a kind of referee to keep the ring between the terrorist and their own Forces, leaning a little to the side of the terrorists..." (H.L. Debs., col. 931, 28/7/75).

The British Army, deployed to Northern Ireland by the Westminster government, entered a politically and bureaucratically complex scene. The constitutional genesis of Northern Ireland, as a political and administrative unit, resulted from a compromise by the British government between the Republican demands and Protestant fears (Weitzer 1990:44). Before 1920, British attempts to grant home rule to Ireland (1886, 1893, 1912-14) met with Protestant resistance.[79] Fearing that home rule in a Catholic Ireland would strip them of their social, economic and political privileges, the Protestant minority lobbied for continued full incorporation within the United Kingdom, and were willing to fight the British government in order to secure their right to remain British (Dewar 1985:99). Home rule (dominion status) was thus granted in the south (Irish Free State), while power was devolved to a local, constitutional government in Ulster, allowing the weary British to withdraw. In order to guarantee Protestant ascendancy, the borders were drawn to include only six of the nine historic provinces of Ulster. The British government armed and trained the Ulster police at its own expense (Farrell 1983), and assisted, in conjunction with police officials, in the institutional foundation of a sectarian security system (Weitzer 1990:57).

The Government of Ireland Act (GOA) 1920 structured the political relationship between Britain and Ireland. Intended as temporary legislation, the GOA continued to function as a constitution in Northern

Ireland for fifty years, while granting supreme authority to the United Kingdom. "For some fifty years, Northern Ireland has existed as a semiautonomous state under a written constitutional document within a larger political community, Great Britain, which does not possess a written constitution" (Finn 1991:8). Westminster retained supreme control over matters concerning the United Kingdom as a whole (defense, trade, etc.), while Stormont controlled the civil service and police. Because border defense was a metropolitan responsibility, three thousand British Army soldiers were permanently garrisoned in Ireland. Although Britain had abrogated its imperial control over political and coercive institutions, it continued to claim sovereignty over the territory and retained its prerogative to legislate on any matter (Cunningham 1991:10-14). Lack of operational control of security matters by Westminster, and the pre-eminence of the executive over the judiciary branches of government, combined with *laissez-faire* political policies throughout the 1930's and 40's of keeping Ireland "at arm's length" led to controversy and confusion (Cunningham 1991:14).

Weitzer (1990) characterizes Northern Ireland as a 'settler state', a communally divided polity, fractured along racial, ethnic or religious lines which has achieved some autonomy from the metropole in the exercise of political authority and coercive power. In a settler state, the security sector, "that cluster of organs with direct responsibility for domestic order and external defense" (3), tends to be repressive. When an overt relationship of domination and subordination exists between political, ethnic or racial groups, the security system retains a partisan orientation on behalf of the dominant sector. Thus, "the paramount function of the security system is to preempt or neutralize opposition rooted in the subordinate group" (Weitzer 1990:9). The security system may appear autonomous in relation to other branches of the state, which "reflects either political subordination of judicial and legislative branches or active collusion with the security establishment" (Weitzer 1990:5). State agencies represented in British democratic tradition as 'neutral' are often partisan in Ulster (Farrell 1983:278). The Catholic population has been subjected to stringent security laws, electoral gerrymandering, police harassment and occasional intimidation by Protestants (Weitzer 1990:53). Catholic opposition and resistance have often been violent. As Lord Salisbury, former prime minister of Ireland said: "the internal history of Ireland has been a continuous tempest of agitation, broken by occasional flashes of insurrection" (cited in Townshend 1986:45).

In the late 1960's, fueled by the success of the non-violent American civil rights movement, the Northern Ireland Civil Rights Association (NICRA) began a protest campaign against the sectarianism of the Stormont government (Purdie 1990). Responding to grievances of the civil rights movement (housing discrimination, electoral gerrymandering, and sectarian police), political reforms were encouraged by Chichester-Clark, the Prime Minister of Northern Ireland. Sensing that these reforms constituted a threat to the Unionist power structure, Protestant groups attacked and burned Catholic ghettos. Both Protestant paramilitary groups and the B-Specials of the Ulster Special Constabulary were involved in the burning and rioting (Dewar 1985:29-33). In conditions of collusion within the settler state "institutions of organized physical coercion are themselves uniquely placed to break the law with impunity since there is no superior coercive power to which they are subject" (Beetham 1991:123). Although the police were acting outside the law, the Government of Ireland Act stipulated that all civil power should be deployed before the Army was introduced. This requirement meant that the B Specials, "the most discredited force in the whole of our Modern history [sic] and we had to commit them before Westminster would agree to commit the army" (civil servant cited in Hamill 1985:21).

The British Army was introduced as a 'peace-keeping force' into Northern Ireland, not to keep the peace between 'Protestants' and 'Catholics', fueled by atavistic ethnic hostility, but between partisan police and Catholic civilians. According to Townshend, the "disciplinary failure in 1969 was the direct cause of British military intervention" and, in this sense, the RUC was the "epicenter of the crisis" (1986:70). The presence of the Army was intended to bring the RUC back under political control.

With no normal police presence in Catholic areas of Belfast, the nearly defunct IRA reorganized and rearmed as a defense force against the Ulster police (deBaroid 1990). The wholesale movement of populations from burned out areas was the largest movement of a civilian population since World War II (Hamill 1985:36). Police participation in the rioting called into question not only the legitimacy of the state security apparatus, but the legitimacy of the state itself. The introduction of a 'foreign', British Army into Northern Ireland to restore order jeopardized the legitimacy of the Stormont government's rule in Northern Ireland (Hamill 1985:6). When James Callaghan MP, announced the deployment of troops into Northern Ireland on behalf of the Labour government, he reassured Unionists that the

Army's presence did not constitute a suspension of Stormont's authority in the North:

> The General Officer Commanding Northern Ireland has been instructed to take all necessary steps, acting impartially... to restore law and order....the troops will remain in direct and exclusive control of the GOC, who will continue to be responsible to the United Kingdom government...The Ireland Act of 1949 affirms that neither Northern Ireland nor any part of it will in any event cease to be part of the United Kingdom without the consent of the Parliament of Northern Ireland... (Cited in Hamill 1985:7).

The first GOC in Northern Ireland was Major-General Ian H. Freeland. Prior to his command in Northern Ireland, Freeland had been the GOC in East Africa. Following the granting of independence in Tanganyika in 1961, Uganda 1962, and Zanzibar 1963, senior British officers were retained until replacements could be found. These hold-over British appointments were unpopular; local African units mutinied against British officers in Tanzania 1961, Uganda 1963, and Zanzibar in 1964. British GOC Freeland suppressed the rebellion, and was rewarded with an appointment to Ireland, considered a quite out of the way sort of place (Blaxland 1971:414, 478). It is no wonder that Freeland saw the turbulence on the streets in Belfast as yet another temporary rebellion.

The Labour Government committed troops to active duty in Northern Ireland reluctantly, hoping that military intervention would make direct political control unnecessary (Bew and Patterson 1985:21). The government's aim was "restabilization with minimal direct involvement in the province" (Weitzer 1990:132). The Home Office, seeking to absolve itself of responsibility for conditions in the province, assumed incorrectly that "questions of law and order are entirely for the Government of Northern Ireland" (Frank Newsam, Permanent Under-Secretary of State at the Home Office, quoted in Bew, Gibbon, and Patterson 1979:177). Harold Wilson stated, "...intervening in Irish affairs: none of us wanted that" (1971:672).

Despite the formation of a Northern Ireland Department at the Home Office in London, the Cabinet's desire to avoid legislative intervention allowed the situation to worsen, "since there was no mechanism for metropolitan judicial intervention" (Weitzer 1990:118). Westminster was not prepared to deploy troops without operational control of security, but was unwilling to disempower Stormont (Cunningham 1991:23), thus the Army was deployed into Northern Ireland without any accompanying changes in the structure of Ulster politics. The British government was now in the awkward position of having deployed its own army 'in aid of the civil power' of an ostensibly foreign government, which it had itself created.

The Hunt Committee Report in 1969 recommended that the Ulster Special Constabulary (B Specials) be dismantled and that the RUC be disarmed and demilitarized: "any police force, military in appearance and equipment, is less acceptable to minority and moderate opinion than if it is clearly civilian in character" (Hunt Report, 1969:21). Two days of Protestant rioting followed its publication, during which sixteen soldiers were wounded, and one police officer killed. Having accepted the proposals but dreading the Protestant response the British government shelved or delayed many of the recommendations.[80]

In curtailing the military functions of the police, responsibility for containing the violence was transferred to the Army. Although between 1969 and 1972, responsibility for internal security in Northern Ireland was theoretically shared between Stormont and Westminster, after the institution of some of the Hunt Report recommendations, the GOC of Northern Ireland became responsible for law and order, including the operational control of police for security operations and the everyday functioning of the Army. Because of the divided constitutional and administrative responsibility for Ulster, the GOC was serving two masters: the Northern Ireland government at Stormont, which made the laws under which security forces functioned, and the United Kingdom, which provided the military forces to enforce the laws. Although technically responsible to the Ministry of Defence in London, divided channels of control between Stormont and London (and in London between MOD and Home Office) meant that the army was aiding two civil powers with conflicting interests and policies.[81]

> It was thought that if we now had the police, the B Specials and the Army totally committed, then we could beat the Micks. The trouble was there were two levels of thinking at that time because the Army was not there to beat the Micks. There was an enormous chasm of understanding between the Stormont and Westminster Governments (cited in Hamill 1986:13).

The continuing separation of Army, police, and civil administration "sometimes led to three different campaigns being waged" (Evelegh 1978:57, 110). While the British deferred political control to the military, stating that events in the North were "entirely a matter for the GOC Northern Ireland" (cited in Hamill 1985:19), the Army continued to insist that, "We are in aid of the civil power and it is at the request of the civil power that we take action" (cited in Hamill 1986:19). Meanwhile, the GOC affirmed that the soldiers were under military control and the police were not giving them orders (Hamill 1986:21). The divided control of the military may have encouraged the UK to impose direct rule.[82]

The executive order for internment, on 9 August 1971 exemplified Westminster's *laissez-faire* attitude towards security policy. Acting under the Special Powers Act, the Stormont government authorized the arrest of four-hundred and fifty Catholics.[83] A military operation in pursuit of political aims (Weitzer 1990:129), internment was introduced against the will of the police and the military (McGuffin 1973:86; Hamill 1985:232), who for the most part were not consulted (Cunningham 1991:58).[84]

With the Stormont government discredited, the British government unwilling to become involved with provincial political problems, the RUC disarmed and the B Specials disbanded, the Army was left "as the only force to keep law and order" (Hamill 1985:34). "Whatever the legal theory, the soldiers were no longer acting 'in aid of the civil power' but in certain areas acting in place of that civil power" (Hamill 1985:19; Fisk 1978:86; Fox 1974:304). Sir Ian Freeland's plea for a statement of aims, and a policy for the security forces went unanswered. A senior officer said: "Politicians and the media often referred to our having 'softly-softly or 'go in hard' instructions. No instructions of that sort ever arrived! We worked on our own...The one firm guideline we had was the law — until we tried to get the lawyers to interpret it!" (cited in Hamill 1985:26).

> There were endless directives on high and low profiles. Directives on whom you could arrest and whom you could not arrest... ad hoc control, depending on rows in Parliament or the papers. There were people in court cases one was not allowed to touch... then suddenly one was allowed to arrest them. You know — one bad bombing and you'd be allowed to arrest everybody (cited in Hamill 1985:106).

With no statement of political aims to guide the deployment of Army, the chain of command in chaos and with very little public confidence in the Stormont government, PIRA established 'no-go' areas in West Belfast and in the Creggan and Bogside districts of [London]Derry, controlling their own roadblocks and patrolling openly on the streets. British Army Intelligence believed that any attempt to retake these areas would result in massive civilian casualties and constitute a violation of the doctrine of minimum force, which at that time was still unchallenged (Dewar 1985:70). Colonel Dewar wrote that the continued existence of the 'no-go' areas meant "permitting gun law to rule for a time in parts of two cities in the United Kingdom" (1985:70).

> In a rather vain attempt to show that the Creggan and Bogside were not 'No Go' areas, the Army occupied a factory...to

guard a little police post. Each day a solitary policeman would be escorted in and would sit in his office, drinking tea, until it was time to go off duty. He was guarded by a full company of soldiers, and never put a foot outside (Hamill 1986:73).

Because the British Army in Northern Ireland was under the control of Westminster, and the government was negotiating with PIRA, the Army was ordered not to interfere with PIRA operations. "Thus for nearly a year...the Army accepted orders not to enforce the law in Londonderry" (1978:17). Without discernible constitutional rules to guide the application of force, the Army appeared to be "acting with a certain aimlessness" (Evelegh 1978:3). Because the Cabinet controlled the Army through informal channels "few limits are set to a system of control which is not acknowledged by the law as even existing" (1978:19). This resulted in flexible, arbitrary and unpredictable law enforcement. According to Captain Lord Richard Cecil: "Daily in the Creggan our troops come face to face with men who are known to have murdered British soldiers. They are not allowed to arrest them because the government has a truce with the Official IRA" (cited in Evelegh 1978:27).

According to Brian Faulkner, a Cabinet member of the Stormont government: "Instead of moving in to support the civil power they dug in as a sort of peace-keeping force on the fringe of the barricaded areas and tacitly accepted the right of various Republicans and known IRA men to rule..." (cited in Hamill 1986:23). Minus metropolitan political control, the Army was forced to act as a non-partisan benign policeman between hostile, warring tribes,[85] a task most soldiers found odious, and for which they were untrained. Criticism produced by British military figures and articles published in 'in-house' Army journals reflected the distaste for internal security duties, and the emerging consensus that military force was being misapplied (Fox 1974; Lunt 1974; Evelegh 1978). According to one officer,

> We were there to try and protect them...But we had to disobey all the logical military principles because of the political forces and because it was the United Kingdom. It was a frightening thing to watch in the Shankill Road...people carrying Union Jacks, singing God Save the Queen, and slinging petrol bombs at us. We'd be standing there with the police in front of us, because we were still sticking to the principles of Internal Security (Senior Officer, 1 Para, cited in Arthur 1987:11-12).

Peacekeeping by troops replaced sectarian, paramilitary policing as the primary means of maintaining civil order; an unpleasant job for the Army, but not an unfamiliar one. Internal security operations and pacification of hostile tribes at the margins of empire had been the primary task of the Army prior to World War I (see Pimlott 1988).[86] These 'small wars' were "expeditions against savages and semi-civilized races by disciplined soldiers" (Callwell 1906:21). The enemy was defined in opposition to the West: "the forces opposing these [regular armies] whether guerrillas, savages or quasi-organized armies, will be regarded as the enemy" (1906:23).[87]

Fighting an uncivilized enemy to whom the protection of the crown did not extend, the Army was not answerable to the military codes of the metropolis: "operations are sometimes limited to committing havoc which the laws of regular warfare do not sanction" (Callwell 1906:42). Military behavior was restrained by the moral component of benevolent colonial despotism (Stocking 1987:81). Although brutality was often sanctioned as a means to induce cooperation, the Army's intention was usually pacification rather than extermination.[88] In Mesopotamia in 1920, for example, aerial bombings of Bedouin tribal groups were intended to 'inconvenience' the population rather than to inflict 'casualties' (Mockaitis 1990:32).[89]

The colonial subject pacified by the Army was an intellectual construct of Victorian anthropology. Not only were the Irish probably inferior and definitely primitive, they were also marked in British political discourse as racially *different* (Szwed 1975). Anthropology articulated the nature of the savage, and justified pacification as a moral duty of white men. Savagery would be replaced with civilization, and the 'violence' of primitive society with consent-based modern forms (Stocking 1987:206; Fitzpatrick 1992).[90] "War is now infiltrating the social sciences" (Virilio and Lotringer 1983:49), but the social sciences infiltrated war long ago.

The Irish provided the Victorians with an example of 'savagery':

> for Englishmen at home and abroad, domestic class and overseas colonial society were linked by the 'internal colonialism' of the Celtic fringe. Thus Ireland, especially, had since Elizabethan times provided a

mediating exemplar for both attitude and
policy in relations with 'savages' overseas
(Stocking 1987:234).

Despite the influx and spread of industrial capitalism,
the population of the Celtic fringe remained an atavistic
survival from a preindustrial world, connected not by the
economy of capital but by tribal genealogy: "tribal and
clan ties were till very lately in full force" (Stocking
1987:219).[91]
The "metaphorical extendibility of socio-cultural
evolutionism" (Stocking 1987:229) slotted criminals,
women and Irish savages into "a subordinate hierarchical
relationship to those who...articulated the cultural
ideology of mid-Victorian Britain" (Stocking 1987:230).
Benevolent Victorian paternalism attributed the rebellion
of colonial subjects to their childish nature.

[T]he British viewed civil disorders as an
expression of an innate human tendency to
rebel against authority. Thus they attached
no great political significance to civil
unrest. Disturbances had to be quelled just
as schoolboys had to be disciplined. The
British tendency to view colonials as
children reinforced this attitude (Mockaitis
1990:65).

In certain ways, the deployment of the British
Army into Ulster in 1969 invoked the Victorian discourse
of tribalism and recapitulated certain structural
conditions of colonial internal security duties. Like a
gentle parent, separating squabbling children, the Army
was required to remain impartial, to act as a neutral, non-

partisan mediator between hostile, warring tribes. In any peace-keeping operation, according to Major-General Perkins, "The security forces must be absolutely impartial between the various factions (1981:31). According to one officer,

> We felt that we had gone out there to protect the Catholics. At that stage it was the Catholics who were giving us cups of tea and the first shots fired at the Army were fired by Protestants. We were there to keep the peace, that was the main role, but when one side is giving you cups of tea and the other side's shooting at you, you do tend to lean towards the faction that is giving you cups of tea (cited in Allen 1990:207).

'Protestants' and 'Catholics' were identified as the primary political agents; political violence was seen as the result of tribal, ethnic conflict,[92] such that "the insurgency thus resembles tribal war more than revolution" (Townshend 1986:68). 'Tribal maps' hung in Operations Rooms around the province used green and orange ink to demarcate ethnic territories.[93]

Genealogical connections were identified as the basis of PIRA organization (see illustration 3.1). According to a Brigadier from 1 Para: "Many of the people of South Armagh...were a tightly knit and inbred group of lawbreakers" (Morton 1989:50). According to Conor Cruise O'Brien: "Since Irish Republicanism — especially the killing strain of it — has a very high propensity to run in families, and since the mother whose sons behave in this way has had something to do with what they believe and how they behave" (cited in Curtis 1984:193) family connections are a vital aspect of intelligence work-ups. According to the Company Commander of the 39 Infantry Brigade, "We would sit down and work out all the various connections between the various families and we would then go visiting those houses..." (cited in Allen 1990:243). Tribal loyalty provided an interpretive framework for soldiers to understand "their hate for us, and their support for the armed youths, [which] was very poignant, an intense tribal loyalty which defied morality" (Morton 1989:169).

The genealogical ordering of paramilitary groups was the basis of a culture of violence. According to a Captain in the UDR, "Terrorism has been here a long time. The people of 1956 are the Godfathers of today's fighters. They haven't changed all that much. In East Tyrone terrorism is a family business — and you'll find that security forces is a family business. So if your father was in it, you probably join" (cited in Arthur 1987:228-9). Security in Northern Ireland is also a family business. According to a member of the UDR: "My father served in the Royal Irish Constabulary in the 1920's and subsequently in the RUC. I had a brother in the 'B' Specials...My ancestor came over with King William's Army in 1689 and it has always been a tradition in the Enniskillen area to provide and join the security forces" (cited in Allen 1990:271).

KEEP IT IN THE FAMILY

3.1 "Keep it in the family". A photograph printed in Visor (1 May 1974), a weekly magazine for British troops serving in Northern Ireland. The photo was apparently found during a house-search, and shows Martin Meehan and two other IRA men posing with local kids.

The dominant motif for this tribal conflict model was cowboys and Indians. During the early period, there was "wild-arsed cowboy stuff on both sides. In '72 you were for ever charging around re-loading your magazine — certain parts of Belfast really were like cowboy towns..." (Captain, 2 Royal Anglian, cited in Arthur 1987:122). Civilization extended no further than the 'bandit country' of South Armagh and the borders of Enniskillen, impenetrable to British Army ground forces. "And then there was the Lower Falls, or 'the Reservation' as we called it..." (Royal Green Jacket, Company Commander, cited in Allen 1987:238), a heavily Republican area where the barracks were known as "forts." The soldiers saw themselves as guardians of the cultural frontier, "paid to keep the forces of barbarism at bay, to form a hard protective outer rim for civilisation..." (Lake 1990:122).

During the early 1970's PIRA and the British Army both conducted themselves as if it were a gunfight at OK corral. PIRA, with no political directive, was simply out of control (Smith 1991). The break down in the British chain of command during the transition to direct rule would seem to suggest that the British Army was also having a bit of a 'shoot 'em up'. While British officers asserted that "purely military victory in Ireland is not possible" (Dewar 1984:159; Townshend 1986:24), in June 1971, General Tuzo[94] said that the Army could "best" PIRA, allowing politicians to find a solution (Hamill 1985:54). Tuzo's plea for increased political direction (Hamill 1985:83) was frustrated by political references to "achieving an acceptable level of violence" (Home Secretary Reginald Maulding cited in Hamill 1985:84) and the government's apparent contentment with a stalemate. Heath's "attitude was that he would not go on carrying the can for decisions made in Stormont" (Hamill 1985:102), and the Army's attitude seems to have been that it would not go on 'carrying the can' for Westminster.

communists do not play golf

In [London]Derry on Sunday, January 30, 1972 the Parachute Regiment was ordered to arrest demonstrators in an illegal anti-internment march. Believing that they were being fired on, soldiers opened fire on a crowd of civil rights demonstrators:

> Acid bottle bombs were being thrown from the top of the flats, and two of our blokes were badly burnt. By that time there were lots of running crowds. It was very busy, very chaotic: panic had stricken. People were running in all directions, and screaming everywhere. Innocent people were being bowled over--we were running past women and children, shouting at them to get out of the way (Sergeant, 1 Para, cited in Arthur 1987:73).

The over-kill in Derry on Bloody Sunday was a startling violation of the most basic principle of counter-insurgency — minimum force. According to the Staff College Counter-Revolutionary Warfare Handbook "there is a direct relationship between the appropriate use of force and successful counterinsurgency" (1986:2). But what happens when commanders misread a situation? "[I]f the nature of the challenging 'force' is misunderstood, then the counter-application of force is likely to be wrong" (Townshend 1986:59). Understanding the situation as a war, the Paras responded to the rioters as if they were an opposing army.

> The Paras are trained to react fast and go in hard. That day they were expecting to have to fight their way in. It was very tense. In those street conditions it is very difficult to tell where a round has come from...that section, quite frankly lost control. For

goodness' sake, you could hear their CO
bellowing at them to cease firing, and only
to fire aimed shots at actual target (Army
officer cited in Hamill 1985:93).

Violations of the minimum force doctrine resulted
from forcing the Army "to act apolitically in a political
arena, and act without the advantages normally flowing
from superior force" (Tugwell 1989:13). Deploying a
combat trained army onto city streets, and expecting
them to keep the peace, inevitably led to violations
(Charters 1989:193; Fisk 1978).[95]

The role of the Army in aid of the civil
power is perfectly clear and definite...It is
not to replace the police. It is not to
supplement the police...It is to act as what it
is: a killing machine, at the moment when
authority in the state judges that order can
no longer be maintained or restored by any
other means. The Army is then brought in
to represent the imminent threat, and if
necessary to perform the act of killing,
albeit minimal, controlled and selective
killing. Having performed this role, it is
instantly withdrawn and the police and
civil powers resume their function (Enoch
Powell, 1977, cited in Hamill 1985:278).

The Paras did what all armies are trained to do:
they killed their enemies. Death is the natural
commodity of any war machine. Bloody Sunday was the
cumulative result of ad hoc control of the military, a
failure to establish the legal parameters for the use of

military force, and unrealistic expectations that combat-trained soldiers would abide by a doctrine of minimum force. Bloody Sunday signified the end of peacekeeping between hostile tribes and the emergence of a new paradigm. What political or legal changes finally allowed the Army to assert autonomy? First, the British Army became a target of the people it had intended to protect. According to one battalion commander:

> The feel of that city [Belfast] had changed dramatically. The atmosphere was harsh and tense. They were all predicting that it was going to go on for very much longer and that bloodbaths were going to occur — and everybody was waiting (cited in Allen 1990:217).

Military non-neutrality resulted from the empowerment of a Tory government in Westminster, where a traditional alliance encouraged Conservatives to yield to Unionist demands (Weitzer 1990:126). Former Prime Minister Harold Wilson explained that "one element in a gravely deteriorating situation is the growing appearance of a British Government departing from its position of neutrality and accepting a state of alliance with a single Ulster faction" (cited in Weitzer 1990:126). Under the Tories, the Army became an "instrument of the Unionist hegemony" (former minister Roy Hattersley cited in Weitzer 1990:127). Essentially, "imperial inertia from 1970-1972 allowed power to revert to the settler elite" (Weitzer 1990:126). Following the power shift back to Stormont, British soldiers "would once again be seen as the traditional enemy" (Hamill 1985:22) doing the bidding of the Unionist Government

(Callaghan 1973:21, 27). In these conditions, the Army could not stay neutral, "it would only be a matter of time before the Catholics had their worst fears confirmed: that it was not themselves who were being protected; but the Unionist government" (Hamill 1985:31).

Once Republicans identified the Army as a prop of the Unionist government, the IRA preferred to shoot rather than accept concessions. In December of 1969, the IRA voted to give token recognition to the parliaments at Westminster, Dublin and Stormont. At Sinn Fein's January, 1970 conference, this motion prompted a walk-out by a militant Republican faction, who saw recognition as an implicit legitimation of either imperialist, collaborationist or sectarian governments (see Foley 1992). The strain of militant Republicanism "that equates political compromise with treachery" (Burton 1978:127) was inherited by the 'Provisional' IRA, while the 'Official IRA'[96] eschewed violence in favor of political compromise (the physical force tradition of Republicanism is highlighted in Chapter Six). Although PIRA favored the use of force, ineffective training and lack of arms prevented a full-scale shooting war against the British Army until 1971.[97]

By 1972, a familiar peace-keeping operation had mutated into something resembling a war. Ulster Prime Minister Major Chichester-Clark announced: "Northern Ireland is at war with the Provisional IRA" (Hamill 1985:45). During the afternoon of August 9, the introduction of internment, Faulkner announced, "We are, quite simply, at war with the terrorist" (cited in Hamill 1985:61).[98] One historian of the British Army in Ireland wrote, "I think it would be naive to say that there was not a 'war' situation in Ulster" (Barzilay 1973: intro.). The Commander Land Forces (CLF), Major General

Farrar-Hockley, announced that the Army was "nearer to the threshold of harder operation" (Hamill 1985:40). In April 1970, the GOC made a public statement that petrol bombers were "liable to be shot" (Hamill 1985:32). At the end of May, Faulkner repeated this warning:

> At this moment any soldier seeing a person with a weapon or acting suspiciously may fire to warn or with effect, depending on the circumstances, without waiting for orders from anyone....this is a firm warning to the whole of the Ulster community as to what can happen at this time (cited in Hamill 1985:53).

As one commander put it: "We tore up...the Aid to Civil Power" (Styles 1975:77).

Despite the belligerent posturing and in-house admissions that Northern Ireland was a 'war' situation, the government's continued credibility came to depend upon maintaining the political fiction that the conflict was not a 'war.' To avoid conferring belligerent status on the IRA, military operations had to be kept under the aegis of the somewhat disorganized and disinterested civil power. While conflict could not be called a 'war,' it could be identified as an 'insurgency', a shift in rhetoric and paradigm which introduced a new set of preconceived assumptions, prescriptions and expectations.

Kitson (1971) defines insurgency as "the use of armed force by a section of the people against the government" in order to overthrow the government or force it to capitulate to insurgent demands (3), a definition later amended to include risings "in active revolt against the constitutional authority of a country" (1977: xii). Beckett and Pimlott (1985) define insurgency as "politico-military campaign waged by guerrillas with the object of overthrowing the government of a state" (1). Sir Robert Thompson defined insurgency as "a form of warfare which enables a ruthless minority to gain control by force over the people of a country, and thereby seize power by violent unconstitutional means" (1970:41). Although insurgency may be instigated by a minority, it entails the support of the populous (Paget 1967b:14).

The roots of counter-insurgency lay in the "expansion of colonial empires in the late nineteenth century" (Beckett 1988:8), colonial policing, "an invaluable source of experience for the British authorities," guerrilla actions of World War II (Pimlott 1988:38) and wars of decolonization. Conventional war was the exception for the British Army (Charters 1977; Jeffery 1987; Mockaitis 1990:147). Commandos units like 2 and 3 Para, for example, had conducted operations in Northern Ireland since 1969, but had not experienced conventional war for twenty-five years.

The axiomatic principles of British imperial policing and 'aid to the civil power' (minimum force, civil-military co-operation and tactical flexibility), formed the basis of counter-insurgency doctrine (Mockaitis 1990:180) and have been the subject of numerous scholarly studies (Beckett and Pimlott 1985; Beckett 1988; Mockaitis 1990; Jones 1991; Townshend 1986). A number of COIN principles distilled from combat experience are being taught as Army doctrine (see Beckett and Pimlott 1985, on which the following passage is based).

Counter-insurgency entails a recognition that the problem, and hence the solution, is basically political (Pimlott 1985; Pimlott 1988:20). Military action must therefore be circumscribed by political considerations as a fundamental matter of strategy (Thompson 1989:4). As Field Marshal Templar explained "The answer lies not in pouring more troops into the jungle, but rests in the hearts and minds of the Malayan people" (cited in Charters 1989:195). During an insurgency, the government's legitimacy becomes a 'center of gravity' target. The "central goal of an insurgency is not to defeat the armed forces, but to subvert of destroy the government's legitimacy, its ability and moral right to govern" (Prisk 1991:69). COIN depends on establishing a civilian-dominated coordinating apparatus at every level of operations. The collection (intelligence) and dissemination (PSYOPS) of information become vitally important. A counterinsurgency cannot be treated as a local problem; public condemnation of Dyer in 1919, leaks about the war atrocities in Burma, and outcry about Bloody Sunday demonstrated that military operations are not immune from public scrutiny. Excessive use of force is inevitably counter-productive.

> To maintain the moral advantage government forces must not only avoid wrongdoing but also the appearance of wrongdoing. In counterinsurgency an atrocity is not necessarily what one actually does but what one is successfully blamed for (Mockaitis 1990:37).

Precision is therefore preferable to overkill: "COIN operates by precise tactics. Two weeks waiting in ambush and one kill to show for it is far better than to bomb a village flat" (Col. Michael Dewar, interview, 1994). Force limits tend to be established by the insurgents and "terror must be dealt with by effective police methods that do not include counter-terror" (Mockaitis 1990:7). COIN conforms to its own internal limiting principles. Because shooting guerrillas in the jungle will not win votes at home (Klare and Kornbluh 1988:16; Barnet 1988:220) "strategy depends upon secrecy" (Barnet 1988:218). COIN tends to disregard the conventional limitations on armed conflict imposed by international law (Barnet 1988:215). A US Army general wrote that, "low-intensity conflicts do not conform to democratic notions of strategy or tactics. Revolution and counterrevolution develop their own morality and ethics that justify any means to achieve success....Survival is the ultimate morality" (Sarkesian 1985:7, 11).

Dependence on local resources and exploitation of local knowledge are hall-marks of counter-insurgency operations (Pimlott 1988:21). Counter-insurgency, characterized by 'unconventional' tactics, informal engagements (Thompson 1989:4) and ad hoc operational procedures, makes troops with conventional training only marginally effective against insurgents (O'Neill 1990:128). Counter-insurgency demands an unconventional approach. According to Kitson:

> At the moment many of these people [officers commanding units combating insurgents] deliberately try to present the situation to their subordinates in terms of conventional war....they are in effect encouraging the development of the characteristics which are unsuited to this

particular type of operation...In other words they are leading their men away from the real battlefield on to fictitious one of their own imagining (1971:201).

The Malayan Emergency (1948-1960) is considered "the textbook counterinsurgency campaign" (Mockaitis 1990:113; see Jackson 1991). During the Malayan Emergency, insurgents were identified in public discourse not merely as anti-imperialist, but specifically as communist guerrillas (Heilbrunn 1954; McCuen 1966; Thompson 1966; Kitson 1971). The British did not see the uprising in Malaya as a nationalist movement with the aim of self-governance. Rather, the majority of 'Communist Terrorists' killed were Chinese, a fact that added credibility to notion that Malaya was not an isolated, local war, but under external direction and influence. According to Gerald Templer, "Communists...seldom go to races, they seldom go to dinner parties, cocktail parties, and they don't play golf" (Speech, April 1952 to the Rotary International Conference in Penang. Templer papers 7410-29-1-9).

Keeping the Peace (Duties in Support of the Civil Power), 1957 was the first official publication to explain anti-imperial rebellion against incumbent regimes as a 'communist phenomena'. Maintaining the 'balance of power' obliged numerous first-world Western states to intervene in order to ward off the 'threat' of communism.[99] The causes of insurgency were defined by their supposed consequence: communist expansion (Shafer 1988:152). Mao's three stage model of insurgency (guerrilla warfare, mobile warfare and positional warfare), stressing civilian support,[100] provided the standard recipe to which American and British counter-insurgency experts reacted (see Thompson 1966): "the

Army did seize upon the communist revolutionary *technique*...almost exclusively as the 'model' for the organisational, politico-military and tactical aspects of the insurgent" (Charters 1989:193; Beckett 1982:207). Whether this model reflected reality is another matter: "It was a tragic day for the French army, the Fourth Republic, and de Gaulle — to say nothing of countless Algerian Moslems — when some do-it-yourself military theoreticians stumbled across the writings of Mao Tse-Tung" (Furniss 1964:41).

The appointment of Frank Kitson to 39 Brigade, Belfast corresponded with the paradigm shift from peace-keeping to counterinsurgency. Kitson, endowed with expertise from low-intensity operations in Cyprus and Kenya, bureaucratized the war effort in Northern Ireland, introduced Special Forces (1987:171), and officially proposed the use of the SAS to the War Office (1987:198). Kitson, perceiving communist insurgency as a global form, placed a neo-Weberian emphasis on ideas as the motivating factor in insurgent violence: "the main characteristic which distinguishes campaigns of insurgency from other forms of war is that they are primarily concerned with the struggle for men's minds" (Kitson 1971:290). Ideological motivation makes communist insurgent violence amorphous, intractable, and permanent (see also Virilio 1986):

> And so for many years the peoples of a hundred tribes fought in hate at the bidding of the Communists' leaders, knowing neither why, nor for whom, they struggled...In each case the Western Powers countered with conventional forces which were totally unsuited to the task. They

called it *the* 'Cold War'...A few people realized that it was not the Cold War but the war (Kitson 1977:63. Italics are his).

While Kitson was certainly instrumental in the development of a COIN paradigm,[101] and changes in praxis in Northern Ireland correspond with his tenure, the conspiratorial fantasies of Faligot (1983), who suggests that Kitson was assigned to Ulster in order to test the theories set out in Low Intensity Operations, are no doubt misguided. Members of the British military establishment have suggested similarly that the public focus on Kitson is misplaced, and that his influence has been overestimated. Nevertheless, in the mid-1970's Kitson's book was required reading for new PIRA recruits (unattributed, 1991 fieldnotes). Violence attributed to an uprising "orchestrated and led by communist agitators" (Styles 1975:88; Sterling 1981; see also Wilkinson 1986:91)[102] also correlates with ascendancy of MI5 and the restructuring of the military intelligence bureaucracy in the mid-1970's.[103] In any case, ideology replaced genealogy as the organizing principle attributed to PIRA violence.

ubermensch boyscouts

Increasing violence, allegations of internment abuses and confusion about security arrangements led Secretary of State Whitelaw to suggest that the division of responsibility between Stormont and Westminster was no longer satisfactory (Cunningham 1991:64; see HC vol. 834, Col. 240, 28/3/72). In 1972, expecting civil war, the British government, acting under Article 75 of GOA, resumed direct rule, suspended the constitution, and abolished Stormont. Security arrangements were also affected by metropolitan intervention.[104]

During the confusion of political collapse, the numerical strength of the Army was decreased, allowing PIRA no-go areas to become "rival mini-states" (Hamill 1985:104) controlled by paramilitaries. Following the institution of direct rule, which re-incorporated Northern Ireland into the United Kingdom, territory assumed new political connotations. In July 1972, a maneuver to reclaim the state's territory from PIRA, Operation Motorman, was the biggest military operation since Suez (Hamill 1985:113). The troop level was boosted to over twenty thousand (Dewar 1985:105): "the battalion was given the task of holding, and dominating, the main part of Andersonstown" (Barzilay 1973:109).

By 1975, "holding and dominating" parts of Belfast was no longer the issue: the war was beginning to go underground.

> What struck me about our '77 tour was the change relative to 1972... By 1977 the whole emphasis was on police primacy....It astonished me...how strait-jacketed the Army had become. We were talking in codes because we didn't have secure radios, giving long elongated sets of orders for comparatively simple tasks; we were tying ourselves in knots, really, over-briefing and de-briefing on every single little thing... (Captain, 2 Royal Anglian, cited in Arthur 1987:122).

This era was characterized by "removing the military from the front line of counterinsurgency and saddling criminal justice agencies with responsibility for internal security" (1990:237). Overt militarism was de-emphasized in favor of legal control. The law was now seen as the basis of legitimacy for military use of force. Conservative Secretary of State Humphrey Atkins (1979-1981) said: "The aim is to defeat the terrorist by use of the law. Generally it means accepting the law of civilized countries" (cited in Weitzer 1990:206). The blue print for police primacy was a report known as 'The Way Ahead,' produced by a committee of senior members of the security establishment, headed by John Bourne from the NIO. The 'Way Ahead' reduced the Army's role, increased the RUC's authority, and expanded local forces such as the UDR. This document established three policy components: 'criminalization', 'normalization', and 'ulsterization.' After referring to this report in the House of Commons, Secretary of State Merlyn Rees explained that although he could not disclose any details, "at the heart of the committee's conclusion is the idea of securing police acceptance and effectiveness. By securing police effectiveness is meant the integration and acceptance of the police in the community to enable them to administer the law effectively...." (cited in Hamill 1985:185).[105] Police primacy meant a return "full circle to the view of fifteen years before, only now the police were able to control the trouble" (Hamill 1985:276).

Global ideological war between communism and capitalism was replaced with local conflict: 'ulsterization'.[106] Locally recruited and trained police grounded the conflict in the province and released Great Britain from the responsibility of law and order. The Army was gradually phased out.[107] The Kitsonian master narrative of communist insurgency was replaced with a micro-narrative of local conflict. "The great narrative of the Total War has crumbled in favor of a fragmented war which doesn't speak its name, an intestinal war in the biological sense" (Virilio and Lotringer 1983:36). The "total war" involving the hearts and minds of the people was displaced by the compartmentalized war of criminalization, requiring the involvement of the very few — namely police and criminals. Regionalization of war drastically increased the value of "intelligence based on local knowledge" (Townshend 1986:33).

When 'crime' replaced 'war' as the dominant metaphor for violence between PIRA and the state, provincial violence came to be seen as a 'normal' result of 'wide spread' crime. PIRA were no longer described in public discourse as communists, but as godfathers and thugs promoting the war through extortion and racketeering. Political acknowledgment of an 'acceptable level of violence' — an implicit admission of military stale-mate— also normalized violence. PIRA appeared to be attacking the security forces while "many other people were suffering apparently at random" (Elliot, cited in Curtis 1984:99). Violence became a systematic and unresolvable social problem. As Booth points out, "if it is 'forgotten,' it is not a 'war': it is merely regarded as a way of life" (1978:145).

As the conditions of war were increasingly seen as an aspect of everyday life in Northern Ireland, the mechanisms of state power became invisible and silent.[108] The new situation reflected

> a campaign that had changed radically since the days when soldiers went out on the streets in full riot gear. Now the emphasis was on undercover work of such secrecy that the policemen involved never

used police stations, never went into court, worked from dead letter boxes... (Hamill 1985:271).

Armed contact between security forces and paramilitaries continued to occur, but mainly as covert action: the center of the war shifted from petrol bombing to information, penetration, and surveillance. Special forces, introduced into Ulster "as a precise cutting tool for political policy" (Geraghty 1980:1), and to counter IRA Active Service Units (Dewar 1985:114), eventually became the most effective counter-force in Northern Ireland. Although a full squadron of the SAS had allegedly been deployed earlier as a covert operation,[109] the public announcement of their presence in Ireland in January, 1976, "signaled a change of role from one of intelligence gathering, control and analysis...to one of combat by the regiments four regular squadrons on rota" (Geraghty 1980:181). Special Forces often mimicked the structure and operations of PIRA: US Lt. Col. McCuen wrote that the success "lie[s] in developing a counter-revolutionary strategy which applies revolutionary strategy and principles in reverse to defeat the enemy with his own weapons on his own battlefield" (1966:78). Accustomed to treating the opposition as co-belligerents subject to wartime rules of engagement (Charters 1989:211; Geraghty 1980:204), these Special Forces often violated the minimum force doctrine, only unlike Bloody Sunday publicity was scarce or sympathetic.

Meanwhile, political violence was no longer being publicly described as 'communist insurgency,' but as 'terrorism' (see for example Priestland 1974; Clutterbuck 1975, 1978; Wilkinson 1986).[110] The 'terrorism' label stuck regardless of whether violence was perpetrated against a civil community or against the state. Terror is a technique; this does not make everyone who uses the technique a "terrorist". Both the British Army, Loyalist paramilitaries, and perhaps most notoriously, the Provisionals use terror strategically. But the British can never be called "terrorists" because they have an air force. Political violence by groups lacking airpower was coded as "terrorism": intimidation by kneecapping figures prominently in media reporting during this period. Anthropologists produced ethnographies to demonstrate that because the Provisionals had a high level of community support, *ipso facto* they were not terrorists (see, for example, Sluka 1989)

As a violence against the state, 'terrorism' favors 'illegitimate' targets (prison wardens, Harrods, Mountbatten) over 'legitimate' military targets (soldiers, barracks, depots, checkpoints), and is not governed by rules, but by the whim to terrorize.[111] The randomness of terrorism makes it seem symbolic and spectacular. According to Brian Jenkins, "terrorism is theater": it is "aimed at the people watching, not at the actual victims" (1975:16). PIRA violence thus appeared as little more the outgrowth of minority strategies to get into the news (Schmid and DeGraf 1982).[112] Media representation became significant,[113] and even the camera shy British Army began to show concern for its public image.[114] "Persuasion," according to Kitson, "will become more important in comparison with armed offensive action" (1971:199). Kitson's criticism of British apathy towards psychological-operations (1971:189) resulted in an expansion of psychological operations (PSYOPS).[115] Impartiality was eventually so thoroughly imparted to the Army that sociologist Philip Eliot, after examining press coverage from autumn of 1974 to spring of 1975, concluded that "the army appeared as almost above the fray...a rather superior kind of Boy Scout Troop" (cited in Curtis 1984:83).

Characterizing PIRA as terrorists benefited law enforcement agencies; INTERPOL, having recently decided to treat terrorism as a crime, was now able to investigate PIRAs activities (see Manwaring-White 1983:48). Denouncing Republican armed action as 'terrorism' was politically expedient and "has proved too great for the temptation ever to be resisted, even when the action is patently military, such as ambushes of security forces or attacks on barracks, or self-directed violence such as hungerstriking" (Townshend 1986:69). This 'terrorism' rhetoric masked a deeper transformation of the war.

> The IRA had become much more professional. Instead of firing off a thousand bullets they were choosing their targets very well, and the ratio of shots to kills had come down dramatically. A wild game had become far more a professional war, which was quite depressing, in the sense that although shootings were few, we felt much more tension, because if there was a shooting there was a real chance of losing somebody (Sergeant, 2 Royal Anglian, cited in Arthur 1987:139-140).

The political rhetoric of counter-terrorism should not be confused with the understandings reached behind closed doors at the Ministry of Defence. Reductionist characterizations of paramilitary violence as pure crime are for public consumption; practitioners of counter-insurgency recognize it intuitively as nonsense:

> The IRA have had eighteen years' practice; I would rate them as probably the most professional terrorists in the world. You've got to have respect for their abilities. You can't go over there and class them as a bunch of clowns: you're dead if you do, you come back in a black box (Corporal, 40 Commando cited in Arthur 1987:250).

If the security forces actually believed that "terrorism" accurately described the violence in Northern Ireland, operations would be conducted quite differently.[116] One cannot fight "terrorists" using the tools of counterinsurgency. The military effectiveness of counter-insurgency *depends* on the depth and breadth and accuracy of knowledge — including cultural knowledge — of the opponent. Army officers are the first to articulate a grudging respect for their adversaries: "So you began to realize that the IRA aren't mindless idiots, that in fact they are quite sophisticated people, using sophisticated techniques and they think about what they are doing" (cited in Hamill 1985:13). Good soldiers recognize each other as soldiers: the bizarre oxymoron "professional terrorist" captures the tension of avoiding the term "soldier" in regard to PIRA.

Anthropologists, like practitioners of counter-insurgency, know that paramilitary violence is not terrorism, only unlike practitioners, they are far too willing to prove it. Burton (1978), Sluka (1989), and other apologists for PIRA have countered the political violence as 'terrorism' model by demonstrating that

Republicanism *as a culture* has historical memory and rational politics. Burton wrote, in a moment of great naiveté, that the reality of the community he studied "exposes the counter-insurgency theorists' writings on Ireland for what they are: bad sociology and political theory...There is no suggestion that the guerrilla is part of his [or her] community or that his causes can be ...grasped and rationally embraced by that community" (1978:118-119).[117] In one sense, Burton internalized the hype and failed to understand the political expediency of the representations.[118] The most radical variety of military theory — counter-insurgency — is actually good anthropology.

the terrible paradox

Spivak questions whether colonial hegemonic discourse allows for subaltern speech (1988). In Northern Ireland, the subaltern is always speaking. The question has become, who is listening? Direct negotiation between the British government and Provisional Sinn Fein regarding truces and treatment of prisoners has been commonplace for years.[119] Rarer, although not unheard of, are communications between PIRA and the Army: "Someone in the battalion sent the IRA a Christmas card in 1972. Their card to us came by return of post, and said, 'We reciprocate your greetings, and hope that by this time next year you'll be back in your own country'" (Captain, 2 Royal Green Jackets, cited in Arthur 1987:93).[120] Despite the public rhetoric of 'terrorism,' counter-insurgents understand the nature, and the 'speech' of their opponents. Do these paradigms, then, represent some sort of misunderstanding of the enemy?

How well do PIRA and the British Army actually 'speak' the same language?

Graveside oratory, another highly symbolic speech act, is designed to be de-coded by two particular audiences. The Army recorded the following speech by Jimmy Drumm at Wolfe Tone's Grave at Bodenstown:

> The British government is committed to stablilising the Six counties and is pouring in vast sums of money...to secure from Loyalists support for a *long haul* against the IRA. We find that a successful war of liberation cannot be fought on the backs of the oppressed in the Six Counties, nor around the physical presence of the British Army. Hatred and resentment of the Army cannot sustain the war (cited in Hamill 1985:238. Italics are mine).

From this speech, the British Army predicted PIRA's transition from a "one more push" to a "long haul" approach (Hamill 1985:238; Urban 1992:31). Until the early 1970's, PIRA assumed that the British government was not fully committed to a protracted war, and that victory would follow from military escalation. Their militarization after the collapse of the 1972 truce rested on this assumption, and was encouraged by a tri-phase Maoist model of progressive increase in popular mobilization during guerrilla war (which would seem to confirm the 'communist insurgent' analogy). PIRA, however, misinterpreted British ambivalence towards the violence in Ireland and therefore "fell victim to the escalation fallacy in limited war" (Smith 1991:233). Circa 1972, at the same time that the British Army's chain of command degenerated into chaos, PIRA's command structure also collapsed under the pressure of escalation, militarization and the 'year of victory' rhetoric: "the natural conclusion that we come to is that PIRA was simply out of control" (Smith 1991:237). By 1977, a younger, Northern, politically radical, hardline faction in PIRA (led by Gerry Adams, Tom Hartley, Danny Morrison, Joe Austin, and Martin McGuinness, some of whom had been interned in Long Kesh) recognized that the British Army was not going to simply withdraw no matter how intense the war became. The assumption of imminent withdrawal was finally exploded by the acknowledgment of 'an acceptable level of violence' by both sides; that level was much higher than PIRA had anticipated. This was the political background to the emergence of the 'long haul' approach.

While Drumm's Bodenstown speech altered British intelligence to a change in PIRA's military strategy, it also functioned as a policy enunciation to other Republicans. That Jimmy Drumm delivered the speech was critical: he was widely respected within the movement, had served more time in prison than any other living Republican, and "carried the credentials of orthodoxy" (Clarke 1987:40). Drumm introduced to Republican listeners the shocking notion of a shift away from pure force: Drumm's comments that "hatred and resentment of the army [the British] cannot sustain the war" rang loud and clear to Republican ears. The primacy of pure force has been an unquestioned principle of the IRA since 1921. Because political compromise was seen as treason, abstentionism characterized Republican politics. Any suggestion of political participation typically led to factional splits, such as that of the Provisionals and Officials in 1971. Sinn

Fein, until Drumm's speech, had been peripheral to the struggle, functioning mainly as a proxy for PIRA. Drumm's speech signaled that PSF was to assume center stage in a joint political-military strategy of 'armalite and ballot box'. The political and military aspects of the war were going to be unified under an umbrella strategy.

Drumm's Bodenstown commemoration speech reiterated and presented for public scrutiny ideas which were already circulating within the Republican movement. In 1977, Gerry Adams had written a report for the General Headquarters Staff of PIRA called the Staff Report which was discovered in Seamas Twomey's flat by Gardai when Twomey was arrested in the weapons bust aboard the MV Towerstream in December, 1977. This Report identified organizational and training deficiencies in PIRA, and outlined a plan for a new military infrastructure not based on commands, brigades, battalions and companies. "The old system with which British and Branch are familiar has to be changed" (cited in Clarke 1987:251). According to one commentator, "Using military terminology to describe its units was part of the Provisional's attempt to root themselves in the tradition of insurrectionary republicanism which had won independence for the south" (Urban 1992:30). Their structural organization, however, was not merely a legitimacy device, although this may have been an added benefit. Rather, PIRA's *existing organisation was loosely based on that of the British army.* There were three battalions in the Belfast Brigade area covering respectively Upper falls, Lower Falls and the rest of the city. Another brigade existed in Derry City and there were units of organizations in Newry, North Antrim, North Armagh, South Armagh, South Fermanagh ..." (Clarke 1987:41. Italics are mine). British intelligence knew that PIRA's organizational structure duplicated their own, and could therefore penetrate and/or second guess PIRA quite easily.

British military intelligence penetrated the Republican network by tapping informers willing to make confessions during seven day detentions. The Staff Report explicitly identified the deleterious effect of detention orders on Republican prisoners undergoing interrogation by the RUC and the Special Branch at Gough Barracks and at Castlereagh. British interrogation methods were extremely effective at breaking PIRA members in detention. PIRA's battalion system, identical to that of the British Army, made them vulnerable to exposure if a prisoner broke under interrogation because everybody knew the chain of command. These interrogation techniques had been used on British soldiers by the Chinese in North Korea and the techniques adapted for future use in other colonial encounters. Because their own military structures had been penetrated and destroyed by interrogation during detention, the British knew that this tactic produced results. Because the British knew that PIRA functioned *just like they did*, they knew interrogation would work. The Staff Report and the switch to ASU's is the actual evidence that it did work.

According to the Staff Report, "it is the IRA's fault for not indoctrinating volunteers with the psychological strength to resist interrogation." The Staff Report recommended that volunteers be "reeducated and given up-dated lectures in combating new interrogation techniques." According to Clarke, "this meant the isolation of the basic interrogation methods used by the British and the RUC...together with reminders that while interrogation was limited to seven days, imprisonment was open ended and the punishment for informing was death" (1987:41). These interrogation techniques were also used by PIRA on suspected informers, and apparently on at least one captured British soldier. PIRA thus redeployed the techniques used against it, exemplifying the mimetic relationship between armies and insurgents.

Informing, which signifies penetration of the cultural interior, can be prevented by indoctrination. Internees can be restrained from breaking under interrogation by instilling in them the Irish culture and the Irish language. The culture and especially the language work like a shield during detention and in prison. According to an INLA man

> A lot go into Castlereagh, and the police have things in common with them...that used to break down their defences. You see guys come in for interrogation and they're as English as the English, except they claim to be Irish. The have a lot in common with the police; they break down and become informers. If they had the culture and the Gaelic stronger, they could hold out. If you don't have any culture, how are you different from the Brits? You're fucking just the same... (cited in Feldman 1991:132).

Learning Irish in prison not only distinguished Republicans from the British warders, but made Republican prisoners psychologically and culturally impenetrable (Feldman 1991:216). Linguistic separatism made it impossible for the 'screws' to disrupt the prisoners' cohesion as a group. "The jails proved that when you become culturally separate it breaks the enemy" (PIRA member, cited in Feldman 1991:213). Cultural separatism as a weapon and as a means of survival was a lesson learned in the prison. Transferred to the outside, Republican cultural separatism not only enlivened and enriched Irish pre-school education for ghetto kids, but made it impossible for the British to "know the enemy."

Other ways to control the damage done by men breaking under interrogation is to limit knowledge. The Staff Report recommended that the vulnerable and too easily penetrated brigade structure be abolished and replaced with operational cells called Active Service Units (ASUs). In urban centers, cells of four volunteers organized by specialty (bombing, sniping, and execution) were to be controlled militarily by Brigade/Command Intelligence Officers. Cells would have no control over weapons or explosives, which would be retained by the Quartermasters, and would operate out of their own areas as often as possible. The basic model for the ASU was probably Latin American guerrilla groups who used the form quite successfully against bigger and better equipped state armies. Being small and autonomous, the ASUs would be fairly invulnerable to penetration or informing. Members of the ASU would know only their own identity, "only the ASU commander would have contact with the next level of authority" (Urban 1992:31). Republican graves serve as artifacts of this shift: after 1977, Republican dead are longer listed by rank, just simply 'Volunteer'.

The British Army, who could so easily penetrate and defeat a brigade structure which mirrored their own, was forced to alter their own military structures to cope with PIRA's reorganization into ASU cells. The introduction of semi-autonomous special forces units by the British Army, which duplicated the ASU structure of PIRA, shows that PIRA's move to ASU's was an effective counter-measure against interrogation, and is scrutinized more closely in Chapter Five. The shift in rhetoric, the new long haul strategy, the emergence of a radical Northern military faction, and the introduction of Sinn Fein as a player in the war, produced some anxiety in Army intelligence.

In 1978, Brigadier James Glover on the General Staff (Intelligence) wrote a classified report called Future Terrorist Trends. This document should be seen in relation to the Staff Report, since it comments on the changes outlined in the Staff Report. PIRA somehow acquired a copy of this classified report and transmitted to various left-wing journalists in the UK. Duncan Campbell wrote that "The terrorists' abilities, their professionalism, and their expertise are all on the increase the army believe. Indeed, substitute "soldier" for "terrorist" in the study and the report might almost be a careful appreciative analysis of an allied Army" (Newstatesman, July 13, 1979).

Future Terrorist Trends is essentially an ethnography produced by the security forces. As such, it is quite an excellent piece of anthropological analysis. It acknowledges the switch to ASU's and that the "Provisional leadership is deeply committed to a long campaign of attrition". Contacts and links with the PLO are specified as a "potentially rich source of weapons and of an exchange of ideas on terrorist technique" — the 'terrorist' network alluded to in Chapter Two. In assessing manpower resources, Future Terrorist Trends notes that "there is a substantial pool of young Fianna aspirants, nurtured in a climate of violence, eagerly

seeking promotion to full gun-carrying terrorist status..."
This suggests that the Security Forces understand how a
culture of violence reproduces itself. The Report states
that "our evidence of the calibre of rank-and-file terrorists
does not support the view that they are merely mindless
hooligans drawn from the unemployed and
unemployable. PIRA now trains and uses its members
with some care. The Active Service Units (ASUs) are for
the most part manned by terrorists tempered by up to ten
years of operational experience". This suggests,
furthermore, that the British Army doesn't believe the
'terrorist' hype any more than do professional
anthropologists. Middle class graduates are deterred
from joining the Republican movement because of "the
Provisionals' muddled political thinking. Nevertheless
there is a strata of intelligent, astute and experienced
terrorists who provide the backbone of the organization."
Financial resources, external support, command and
control, communication networks and strategy and
targeting are analyzed in the document and mock
budgets drawn up for weekly operating costs.

It was not only references to "muddled political
thinking" in Future Terrorist Trends which irritated PIRA.
The report also stated that PIRA's anti-aircraft capacities
were ineffective, small arms fire and the RPG-7 having
been the main methods. "We know that the PIRA has
long wished to obtain hand-held anti-aircraft missiles.
The black market price for the SA-7 in 1976 was £7,000,
seemingly within the Provisionals' grasp." This taunt so
offended certain members of PIRA that they videotaped
their use of an SA-7 in a helicopter attack in Crossmaglen
and allegedly forwarded the video to the Army
(informant statement, 1991).

Future Terrorist Trends is excellent ethnography because it explicitly acknowledges that the war is being fought according to certain codes: "PIRA still sees itself as an 'Army' and clings to the remnants of what they believe to be a military code of ethics." Furthermore, the report identifies an autonomous force tradition of "violence for its own sake", and what is perhaps a very significant omission, Future Terrorist Trends only mentions Sinn Fein once. Interpreting this omission in light of the Staff Report is quite illuminating. The Staff Report recommends that "Sinn Fein should come under Army organisers at all levels.... Sinn Fein should be radicalised (under Army direction) and should agitate about social and economic issues..." (cited in Clarke 1987:253). PIRA and PSF follow a pattern of civil-military relations which derives more from Ludendorff than from Clausewitz: PIRA has total military autonomy from Sinn Fein. At best, Sinn Fein exerts a restraining influence, but they have no real authority over the military cadre. Furthermore, PIRA does not derive its legitimacy from Sinn Fein — on the contrary. According to Gerry Adams, "on mandate, I think, very, very simply, if you are talking about armed struggle, that the very, the presence of armed forces and of a British presence in this country is where people derive their mandate from" (interview, cited in Clarke 1987:11fn.). PIRA needs no political mandate because the Army Council believes that it is the legitimate government of Ireland. This is the first principle of the Republican movement, subject to neither negotiation nor question. Force is at the center of the Republican movement and the British Army knows this.

Understanding the enemy is, of course, the basis of all effective military strategy: "it could be maintained that *it is not hatred* so much as understanding of the enemy, and not a conscienceless squandering but a humane conserving of his own forces which are the hallmarks of an efficient commander" (Dixon 1976:275. Italics are mine). Even hatred is a form of subjective involvement by which we are bound to the hated object. Remember Booth's "terrible paradox" at the heart of military life (1979:100): the aggression necessary in war is the antithesis of the cultural understanding necessary to construct strategy and fight effectively. According to Booth, ethnocentrism interferes with "knowing the enemy," which is the foundation of all strategic analysis (1979:101). Effective war-making demands empathy at the same time that it demands suspension of empathy. Successful proto-counter-insurgents, such as TE Lawrence (and Frank Kitson), were adept amateur anthropologists. But if bad anthropology makes for botched war, does 'good' anthropology make for better killing? Future Terrorist Trends and the Staff Report suggest that it does. An empathetic, compassionate and accurate representation of their opponents as politically motivated, communally oriented, and so on, keeps the war running smoothly. To know is to love, but during wartime to know is also to kill. To what degree, then, is elegant theory culpable for a war efficiently fought?

Chapter Four:

Permanent Emergency and the Control of Force

"naught better can betide a martial soul than lawful war"
— Lord Krishna to Arjuna

In 1893, Durkheim wrote that, "it is an error to believe that vengeance is but useless cruelty" (1933:87). In Northern Ireland, political violence often appears to be little more than grotesque bloodshed: a Catholic taxi-driver is doused with petrol and set on fire by Protestant paramilitaries, a British soldier has his left leg blown off by a PIRA land mine, a ten year old girl is killed with a plastic bullet from a British gun. Although gruesome, these events are neither random, indiscriminate, nor irrational. Long-term political violence adheres to certain predictable patterns. While some of the patterns reflect informal norms, other patterns of political violence have been established via legal codes and frameworks.

Contrary to Cicero's dictum *inter armes silent leges*, in time of war the law is hardly silent. The limit of absolute force is not the only restraint on war (Howard 1979); logistic, political, and legal considerations constrain the armed conflict in Belfast and South Armagh. Law limits, contains, and structures conflict (Hattendorff and Murfett 1990). In Northern Ireland, two formal legal systems interact to limit conflict. International humanitarian law applied to armed conflict, also known as the laws of war, and internal emergency legislation, or law as a security policy, both impose limits on the scope, intensity and conduct of violence. The laws of war and security legislation, however, may have less connection with actual on-the-ground military praxis than the informal codes between combatants, as we saw in the previous chapter. Formal legal structures thus may function as ideal, imaginary norms, while informal quasi-legal codes may exert a greater actual control on behavior. This chapter explores the influence of formal, legal frameworks on the practice and definition of armed conflict.

Formal and informal legal codes interact, intersect and contradict each other at different points, forming a series of competing yet related systems of social control, in which "different levels of legal control act upon each other, i.e., are interdependent" (Nader and Metzger 1963:585). Anthropological theory provides a concept of a 'multiplicity' of legal systems to describe the complexities of this legal jumble. Postulating a multiplicity of legal systems entails an assumption that the state has no monopoly on political authority or legal organization. Hoebel writes that "where there are subgroups that are discrete entities within the social entirety there is political organization — a system of relations between groups, or members of different groups within the society at large" (1949:367). Political subgroups establish binding, normative rules governing the conduct of members. Such rules can be interpreted either as a sign of a fully developed 'legal system', or as a rudimentary form of social control, depending on where one wishes to draw the analytical line. As Durkheim (1933), Mauss (1906), and Malinowski (1932) demonstrated, within any society several systems of rules may compete, conflict, or complement one another and may exercise different degrees of social control on the subgroup members. Thus, law cannot be assumed to be the property of society as a whole, for "any human society does not possess a single consistent legal system, but as many such systems as there are functioning subgroups" (Pospisil 1974:98). The creation of legal order is, then, not necessarily the province only of the state: "it is not an essential element of the concept of law that it be created by the state, nor that it constitute the basis for the decisions of the courts or other tribunals, nor that it be the basis of a legal compulsion consequent upon such a decision" (Ehrlich 1936:24).

The construction of laws governing the conduct of human violence is not solely the prerogative of states. Tribes, bands, chiefdoms, kingships — not to mention guerrilla groups — have rules and proscriptions about who may and may not be killed, by whom, with what weapon, and when. The laws of war, like other legal systems, are multiplicitious and polyvalent. Numerous interdependent, complementary and contradictory codes may be brought to bear on the conduct of violence between subaltern groups, and especially when 'primitive' groups engage nation-states in armed conflict. The laws of war predominating amongst states can, in this sense, be read as a cultural artifact which entails a set of assumptions about the nature of organized human violence. For example, which parties to the dispute are accorded the legitimate right to make war? Do primitive, guerrilla or 'non-civilized' actors have the capacity to attack states as co-belligerents? Can states be the 'victim' of aggression from non-state actors? Do terrorists have law, or are they beyond the pale of international legal codes? The international humanitarian law of armed conflict is rife with assumptions about power, order, justice and legitimacy.

"The development of the international humanitarian law of armed conflict was in essence one of limiting, by recourse to international law, the scope of armed conflict" (Suter 1984:11). International humanitarian law of armed conflict concerns the ways in which war should be conducted (essentially the *jus in bello*), and was traditionally divided between the Law of the Hague, which concerns the rights and duties of belligerents during the conduct of operations and limitations on warfare, and the Geneva Conventions, which are designed to protect casualties and non-combatants. At one time, the laws of war only limited the conduct of war between states; what states did in their own territory was their business: "the law of nations, or international law, may be defined as the body of rules and principles of action which are binding upon *civilized states* in their relations with one another" (Brierly cited in Suter 1984:11. Italics are mine). While the laws of war have been periodically updated following major wars, they have had only a minor influence on contemporary armed conflict, partially because large-scale violence now occurs more frequently within states than between them. Outlawing 'war' *per se*, with the Kellogg-Briand Pact of 1928, and substituting the category 'armed conflict' in its stead, was meant to limit the recourse of states to the use of armed force as a means of dispute resolution *between states*. Similarly, under Article 2(4) of the UN Charter, states agree to refrain from force or the use of force *against other states*. Although Article 3 of the 1949 Geneva Convention was intended to apply to non-international armed conflicts (that is, to domestic conflicts such as civil wars), the failure to adequately define the level at which internal strife becomes 'armed conflict' merely blurred the distinction between treason and civil war (Suter 1984). None of these legal documents proved totally relevant for the most common type of war after 1945: guerrilla war.

Prior to World War II, war was considered a matter for states and their armies. "War," according to Rousseau, "is something that occurs not between man and man, but between states" (1960:176): private individuals had no business fighting unless they joined armies and became soldiers. Carefully drawn legal distinctions between combatant and non-combatant reflect a traditional view of warfare as the political prerogative of states, and a concern to protect individual civilians from the violence of war. According to the laws of war prior to 1977, combatants, as members of regular armed forces, could wage war, could be attacked by the enemy, and upon capture could not be tried as individual war criminals, but had to be accorded the status of prisoners of war. Non-combatants, on the other hand, could not be attacked, nor could they commit acts of war against an opposing army. If non-combatants did commit such acts, they were liable to be prosecuted as war criminals. Whether belligerent status ought to be extended to categories of persons such as *francs-tireurs*, partisans or *levees en masse* was a central issue at the 1899 and 1907 Hague Conventions, since such irregular combatants hovered betwixt and between the neat categories provided by law. *Francs-tireurs*, individuals voluntarily taking part in military operations without being members of regular armed forces, were condemned to death upon capture for acts of unlawful belligerency as recently as World War II (Clarke, et al. 1989), while uniformed partisans acting in groups were often accorded POW status.

Whatever distinctions once existed between combatant and civilian collapsed in the guerrilla wars following World War II. Guerrilla warfare simply does not correspond to the classic model of conventional warfare in the Hague law and the Geneva Conventions. Guerrilla forces are often composed on a mixture of civilians and soldiers conducting unconventional operations, ranging from clandestine operations (such as sabotage and assassination) to open combat. Such wars raised questions about the application of the international humanitarian laws of armed conflict, especially giving combatant or belligerent status to guerrilla forces (Vetschera 1993; Suter 1984). Belligerent status accords certain privileges for the use of armed force and certain obligations under the laws of war — perhaps most fundamentally, the right to be treated as a prisoner of war rather than immediately tortured or executed upon capture. Are guerrillas legitimate belligerents, entitled to be treated as prisoners of war? Or are guerrillas merely terrorists, fit to be treated as criminals? The answer to these questions depends very much on whether you ask Gerry Adams or Margaret Thatcher.

The 1977 Protocols attempted to rectify the gap between the actual conditions of combat and legal protective frameworks offered by international law. Under Protocol I, non-international conflicts are brought within the ambit of international law, and the distinctions between regular, irregular and volunteer armed forces found in the Hague laws and Geneva Conventions are abolished (Clarke, et al. 1984). Simultaneously, the Protocols attempted to preserve the distinction between combatants and non-combatants under Article 44(3) of Protocol I by obliging combatants to distinguish themselves from the civilian population during military operations preceding an attack, and during any actual attack. While Protocol I makes contra state guerrilla warfare an international conflict by definition, Protocol II makes internal or domestic disturbances *above a certain threshold* subject to international law. Protocol II is similar to Article 3 insofar as both have an implicit level of violence threshold in their material field of application.

The laws of war are intended to function as a limiting *device* in inter-state, and especially in contra-state, armed conflicts. On the other hand, the notion that law can be consciously *used* by a state to limit non-international armed conflicts within its borders should be identified at the outset as a security strategy. Law at the margins of empire is often used to control political violence (Burman and Harrell-Bond 1979). British security policy in Northern Ireland, similarly, is based on the concept that social disorder resulting from political violence is best controlled through legal means (for example, Evelegh 1978; Hogan and Walker 1989:127). Both international humanitarian law of armed conflict and domestic security legislation may reproduce the very conditions of disorder they seek to contain, limit or control. In this sense, the imposition of law whether domestic or international creates and sustains social catastrophe. In Northern Ireland, attempts to contain and limit armed conflict through security legislation may be counterproductive (Rolston 1991), and may actually result in an eventual and inevitable displacement of the 'rule of law' with 'the rule of the gun'.

'Neutral' jurisprudence often conceals the ideological component of law (Kerruish 1991). We should therefore investigate not only 'the rule of law', but *what kind* of law rules. The domestic security legislation that currently organizes legal praxis in Northern Ireland is a direct inheritance from war-time regulations and as such contributes to a 'garrison state.' Law used to impose political order reproduces the social pre-conditions of violence it is meant to alleviate. This is the legal basis of the self-sustaining "culture of violence" (Apter 1991). As Burton (1978) succinctly points out, to argue for the existence of a culture of violence is not to suggest that the

Irish Republican movement "lends itself to the formation of a war machine" (1978:36). Rather, the tight-knit social organization of this paramilitary support community, combined with the entrenched economic interests of the British defence industry, and the limiting effects of domestic and international types of law, dissipate enough social chaos to sustain war.

In the current conditions in Northern Ireland, to what degree does law, as an element in a counterinsurgency program, and as an international, external limiting structure, contribute to the cultural permanence of violence? That is the central question which this chapter addresses. Related questions are: to what degree emergency legislation, used to contain, control and punish political violence, contribute to a culture of violence? And, to what degree does international law of armed conflict also contribute to sustaining the war?

savagery, disorder, terrorism

Savagery, disorder and terrorism, are conceptually linked in international law and in British security policy. .These concepts are aspects of the epistemological basis of legal counter-insurgency. Law backed by force has been the preferred means to civilize the barbarism of colonial subjects, to render them susceptible to law (see MacLeod 1967). Violence, disorder, savagery are nomadic principles which law seeks to control, co-opt and tame. In this discursive formation, the disorder external to the realm of law is the province of the savage. "Colonization is always associated with the bringing of order into a disordered situation" (Fitzpatrick 1992:108; Faris 1973). In most colonial regimes, the application of law is linked to the civilizing process. In the rhetoric of modern nation-states, the cultural disorder of 'savagery' has been replaced by the political disorder of 'terrorism.' Terrorism is perceived by the state as random, indiscriminate, sporadic political violence beyond the pale of the law. Terrorism, although prohibited, is nowhere defined in international legal instruments.[121] The Prevention of Terrorism Act (1978) asserts that "terrorism means the use of violence for political ends" (section 31). This definition contradicts the characterization of 'terrorism' as a non-political, 'criminal' act.

Terrorists fill the discursive position of the nomad, outside the law of states, guided only by the informal, un-codified custom of the tribe. Custom, says Bentham, is "for brutes...written law [being] the law for civilized nations" (1970:153). Terrorism appears as a reversion to a prior evolutionary legal form: un-codified, customary, familial, and nomadic. According to Austin, in a state of nature "men...have no legal rights" (1961-3:9-II; Fitzpatrick 1992:82) because they have no property. In legal, colonial discourse, the savage has no fixity, no place. Terrorism also lacks geo-political fixity, and appears as trans-national, a network of shady figures toting AK-47's in anonymous desert training camps. *Lacking territory, terrorists appear to lack law.* The pervasive, omnipresent paranoid threat of terrorist violence, what Der Derian calls "pure terrorism," is not geo-specific. "[T]he critical production and distribution of the terrorist threat is not territorial, as is the case in conventional war, but temporal: its power is increasingly derived from the instantaneous representation and diffusion of violence by a global communication network" (1992:116). The Republican movement, on the other hand, has traditionally pursued its nationalist and paramilitary projects in terms of a geo-strategic occupation of territory. If anything, the Provisionals have been obsessed with spatial and territorial control, for example: 'no-go' areas, banishing recalcitrant joyriders from West Belfast, maintaining Republican cemetery plots, and the concept of "Free Derry." These artifacts localize the political nation in actual, concrete spatial coordinates.

In Britain, terrorist violence has a strong, historic association with the Irish. Bombs in the City and in Mayfair are not militarily, or qualitatively dissimilar from Fenian bombings of Victorian London. The killing and maiming of civilians by bomb blast is seen as indiscriminate and brutal; the Irish who place the bombs little better than savages. The conceptual opposition between Irish terrorism and the British legal system forms one aspect of the epistemological basis of counter-terrorism: "civilization is the unifying term which legitimates and justifies the counter-terrorists' position" (Leeman 1991:73). 'Terrorism' as praxis exists in opposition to 'civilized' law. Conservative Secretary of State Humphrey Atkins (1979-1981) said: "The aim is to defeat the terrorist by use of the law. Generally it means accepting the law of civilized countries" (Belfast Telegraph, 26 September 1979, cited in Weitzer 1990:206). British soldiers conducting counter-terrorist operations believe they are responsible for securing the cultural frontier, for guarding the perimeter fence of culture, "paid to keep the forces of barbarism at bay, to form a hard protective outer rim for civilisation" (Lake 1990:122).

While terror permeates the London rush hour, the production of a 'terrorism' discourse and the construction of a 'terrorist' image for PIRA are elements of counter-insurgency strategy. 'Terrorism,' hardly a neutral term, encrypts a number of assumptions about violence and about people who perpetrate violence. In describing certain types of violence as 'terrorism', one assumes that the state represents legitimate political authority, and that 'terrorists' are irresponsible renegades from democratic political processes. Terror can better be understood as a military *tactic*, used by states as well as paramilitaries. In certain communities in Northern Ireland, the center of power is not the state, but the Provisional Irish Republican Army.[122] The level of community support clearly demonstrates that PIRA is the armed and dangerous tip of the nationalist Republican iceberg, which can be fairly described as a self-sustaining culture of violence (or 'culture of struggle', the term Republicans seem to prefer). Simply put, terrorists don't have communities. Nor do they have law: because their own law has been stripped from them, legitimacy claims are framed in terms of *justice*. For Republicans, law is contestable and partisan. Seen from the Republican point of view, 'terror' is a necessary and inevitable by-product of military operations, civilian casualties at Harrod's merely collateral damage resulting from the destruction of an economic target.

Laws of war, which establish civil norms of armed conflict between states, do not apply during war with savages or 'terrorists,' since these 'illegal' types of violence exist outside the province of civilized law. Armed conflicts between states and tribes at the margins of empire eschewed formal laws of war (Ferguson and Whitehead 1992). The doctrine of minimum force did not apply to 'un-civilized' opponents, whether 'terrorists' or 'savages', thereby absolving the state all normal ethical restraints (Canny 1976:122). To this end, the applicability of the laws of war to domestic political violence is essentially an interpretive act (compare Wilkinson 1977, 1986:15-17 with Detter DeLupus 1987:19-23). The application of laws of war depends on how terrorism is defined by the state, and entails a great many assumptions about savagery, civilization and the right to perpetrate legitimate violence:

> What is the proper standard against which the performance of the security forces and the courts in Northern Ireland--and perhaps also of those involved in terrorist activity--should be judged? Is it that of the ordinary common law as developed to deal with peaceful conditions in Britain? Is it some lesser standard applicable to states of emergency? Or is it the law of war, as is occasionally claimed on behalf both of the security forces and of their paramilitary opponents? The answer to those questions depends in turn on how the conflict in Northern Ireland is categorized (Hadden, et al. 1988:18-19).

Defining the situation in Northern Ireland as 'war' would have inadvertently conferred legitimacy on PIRA. For political reasons, Great Britain has sought to maintain the rhetorical, if not actual, division between 'emergency conditions' and 'war'. As we will see, "the line between a permanent state of emergency and the declaration of a war situation is very fine" (Bunyan 1976:56).

British experience of colonizing Ireland provides the primary trope of attempts "to civilize this barbarism, to render it susceptible to laws" (Axtell 1985:50), later to become so important in the fight against the 'savage' terrorist. 'Rendering susceptible to law' created bloody conflicts between natives and colonizers, and was often the source of the disorder the law sought to alleviate: "The colonial situation provides another monumental instance of law initiating and sustaining pervasive disorder, even in the pursuit of its pretense to secure order" (Fitzpatrick 1992:81). Perhaps "the imposition of British law in Ireland actually provoked disorder?" (Townshend 1986:55).

the status of war

The obsolescence of major war corresponds with the new status of 'war' as a non-existent *legal* condition, and with the emergence of an omnipresent, perpetual terrorist threat. War has disappeared, but terrorism is everywhere. "In modern international law, the straight answer to the question 'what is the basis of the institution of war?' is: 'nothing'. That is to say that developments in the law since the First World War, and more particularly since the coming into existence of the Charter of the UN in 1945, have outlawed the resort to 'war' as such" (Collier 1991:121), except for the UN's right to armed engagement (e.g., Korea in 1951, Iraq in 1991). Article 2(4) of the UN Charter forbids states to employ force or threat of force except in self-defense, and was expressly framed to be applicable whether or not a war had been declared. A state which resorts to force other than in self-defense violates Article 2(4), even if it declares war. Furthermore, self-defense only justifies the force necessary for defence: excessive use of force is also a violation of Article 2(4) (see Greenwood 1991:136-9). The US justifications for bombing Libya under Article 51 as a "pre-emptive action against [Libya's] terrorist installations" establishes a legal precedent for pre-emptive anti-terrorist actions (Collier 1991:128). The *threat* of future terrorist acts now legitimate pre-emptive strikes (O'Brien 1978; O'Brien 1979).

War still occurs and the law still seeks to control or limit it, but the application of law is no longer dependent on the existence of a formal, legal condition of war (Greenwood 1991:139). The issue of whether or not, in fact, there is a war is irrelevant. Historically, numerous states have eschewed the label 'war' for political reasons, and the refusal to recognize 'war' has prevented application of the *jus in bello*, allowing sneaky states to get away with what might otherwise be called murder. The 1949 Convention's avoidance of the 'war' concept circumvents this problem, by using the objective criteria of armed conflict: "either there is, or there is not, an armed conflict taking place and the views of the individual States concerned are certainly not decisive" (Rowe 1987:141).

In the modern world, fighting happens without declarations, and declarations are made without fighting. The concern is to determine under what conditions in the absence of declarations, armed conflict ('war') exists? Should the criteria be scale, intensity, casualty figures or some other measure? (Greenwood 1991:136). What underlying cultural assumptions are brought to bear in such a decision, and who has the power to decide? This is especially a problem in civil wars. Article 3 of the 1949 Conventions is concerned with armed conflicts of a non-international character, and establishes minimum standards of conduct. States, however, may characterize uprisings or rebellion against their authority by armed groups as merely criminal. If political violence is criminal, it is not an armed conflict. "This is the view taken by the United Kingdom in relation to the activities of the I.R.A. in Northern Ireland. The International Committee of the Red Cross was permitted to enter the 'H' block prison in Northern Ireland [Maze], but only on the express understanding that permission to do so did not carry with it any admission that the United Kingdom government admitted the applicability of common Article 3" (Rowe 1987:142).

The case of the will of Lance Corporal David Jones was mentioned in Chapter One. Sir John Arnold, who decided the case of Re: Jones (Deceased), ruled that Jones, by mumbling his dying wishes to a fellow officer, had indeed produced a valid soldier's will. This decision was based on an Australian precedent (In the Will of Anderson (1958) 75 WN (NSW) 334):

> In the present case there was no state of war and it is difficult to see how there could have been, for there was no nation or state with which a state of war could have been proclaimed to exist, but in all other respects *there was no difference between the situation of a member of this force and that of a member of any military force in time of war*. In my opinion the deceased was in actual military service... (5 at H. Italics are mine).

The character of opposing force was irrelevant to a determination of active military service. "The fact that the enemy was not a uniformed force engaged in regular warfare, or even an insurgent force organised on conventional military lines, but rather a conjuration of clandestine assassins and arsonists, cannot in my judgment affect any of those questions...It is not the state of the opponent, or the character of the opponent's operation...which affect the answers to the questions which arise" (5 at J). The troops are at war: no matter with whom, no matter whether or not war is declared. In the absence of a legal construct to describe armed conflict between high-contracting parties, *battlefield conditions* rather than declarations define armed conflicts. What do battlefield conditions mean during a permanent state of emergency when the normal level of violence is very high? Although battlefield conditions as a means of defining war would seem to broaden the scope of applicability of the laws of war, in fact they create a whole new set of problem.

Although war no longer exists, sovereign states no longer have a monopoly on the legitimate use of force: national liberation movements (NLMs) now may use force to secure the right of their peoples to self-determination (Wilson 1988:91). Following the post-World War II wave of decolonization, "the widespread resort to guerrilla warfare raised questions concerning the application of the law, because in most cases the activities of guerrillas challenged the existing legal conditions for combatant status" (Roberts and Guelff 1989:387). The 1977 Protocol Addendums to the Geneva Convention redefined 'international conflict' as: "armed conflicts in which peoples are fighting against colonial domination and alien occupation and against racist regimes in the exercise of the right of self-determination..." (Roberts and Guelff 1989:390). According to British legal scholar Peter Rowe in <u>Defence: The Legal Implications</u>, "it is not at all clear whether the IRA, or any other such organisation, represents a 'people', as opposed to a group of defined religious (or other) sect. In any event, the whole population of Northern Ireland has, for many years, had the right, through the ballot box, to exercise its right of self-determination and has expressed its wish to remain part of the United Kingdom" (1987:60). Culture, which from the outside seems extremely amorphous and relative, seems from the inside quite absolute: Republicans had no doubt that they were 'a people' and explicitly framed their public legitimacy claims in the language of the Protocols. While on hungerstrike, Bobby Sands wrote: "I am a political prisoner because I am a casualty of a perennial war that is being fought between the oppressed Irish people and an alien, oppressive, unwanted regime that refuses to withdraw..." (1981: Day 1).[123] This rhetoric of 'oppressed people' and 'alien regimes' can be read as an attempt to position the Republican movement as a NLM for inclusion under Protocol I.

The inclusion of national liberation movements under the auspices of Protocol I dealing with international conflict, "seems to imply that there is a new status in international law" (Wilson 1988:105). This *de facto* status of national liberation movements outside of the category of 'internal strife' is only possible if the concept of 'a people' is separated from the concept of 'a state':

> If a 'people' is separate and distinct from the power administering it, then the armed conflict between the administering power and a liberation movement...must be international by definition. The implication...is that wars between States are not the only kind of international war (Wilson 1988:94).

Although armed factions of every hue have been asserting their military legitimacy loud and clear, the elevation of national liberation movements to belligerent status in an international conflict gives that self-proclaimed legitimacy an official stamp.

Protocol II offers protection to victims of 'internal' or 'civil wars' and applies to armed conflicts not covered by Protocol I

> which take place in the territory of a High Contracting Party between its armed forces and dissident armed forces or other organized armed groups which, under responsible command, exercise such control over a part of its territory as to enable them to carry out sustained and concerted military operations and to implement this Protocol (cited in Roberts and Guelff 1989:449-50).

Article 43 of Protocol I, defines combatants as armed forces which must be "under a command responsible to that Party" and "subject to an internal disciplinary system which, *inter alia*, shall enforce compliance with the rules of international law." Combatants must distinguish themselves from civilians by carrying arms openly during a military engagement, and during military deployment preceding the launching of an attack. Holding 'territory' is an explicit requirement for the inclusion of guerrilla groups within the ambit of Protocol I.[124] The operations of Francis Hughes discussed in Chapter One, especially the wearing of a uniform and the carrying arms openly preceding an attack, could probably be interpreted as an attempt to position PIRA as an NLM for inclusion under Protocol I, although informant statements were inconclusive on this point.

Applicability of combatant status outlined by the Protocols to the conflict in Northern Ireland is a sticky political issue. While the legitimate use of force by national liberation movements, and the definition of wars of national liberation as international in character may be recognized by the Protocols, "no State confronted by a NLM has ever acknowledged that it was fighting an international conflict, nor has any state since the end of World War II accepted the authority of a liberation movement to use force against it" (Wilson 1988:124). The Protocols legislate what no state actually practices.

Although they may not have explicit combatant status, captured guerrillas are commonly given a *de facto* POW status; this was the case in the UK until 1975. British Army lawyers have since then been at pains to point out the distinction between POW status and lawful belligerent status (Clarke, et al. 1989). Great Britain, in any event, refused to ratify the 1977 Protocols. The British delegate explained their unwillingness to sign on the grounds that

> in relation to Article 1, that the term 'armed conflict' of itself and in its context implies a certain level of intensity of military operations which must be present before the Conventions or the Protocols are to apply to any given situation, and that this level of intensity cannot be less than that required for the application of Protocol II, by virtue of Article 1 of that Protocol, to internal conflicts (cited in Wilson 1988:130).

The British jurist GIAD Draper objected to the possible inclusion of PIRA in the Protocols. Protocol II, according to Draper,

> does not apply to fights at football matches...or to the 'kill and run' types of armed activity such as have been experienced in Northern Ireland. Such activities are not 'sustained and concerted military operation' however much it may be claimed that 'organized armed' groups are committed. Neither is there the exercise of such control over a part of the State's territory as to enable such dissident armed groups to carry out such sustained and concerted military operations (1979:149).

Furthermore, "if the capacity of such 'peoples' is limited to the 'kill and run' type of activity, their struggle should properly be excluded from the ambit of both Protocols" (1979:150). British objections to the Protocols centered on the link between combatant status and the level of intensity of the armed conflict, what is known as 'threshold of application'. The intricacies of the interaction of domestic British military law and the Geneva conventions are beyond the scope of this essay. Suffice to say that the British Government has, since 1916, sought to avoid conferring belligerent status or any *de facto* or *de jure* military legitimacy on PIRA.

In the absence of a legal construct to describe armed conflict between high-contracting parties, *battlefield conditions* rather than declarations define armed conflicts. Under Article 3 of the 1949 Geneva Convention, belligerent status depends on demonstrating "a threat to the very survival of the government. The most effective way of manifesting that threat is violence; the guerrilla needs to force up the level of violence from a situation of civil disturbance to a clear Article 3 situation" (Suter 1984:16). In order to gain belligerent status as a High Contracting Party, PIRA would have to boost the activity levels of its operations to a higher 'intensity' (Boyle, et al. 1980:271). There is some evidence to suggest that this was exactly what PIRA was intending to do in the early 1970's in the form of bombing campaigns directed at military targets (unattributed source 1991). In 1978, PIRA attempted to gain recognition at the UN as a NLM. According to a British Army Captain, writing in 1974,

> the government has provided a great deal of the substance of martial law in Northern Ireland while avoiding its form. By so doing, the government has affirmed that the conflict has political, economic and cultural, as well as military form and... *to narrow the conflict to a one-dimensional military form would be playing into the hands of the terrorists. The concept of militarization may be associated with the late Carlos Marighella, but the practice has long been associated with the IRA* (Fox 1974:305. Italics are mine).

Criminalization of political violence can be read as a reaction to the militarization of PIRA operations; a means of de-escalating and controlling the conflict. In other words, criminalization was a means of finding an acceptable level of violence, and preventing escalation which would have jeopardized the well-established non-belligerent status of PIRA in international instruments.

For Britain, "the legal framework for military actions remains blurred, both with regard to its justification as acts of self-defence against an act of armed aggression and with regard to the legal status of those participating in strike operations as long as they do not constitute an "armed conflict" in the meaning of the Additional Protocols, or as long as the state for which they operate has not ratified the protocols" (Vetschera 1993:1581). In other words, to maintain legality of its counter-terrorism measures, Britain must maintain a precarious position as the victim of PIRA aggression; counter operations must appear as self-defense. The appearance of an 'armed conflict' with PIRA must be avoided at all costs, thus the discourse of criminality and 'law and order'. By non-ratification of the Protocols, Britain is unfettered by international legal instrument. As of January, 1994, however, Britain has signaled its intention to ratify the Protocols as soon as implementing amendments can be enacted; criminalization is now so firmly entrenched that inadvertently conferring belligerent status on PIRA by ratification is no longer considered as a serious threat.

Do states have the right to claim justification for counter terrorist activities as self-defense under the war-decision law set down in the United Nations Charter? According to O'Brien, they do. "Terror is an instrument of violence that emphasizes attacks on targets that would not normally be chosen as a matter of legitimate military necessity" (1989:189). Because the Charter is primarily concerned with conventional war, a shift in interpretations of the UN war-decision law would limit Article 51 self-defense measures to defensive reactions to conventional armed attacks (O'Brien 1989:193). The United Nations Charter therefore reduces *competence de guerre*, or the right of states to use armed force, by prohibiting the threat or use of force (Article 2 (4)) except collective security enforcement ordered by the Security Council (Chapter VII, Article 42, Chapter VIII, Article 53), and in self-defense (Article 51). Article 2(4) of the UN Charter was expressly framed to be applicable whether or not a war had been declared. A state which resorts to force other than in self-defense violates Article 2(4), even if it declares war. Furthermore, self-defense only justifies the force necessary for defence: excessive use of force is also a violation of Article 2(4). Early interpretations of the war-decision law limited self-defense measures to those actions taken against continuing hostilities of some duration. "Occasional discrete instances of recourse force — typical of counterterror deterrent-preventive-attrition strikes — were treated as reprisals, and the prevailing UN view was that reprisals were not legally permissible" (O'Brien 1989:193). States, however, may invoke self-defense in sporadic use of armed force against sources of *persistent* terrorist attacks (1989:194). Just where 'persistent terrorist attacks' of Article II (4) dovetail with 'sustained and concerted military operations' of Protocol II (1) is very murky. The US justification for bombing Libya under Article 51 as a 'pre-emptive action against terrorist installations' establishes a legal precedent for pre-emptive anti-terrorist strikes. The *threat* of future terrorist acts now legitimates pre-emptive strikes; terrorists need not be caught with a smoking gun (O'Brien 1978; Suter 1984). Precedents also exist in domestic law, which may justify the application of lethal force by soldiers to prevent the commission of *future crimes* (Rowe 1987:92).

cultural permanence

For inclusion under Protocol II, PIRA would have to boost the level of violence; it would have to militarize. But criminalization and containment of violence in the form of a permanent state of emergency — a sub-war — prevents escalation of the conflict. The Provisionals cannot escalate, but they will not back down. And even a permanent state of emergency is livable. As Darby (1986) and others (Feldman 1991; Burton 1978) have pointed out, the fact that conditions are not intolerable makes the conflict manageable, and contributes to its longevity. Violence can be contained by imprisoning members of proscribed organizations, but violence cannot be eliminated as long as the conditions for continued recruitment and regeneration remain (Boyle, et al. 1980:271). No one doubts that the British Army could destroy PIRA with one smashing blow. According to Myles Shelvin, a PIRA member who negotiated a truce with Whitelaw in the early 1970's: "They [PIRA] can, of course, be beaten. If the British Army puts the boot in they could be flattened. But will they do it?" (cited in McKnight 1974, cited in Smith 1991). Although the Army might prefer this direct approach, the fact is that the military fist is restrained by law. "I have to say that I am not a great advocate of 'economy of effort'. As a soldier I would, as you say, prefer to unleash every conceivable weapon at my disposal to ensure victory" (unattributed 1994). In an internal security scenario, outright war is not feasible. Because escalation is legally prohibited, the conflict settles into a condition resembling a stalemate. Law thus contains the conflict, and makes it manageable. The laws create and sustain the structural conditions of the permanence of violence, within which martial cultures flourish.

This permanence of war is true on either side of the Irish Sea. As of 1988, Northern Ireland became the longest-running active-service commitment of the British Army in this century. For the generation of British kids born after 1969, the Army has always been in Ulster. Life has never been otherwise. In 1987, an 18 year old Private, Mark Drummond, of The Duke of Edinburgh's Regiment, was serving in Northern Ireland. He was born on 14 August, 1969, first day of the war (Army Magazine 1988:9). According to the Diplock Report, in the Republican strongholds of Ardoyne and Lenadoon, children are "growing up in an environment of violence and destruction. To them a battlefield is always at their door" (1972:35).

For Republicans, as long as they are fighting they are winning (Smith 1991). Republicanism has a very strong military force tradition, combined with skepticism regarding political solutions. Militarism obviously has tremendous political utility; the foundation of the Republic in 1921 is convincing evidence of the efficacy of arms. Violence, however, is not merely a symptom of an experience of deprivation for Republicans, but an autonomous, institutionalized practice in its own right (Feldman 1991:21). There is a tendency in PIRA, as in most military organizations, to revere violence for its own sake; an embryonic doctrine of *kreig an sich*. The autonomous force tradition in Republicanism uses the rhetorical ammunition of Pearce's 'cult of violence,' blood sacrifice. The sacrifice metaphor sustains the culture of violence, by giving even military failure a certain mystique (Smith 1991:68).

The institution of 'blood sacrifice', no matter how much certain Republicans scoff at it, is the basis of a strategy of attrition and makes a 'long war' feasible. According to Ruari O'Bradaigh in 1971: "I cannot imagine the IRA driving the British Army into the sea, or anything like that" (cited in Smith 1991:72). Following the cease-fire 1975, the beginning of Ulsterization, and an internal shift of power in PIRA to Northern political factions (of which Adams is representative), a strategy of besting the British Army militarily was abandoned in favor of guerrilla war based on attrition. As we saw in the previous chapter, Drumm's speech at Bodenstown (15 June 1977) signaled the switch. The "Year of Victory" slogan was displaced by drab, seriously grim 'long haul' statements: "We are committed to and more importantly geared to a long term war" (PIRA statement in Republican News 9 Dec. 1978 cited in Smith 1991:304). Organizationally, this strategy manifested in a switch to ASUs, a refinement in targeting, a decrease in the number of civilian killings, and a violence that was no longer based on questionable economic targets (e.g., Birmingham and Mayfair bombs), but on "armed propaganda" (e.g., the sinking of Mountbatten) (Smith 1991:285-296).

An ideology of the desirability of sacrifice and martyrdom makes war sustainable. Rather than damaging or destroying social cohesion, protracted violence may produce a kind of social order. Attrition-based military strategies may sustain war by normalizing it (attrition, in fact, is not a strategy for victory, but continuation of conflict): "the repetitious nature of the terrorist attacks against a narrow range of targets will make the pattern of violence increasingly predictable" (Smith 1991:256). The British public and the

Republican community have come to expect a certain level of violent death in Northern Ireland: "the routinization of imminent violent death is a determining condition of paramilitary life" (Feldman 1991:106). But as Feldman points out, the paradox of the ideology of an acceptable level of violence is that once this belief is internalized, violent death is "received with decreasing levels of collective disturbance" (1991:106). In these special conditions, the more violence occurs, the less it disrupts culture. Death simply becomes the foundation of a certain kind of military culture, ritualized in prison protests, paramilitary funerals, and jurisprudence:

> The dialectic of doing/being done, as a dominant symbolic logic of the culture of war, operates within its own self-directed circuits and economies. The closed circuit of violent reciprocity facilitates an acceptable level of distancing by communities in which the paramilitary lives and for whom he claims to die....The political code of "acceptable levels of violence" indicates that both the state and paramilitaries share a sacrificial logic of political hegemony based on the subtraction of parts from the whole (Feldman 1991:106).

Military cultures welcome violent death as a central aspect of their cultural logic. This is hardly a revelation: death is at the center of any war machine. For any weapon system to function efficiently, it must be capable of surviving attack. Furthermore, it must be able to maintain its ability to deliver firepower during a counter-attack. A culture of war (such as Republicanism)

— basically a weapon-system writ large — must be capable not only of surviving the death of its individual members, but of making those deaths productive. It is, of course, "the nature of a military organization to accept casualties in furtherance of its objectives" (Clarke 1987:172). The ability of a military organization to do so is referred to in NATO doctrine as *survivability*.

Military doctrines can be used as metaphors to explain how military cultures use ideologies of death to promote their own survival. Actually, this approach is not so farfetched: the capacity of a nation or culture to continue fighting over a prolonged period is incorporated in NATO doctrine as "strategic sustainability", a concept focused primarily on logistics and industrial capacity. At the tactical level, the principle of sustainability refers to the ability of a military force to continue operations despite combat losses, and seems to be largely a subjective evaluation by a field commander (Murphy and Blandy 1993:2636). A more precise concept related to sustainability is the Soviet doctrine of *zhivuchest* or "viability," based on the Soviet combat experience in World War II. "The viability of troops (forces) describes the 'capability of troops, aviation and naval forces, weapons, military equipment, rear installations, command and control systems, and logistic rear services to preserve or speedily restore their combat effectiveness under conditions of enemy action'" (Murphy and Blandy 1993:2636). As long as command and control systems remain intact, combat effectiveness can be maintained even with 50-60 percent casualty rates, especially if the casualties are sustained over period of time and are concentrated within one brigade or unit. Viability, which is related to survivability or the ability to engage in combat despite the shock of loss, "includes almost everything that contributes to combat effectiveness" (Murphy and Blandy 1993:2637). The implication here is that the ideology of an 'acceptable level of violence' contributes to the viability of a culture of violence. If death of a certain type (martyrs, combat casualties, and so on) can be welcomed and incorporated into the very structure of a military culture, combat effectiveness can be maintained despite high losses of human life.

The contribution of death to the viability of military systems is only possible because such systems function on the principle of synergy: that is, the whole is greater than the sum of the parts. The subtraction of parts through individual death does not interfere with the synergy of military organizations, but increases it. This is what Feldman approximates with the concept of "a sacrificial logic of political hegemony based on the subtraction of parts from the whole" (1991:106). The 'acceptable level of violence' in this mock-up is simply the threshold of systemic sustainability above which death reduces the synergy of the military system.

The neo-functionalist view adopted herein leads to the perception that systemic equilibrium of military cultures is not disturbed, *but maintained* through the death of soldiers during combat, as long as they do not exceed the limits of the system but remain within acceptable levels. Anything that maintains war at an acceptable level sustains conflict. Coded limitations on violence — strategic, cultural and ideological — prevent conclusion. Politics prevents complete release, to paraphrase Clausewitz. Similarly, legal restraint on armed force controls conflict, but does not eradicate it. Law merely represents the distillation of a code for appropriate killing which keeps violence in bounds. Like the laws of war, internal legal structures perpetuate a cultural permanence of violence, which is examined in the following section.

martial law/military culture

Neither the situation of permanent emergency nor the perseverance or survival of military cultures is a new development in Ireland. In 1787, the Commander-in-Chief of the British armed forces in Ireland stated: "...but for the military there would be no government at all" (cited in Boyle, et al. 1975:165). The common law was never sufficient to pacify Ireland; for that, a standing army was necessary. In 1603, the nine thousand British troops in Ireland guaranteed internal pacification (Pawslisch 1985:5). In 1994, nine thousand British troops are still garrisoned in Ireland. Law, in Ireland, has always been backed by armed force. If law needs to be imposed by an army, does law 'rule' in Northern Ireland and if so, what kind of law?

During states of emergency, civil legal apparatuses cease to function normally and martial law may be used to reestablish control. Martial law signifies the failure of normal civil control, a "point where power surmounts the rules of right which organize and delimit it and extends itself beyond them...and equips itself with ...violent means of material intervention" (Foucault 1984:211). Martial law, by superseding the limits of civil law, is a form of order which exceeds order. In other words, "real martial law is not law. It is rule by the order of someone with enough force behind him to ensure that his orders will be obeyed" (Simpson 1937:99). Martial law is justice applied with a gun.

Although martial law was a regular practice in the colonies (Townshend 1982:167-8),[125] no special legal framework for emergency law, or *Notrecht* existed in Great Britain until the early twentieth century (Vogler 1991:83, cf. Clode 1869:163-7; Brewer 1989:48).[126] By transferring supreme authority to the military, martial law frustrated the British civil authorities' attempts to bring the Army under the administrative and legal auspices of the civil power, and threatened civilian control of the military. After the historical experience of Cromwell, martial law made everyone very nervous.[127] English jurists hesitated to admit that martial law exceeded the rule of ordinary law,[128] and so the pre-conditions for the implementation of martial law within the United Kingdom remained largely unspecified in either common or statute law (Townshend 1986:21-23). Simply put, nobody wanted to talk about it. The Crown Law Officers in 1838 ruled that martial law "cannot be said in strictness to supersede the ordinary tribunals, inasmuch as it only exists by reason of those tribunals having been already practically superseded" (Townshend 1986:21). "Legally, martial law was simply a state of affairs which occurred when the legal system became inoperative" (Townshend 1986:21). But "the problem is to know when these 'states' are deemed to exist" (Evelegh 1978:32).

States of emergency could simply be proclaimed. But a weird legal quirk in common law tradition (*de facto* functioning of marital law through supersession of the common law) meant that a state of emergency could exist regardless of proclamation. "Hence martial law did not have to be 'proclaimed': proclamation did not create an altered legal state but merely gave notice that the courts could not function" (Attorney General and solicitor

General, 16 Jan. 1838. Jamaica rebellion file, WO 32/6235). The "sharp disjunction at the moment of proclamation" made the use of martial law look like a "moment of political crisis" (Townshend 1986:21), which in the interest of maintaining legitimacy, had to be avoided.

Alternately, flexible laws could be developed to cope with emergencies. In such cases, military authority would not supersede the civil authority — rather, the military would become the civil authority. "It may be that the system of 'flexible law' remarked on above is no more than martial law exercised by the Government at it sees fit..." (Evelegh 1978:33). Martial law would function informally, making either a definition of 'disorder' or a proclamation unnecessary and redundant. On the surface, legal structures would look more or less normal.

Although various forms of martial law have been declared in Ireland since the Union of 1801,[129] no equivalent system of statutory or non-statutory martial law has existed in Northern Ireland (Hadden, et al. 1988:16). Because of the militarization of the state, any law whatsoever was already a form of martial law. Acts giving powers to the Army[130] incorporated regulations that "formed the substance of statutory martial law" (Hadden, et al. 1988:16), doing in fact what could not be done in name. The Army had war-time powers without declaring martial law, including the imposition of curfews, conducting internment operations, and shooting in self-defense. "All the powers exercised by the army have been exercised on the basis of express statutory authority...To this extent the army in Northern Ireland may claim to be acting within the ordinary law" (Hadden, et al. 1988:17). Under Northern Ireland's militarized conditions, 'ordinary law' is, in fact, martial law.

Flexible emergency legislation was a central part of a strategic framework for counterinsurgency. According to Sir Robert Thompson (1966), counter-insurgency should be conducted in accordance with the law (1966:52) although special legal powers could be introduced during an emergency (Paget 1967). Collective fines, deportation and detention can be employed, and "statute law can be modified by emergency law" (Thompson 1966:53). "In the long term, adherence to the law....puts the government in a position in which it is represented as a protector of those who are innocent, and it puts the terrorists in the position of criminals" (Thompson 1966:54). Criminalization of guerrillas through the application of flexible law during insurgencies not a new idea: Gwynn pointed out in 1934, "Other advantages of military law are that actions not normally offenses can be made criminal" (1934:17).

In <u>Low Intensity Operations</u>, Kitson suggested that there were actually two possible legal frameworks which could be employed during an insurgency: either increased malleability of law, such that law becomes another "weapon in governments arsenal," or incremental modifications to law, equitably applied to all cases brought before the courts. The latter was intended to conserve the government's legitimacy:

> No country which relies on the law of the land to regulate the lives of its citizens can afford to see that law flouted by its own government, even in an insurgency situation, so everything done by a government and its agents must be legal.

But this does not mean that the government must work within the same set of laws during an insurgency as existed beforehand, because it is a function of government to make new laws when necessary. Nor does it mean that the law must be administered in the same way during an uprising... (1989:55; see an almost identical passage 1977:289).

The development of a "legal system adequate to the needs of the moment" (Kitson 1977:290-1) was the inevitable consequence of the failure of ordinary law to cope with insurgency. Flexible emergency legislation is a hybrid form of law,[131] neither civil nor martial, which complements the structural indeterminacy of counter-insurgency (Townshend 1986; Clutterbuck 1974).[132] When ordinary law could not cope, "...the law should be used as just another weapon in the government's arsenal, and in this case it becomes little more than a propaganda cover for the disposal of unwanted members of the public. For this to happen the legal services have to be tied into the war effort in as discreet a way as possible" (Kitson 1971:69).[133]

Although flexible law may have been a useful counter-insurgency tool in Dhofar or Malaya, expedient legal maneuvering was diametrically opposed to the mind-set of the regular Army. Robin Evelegh's Peace-keeping in a Democratic Society can be read as a criticism of Kitsonian legal flexibility, and of counter-insurgency policy in general. Evelegh identified the constitutional vacuum for controlling the military during the suppression of civil disorder within the United Kingdom (1978:2) as a cause of jurisdictional and operational conflict between the military and the police, and advocated both strengthening and clarifying the law (1978:96-7).[134] In constitutional theory, the basis for military intervention is the common law, soldiers having the same duties as private citizens to aid the civil power in the suppression of disorder. <u>The Manual of Military Law</u> makes it "clear that a soldier must come to the assistance of the civil authority where it is necessary for him to do so, but not otherwise." Because the <u>Manual</u> fails to elaborate on civil authority, "the soldier is left to decide for himself who are the civil authorities he should support" (Evelegh 1978:12).

This vagueness in the legal framework of counter-insurgency complements certain aspects of British legal culture: "the very imprecision of the rules, so disturbing to soldiers was part of the British way. Vagueness was a barrier to despotism" (Townshend 1986:43). But vagueness can also mask the politically expedient use of law, which both Major Evelegh and liberal British jurists found objectionable: "the application of extra-legal military force will...signal a defeat for constitutional democracy" (Walker 1992:12). Expediency should therefore not override legality. Similarly, legal procedure in Northern Ireland for terrorist suspects indicated "the extent to which traditional legal values were eroded by the pursuit of the 'military security' approach" (Boyle, et al. 1975:71).

The concept of 'traditional legal values' existing in Northern Ireland is not so straightforward. All jurisprudence contains an implicit ideology: "jurisprudence has the overall purpose of persuading people of law's innocence" (Kerruish 1991:2).[135] In Northern Ireland, the 'rule of law' — specifically the rule of a non-partisan law — has been repeatedly proclaimed. But what does the rule of law actually mean in a context where law, far from innocent, has contributed to the reproduction of a culture of violence?

MP Arthur Davidson declared, regarding the Northern Ireland (Emergency Provisions) Acts of 1978, "we are not tinkering around with minor evidentiary points...We are altering the whole fundamental criminal process in Northern Ireland..." (cited in Finn 1991:87). In fact, the Emergency Provision Acts and the Prevention of Terrorism Acts are remarkably similar to Coercion Acts of the eighteenth and nineteenth centuries.[136] Most of the current emergency legislation in Northern Ireland derives directly from wartime special powers. "The current powers of arrest, search, temporary detention (for interrogation), and internment...are broadly the same as those provided throughout Britain and Ireland under the Defence of the Realm Acts passed at the outbreak and during the course of the First World War..." (Hadden, et al. 1988:14). Thus, "[t]he current emergency legislation...may be traced back to earlier legal precedents" (Hadden, et al. 1988:13), most of which derive from military statute law.[137]

The Government of Ireland Act 1920, which created Northern Ireland as a legislative unit, empowered Stormont (the Northern Ireland Parliament) to make security law. "Consequently, the main legal weapon fashioned by the Unionist governments, the Civil Authorities (Special Powers) Acts (Northern Ireland) 1922-43 (the 'Special Powers Act') was highly reminiscent of the Restoration of Order in Ireland Act 1920 both in form and substance" (Hogan and Walker 1989:14). This Act, which became permanent in 1933, formed the basis of joint security forces operations in Northern Ireland until 1973, conferring powers of arrest, search, detention and internment on the Army and the RUC.

In 1971, the British executive acknowledged that most military activities authorized by the Special Powers Act were, in fact, illegal. The Special Powers Act 1926 had empowered Stormont to deal with unrest and was the basis for military and policing activities. The Government of Ireland Act 1920, however, restricted the power of Stormont to legally deploy an army. In other words, the legal jurisdiction for deployment did not, and could not, extend to Northern Ireland because the province was not within the metropolitan national boundaries.

This perplexing legal double-bind came to light through a test case, brought by John Hume, MP, who is a somewhat infamous political figure in Northern Ireland. When ordered to disperse during an illegal public assembly, Hume refused to obey a British Army officer who claimed to be acting under regulation 38 of the Special Powers Act. Hume was subsequently brought up on charges in 1971 for failing to disperse. The defendants, convicted by a local magistrate, later appealed to the Northern Ireland High Court on the grounds that powers given to the Army under the SPA were unconstitutional. The Lord Chief Justice of Northern Ireland, Sir Robert Lowry, upheld the appeal and quashed the conviction (Hamill 1985:96), stating that in purporting to confer certain powers on the British Army, the regulation sought to achieve a lawful object by unlawful means (R v. J.P.s for the City of Londonderry, ex parte Hume and Others, 23 February 1972).

Lowry's decision was tantamount to acknowledging that the British government had no legal basis for the deployment and conferral of power on the

212

Army. The basis for Lowry's decision, however, was removed within hours by the enactment of the Northern Ireland Emergency Provisions Act 1972 (Boyle, et al. 1975:132), authorizing the Northern Ireland Parliament to legislate in respect of armed forces when necessary to maintain peace. As Maulding pointed out

> ...it is also necessary to ensure that members of the Armed Forces would not be liable to any legal proceedings based on this technical point in respect of the exercise of power since 1969...The side effect of the bill will be to declare that the law, so far as the powers of the Armed Forces are concerned is, and always has been, what it has hitherto been believed to be (cited in Hamill 1985:97).

The Act conferred retroactive validity on military actions otherwise illegal. If the EPA had not been introduced, the Army would have had to act under common law, an unpopular solution in military circles, since the common law gave "no dispensation for the military in their activities" (Hamill 1985:97).[138]

In 1972, after Westminster assumed direct rule over Northern Ireland (including control over internal security) the Special Powers Act was replaced with the Northern Ireland (Emergency Provisions) Act 1973. Suspension of trial by jury, introduction of amended rules of evidence and continuing administrative detention were introduced. The EPA enacted a list of 'scheduled offenses', "broadly all those crimes regularly committed by terrorists and their supporters" (Boyle, et al. 1975:95), and relaxed common law rules on the admissibility of confession. "[I]ntense political pressure...to apprehend and secure conviction" (Walsh 1988:42) meant that a blind eye was turned to hard-core police interrogation. By 1988, about ten thousand people had been convicted in non-jury courts mainly on confession evidence (Walsh 1988:40), some of it obtained through torture.[139] Statements inadmissible under common law rules — that is, statements obtained by torture — were accepted as evidence in Diplock courts (Boyle, et al. 1975:102),[140] and "the judges ...gave some measure of judicial approval to robust and persistent interrogation" (Boyle, et al. 1975:103). In 1971, allegations of torture resulted in the Irish Republic's case against the British Government for violation of the Convention on Human Rights. The European Commission on Human Rights found that Britain had violated torture provisions in Article 3, citing evidence of "intense suffering and physical injury which on occasion was substantial" (European Human Rights Report 1979:81; Taylor 1980).[141]

The EPA introduced the Supergrass system,[142] which by relaxing evidentiary and hearsay rules, made accomplice evidence acceptable. The Supergrass system was "a distinct counterinsurgency initiative...In essence, it was a method of securing the prosecution and conviction of large numbers of allegedly key members of paramilitary organizations for a huge range of offenses on the basis of evidence given by alleged accomplices in a series of mass trials" (Greer 1988:4). The EPA also instigated detention without trial, which although meant to be a judicial rather than executive system (Boyle, et al. 1975:61),[143] gave "the security forces and in particular ...the army, the power to arrest and imprison a suspect for around six months without any independent judicial consideration of the case" (Boyle, et al. 1975:63).

In response to bombings in Birmingham, Parliament passed the Prevention of Terrorism Act in one day. Introducing the Bill in the Commons, Labour Home Secretary Roy Jenkins said "these powers are draconian. In combination they are unprecedented in peacetime. I believe they are fully justified to meet the clear and present dangers" (H.C. Debs., vol. 882, 25 November 1974). The PTA gave security forces substantial new powers (Hall 1988:183), proscribed membership in paramilitary organizations, outlined the use of exclusion orders (permitting the Secretary of State to exclude suspected terrorists from either Great Britain or Northern Ireland, sections 3-9), and extended the powers of the police to arrest and detain suspects.[144] The PTA is the jewel in the crown of counter-insurgency legal strategy. The executive nature of exclusion orders (to which there is no appeal through the courts, only representations through the Secretary of State) made the law beyond reproach. Exclusion can be employed when there is insufficient evidence to bring a charge for terrorist offenses, although "criminal charges are preferable...for imprisonment is a more effective method of prevention" (Walker 1992:87). Exclusion creates a system of internal political exile (Hall 1988:165-170), and the predominant use of Part III of the Act — to search, arrest and detain — has contributed to the intelligence-gathering power of the police and the Army (Bunyan 1976:55).

Law, in Northern Ireland, is dominated by security considerations. Derrida writes, "...today the police are no longer content to enforce the law and thus to conserve it; they invent it...they intervene whenever the legal situation isn't clear to guarantee security" (1992:42-43). The power relations between security policies and the law are reversed. "The law appears only to have imposed certain minor procedural requirements on security policies which had been decided by the government and the security forces in advance" (Boyle, et al. 1975:52). Because the executive demands that 'terrorism' be controlled, and because normal legal controls on the exercise of powers of arrest and search (Boyle, et al. 1975:41), the security forces are to some degree "allowed to dictate its terms to the criminal justice process..." (Walker 1992:44).[145]

The lack of accountability for suspects killed under disputed circumstances suggests "that security forces have...been granted the power to decide the guilt or innocence of suspected Republican activists without recourse to the courts" (Jennings 1988:105; see also Asmal 1985). This alleged 'shoot-to-kill' policy against suspected PIRA members results from "the security forces...operating in conditions with which the ordinary law was not designed to cope, and in regard to which there are no legal precedents" (Lord Lowry C.J., R.v. McNaughton (1975) N.I. 203 at 208). Because the criminal law fails to indicate when deadly force may be used, it may be used whenever the security forces deem it appropriate to do so. The flexibility of legislation and jurisprudence allows the domination of law by security policy. When the security forces are made to respond to political pressure then the result is "law enforcement in terms of the most recent atrocity" (Evelegh 1978:93). Nobody is happy with this situation, least of all the Army.

This is not to suggest that the Army is totally outside the law. Theoretically, soldiers must always be subject to law. According to an Army lawyer: "...the fact of the matter is that you can't use your troops unless you subject them to the law. The moment you start doing anything else you are producing military anarchy. You must have some regime otherwise it would just be a case of bang, bang, bang--hooray, hooray let's go out and shoot people" (cited in Hammil 1985:49). In a war situation, however, it is difficult to judge soldiers by civilian standards: "the principle of law which entitled a person to use force in self-defence was essentially a domestic or civilian concept. It implied an unwillingness to act aggressively, giving the initiative to the other side and producing a mentality of defense against an assailant rather than an attack on an enemy" (Hamill 1985:167).

When reasonable force is exceeded, soldiers are tried according to common law in a criminal rather than military court. An Army lawyer stated that: "We had as great an interest as anyone in seeing unruly, licentious soldiery being punished. In order to keep discipline every Army has to keep that an objective..." (cited in Hamill 1985:165). Under British common law, agents of the state have the same powers and duties as ordinary citizens. Section 3 of the Criminal Law Act of 1967 allows private citizens and soldiers to use reasonable force in the prevention of a crime. "Soldiers, though they are armed with lethal weapons and controlled by military law are merely a collection of private individuals in the eyes of the common law" (Townshend 1986:19; see also Dicey 1959, ch. VIII). According to one Army lawyer, "To have to justify any violent act a soldier might have to take by use of Section 3 of the Criminal law Act 1967 is ludicrous! The whole basis of Section 3 is that you are talking about a private citizen...Now this did not even begin to meet the situation where the state, as a matter of state policy, decided to use on of the state's arms — the Army — to preserve and maintain law and order" (cited in Hammil 1985:49).

In previous counterinsurgencies declaring a State of Emergency exempted the military from civil jurisdiction, and gave them some measure of freedom to use force. In Northern Ireland such a declaration was politically unacceptable: "All this would have meant the authorities accepting that we were fighting a war, which was never accepted — never!" (Army lawyer cited in Hamill 1985:166).[146] The problem of how to justify the use of deadly force without declaring a war led to the proposal a "hybrid force, matching the hybrid legal state of emergency" (Townshend 1986:25).[147]

permanent emergency

Northern Ireland has been described as a "garrison state" in which emergency laws have been used as a solution to political violence (Hillyard 1988:192).[148] The utility of emergency legislation as a means of controlling disorder has gone mostly unquestioned. Even the Gardiner Committee held to the basic assumption that "temporary powers are necessary to combat sustained terrorist violence" (Review of the Operation of the Northern Ireland (Emergency Provisions) Act 1978 (Cmnd. 9222, 1984)). Critics of emergency legislation have argued that special laws should be eschewed since they "breach the limiting principle of minimum derogation from regular laws" (Hogan and Walker 1989:29). Ordinary law should thus be used in preference to emergency legislation (Hogan and Walker 1989:16; Walker 1992).[149] The recent history of special powers shows that "the legislation has attracted a worrying degree of de facto entrenchment" (Hogan and Walker 1989:29).[150] In other words, temporary emergency legislation is not being phased out, but normalized. Such a state of emergency may very well be a general condition of nation-states: "...what does it mean to define such a situation...as chaotic, given that the chaos is everyday, not a deviation from the norm...?" (Taussig 1992:17).

That political violence is largely generated by the iniquities of special powers laws has been suggested by critics of the law (Hillyard 1983:46; Rolston 1991), official reports (the Baker Report (1978), para. 440, the Philips Report (1986), para. 41, 42), as well as the counter-insurgency literature (Townshend 1986). Nevertheless, the law continues as the primary tool for containing violence: the International Lawyer's Inquiry found that the failure of the British government to curb civilian killings by security forces "supports the view that a certain level of death, violence and public resentment is officially regarded as acceptable, on condition that it is primarily confined to one section of the community in Northern Ireland" (Asmal 1985:125).

The European Convention on Human Rights (Article 15) permits member states to derogate from rights protected under the convention "in time of war or other public emergency threatening the life of the nation." Certain rights, however, are non-derogable and must be respected regardless of national security considerations (see Hadden, et al. 1988:21).[151] The United Kingdom justified its derogation on the following grounds: "the existence of organized terrorism and violent civil disturbance constituting a public emergency in Northern Ireland and the exercise therein of certain emergency powers" (Note Verbale, from the Government of the United Kingdom to the Secretary General of the Council of Europe, 18 December 1978. See Asmal 1985:69). Until 1984, political violence had been formally treated as a public emergency, and excluded from the Convention's protections due to a 'public emergency affecting the life of the nation' (human rights are also protected during 'internal strife', see Meron 1987). There is, however, a territorial scope to the definition of a public emergency (Oraa 1992:28-9). If Northern Ireland is viewed constitutionally as an integral part of the United Kingdom, "it is very hard to see that the situation really threatens the life of the whole nation; the reality seems to be that for the purposes of article 15 of the European Convention, the 'whole nation' is simply Northern Ireland" (Chowdhury 1989:25 fn., cf. Higgins, supra note 14, at 302 in (1978) International Legal Materials (ILM) para. 205:707).

In October of 1984, despite the fact that conditions on-the-ground had not changed, all derogations were withdrawn. This signified full compliance with the European Convention: "the United Kingdom withdrew all its derogations under the Convention on the (disputable ground) that none of the current provisions of the Northern Ireland (Emergency Provisions) Act and related legislation infringe the Convention" (Hadden, et al. 1988:22). By withdrawing the derogations, the state claimed two things about the status of Northern Ireland. First, Northern Ireland is no longer to be formally designated under international instruments of law as a territory in which a 'public' emergency exists. As the International Lawyer's Inquiry pointed out, "the Government's official position thus ruled out 'lawful acts of war' as a possible justification for killings by the security forces" (Asmal 1985:69). Second, the withdrawal of derogations signals that the level and nature of violence in Northern Ireland was 'normal.' As Chowdhury points out, "a de facto state of emergency very often transforms itself into another equally serious aberration known as a permanent state of emergency" (1989:45).

Legal imperialism, or law backed by force, has been displaced by a normalized form of martial law, or pure force made legal. Protections normally offered under the 'rule of law' are absent. Protections offered by international laws of war are useless. 'Counter-terrorist' violence can be justified as self-defense according to the United Nations Charter. States are now the 'victims' of terrorist violence. Permanent emergency clearly requires a reanalysis of norms within 'ordinary law' and, in answer to Taussig's question, an acceptance of the banality of chaos. Arguing for a realistic appraisal of the unintended social consequences of legal functions is not meant to suggest that any active malevolent intent guides the formation and application of security legislation or laws of war. Rather, the contribution of law to a culture of violence, or to a permanent social condition of emergency, is an unintended consequence of the system itself. Neither is it necessary to postulate the existence of an evil genie to explain how the system reproduces itself, nor is it necessary to ask the question, *cui bono*? It may be that *no one* benefits economically from maintaining a permanent state of emergency in Northern Ireland, except perhaps British defence contractors, war correspondents, the Europa Hotel's interior decorators, and provincial glaziers — a far-fetched conspiracy.

The express intent of the British government is not to eradicate violence, but to contain it:

Given the prevailing government and paramilitary strategies, the expectation is that political violence will be contained rather than eradicated (the U.K. government professes the latter aim. See H.C. Debs, vol. 70, col. 575, 20 December 1974, Mr. Hurd)....thus,...the conflict may be sustained almost indefinitely at the present limited level, just sufficiently to maintain an abnormal and disturbed society" (Hogan and Walker 1989:171).

Law, because it cannot ever eradicate terrorism, can only perpetuate the conditions of emergency. The exceptional measures themselves seem to create the problems (Hillyard 1988:191). Law's "interest in a monopoly of violence" (Benjamin 1921:281) means that in order to sustain the rule of law, the chaos of terrorism must be held at bay, even if the establishment of political order reproduces the disorder it seeks to alleviate.

Baudrillard asks, "What kind of state would be capable of dissuading and annihilating all terrorism...? (1990:22). The fact is, no state can maintain a total monopoly on the use of violence. Terrorists make it their business to steal, invent or hi-jack the means of production of violence. While terrorists may be *against the state* (as Pierre Clastres would phrase it), they are not *anti-structure*. Like states, cultures and societies beyond the pale of metropolitan legal authority seek to order and delimit expressions of violence. Like states, they often develop particular, professional classes of people whose task it is to manage the application of violence. Not just anyone, for example, can join the Provisionals; recruits are subjected to a rigorous vetting procedure. Political motivation is a pre-requisite for joining. In the words of Dominic McGlinchey (assassinated by Loyalists in Feb., 1994), "they don't know anything about Mandela but they see the Brits in their fields and they don't like it." Although disliking the British Army is not enough to sustain the war, nineteen year old boys are more easily motivated by adventure than politics. New recruits, frustrated by the prospect of weapons classes on video and years of mundane scouting work, preferred to join the INLA instead, which promised active military operations; "a haemorrhage of trained men was only prevented when, in 1983, a former Belfast operations officer was kneecapped for defecting" to the INLA (Clarke 1987:227). Apparently, not even a well-established paramilitary organization can maintain its own monopoly on violence!

The Provisional Irish Republican Army maintains a system of normative codes which frame and organize the practice of violence. Kneecapping is not simply a grotesque, savage paramilitary praxis, but also serves and is recognized as a sanction within a subaltern legal system (whether or not that legal system is recognized, in turn, as legitimate is another issue).[152]

Similarly, for the British Army, normative codes, unstated assumptions, and implicit understandings, are nested within more formal structures. The independent legal orders and informal combatant codes within and between groups show that military organizations tend to restrict or limit the deployment of violence. War machines self-regulate. The laws of war are the highest, formal expression of this tendency, as it has manifested for the past few hundred years in western European nation-states.

Externally imposed legal codes such as the international humanitarian law of armed conflict are redundant for the control of conflict in Northern Ireland. This is true for two related reasons. First, the laws of war represent and respond to *political pressures* when in fact the issues and actors in the situation are *military*. The laws of war, partially because they legislate what no state practices, and partially because they carry no sanctions, can be seen as a political device. Not everyone is equally innocent before the law; practitioners of revolutionary violence are handicapped. This political bias in the law against paramilitaries, while completely explicable and hardly surprising, does not reflect the reality of the *military* situation where the combatants are more or less evenly matched. As Republicans say, "God made us Catholics, but the Armalite made us equal."

Second, despite the attempts to bring guerrilla warfare within the ambit of the Protocols, jurisprudence lags behind practice. Law not only *normalizes* violence by limiting it, but creates a map of norms for the practice of armed conflict. This map is never quite exact. Because law is continually distilled from custom, reinstitutionalized in legal institutions and negotiated by practitioners, law is always out of phase with social life (Bohannon 1965:37; Falk Moore 1978:39; Kantorowicz 1958). "The discrepancy between the ideal of law and its realization, between the orthodox version and the practice of actual life" (Malinowski 1932:107) exposes the ideological project of law and the contradictions inherent in the legal system. The 1977 Protocols, for example, "reflect a common human trait of making laws for the conduct of war during times of peace and thereby wandering into unreality" (Rowe 1987:150). As Draper points out "the essential balance in the Law of War has probably swung too far in the direction of humanitarianism. Insufficient attention has been paid to the nature of warfare and what commanders are trained to do" (cited in Rowe 1987:151). Lest we forget: armies kill. But remember the most important point: armies also regulate *how* they kill.

Chapter Five:

Systems Intelligence/Intelligent Systems

*"War is beautiful because it establishes man's
dominion over the subjugated machinery by
means of gas masks, terrifying megaphones,
flame throwers, and small tanks. War is
beautiful because it initiates the dreamt-of
metalization of the human body..."*
—Marinetti, Futurist Manifesto (1909)

War has become a permanent condition (Durrant 1978), and Northern Ireland is a permanent state of emergency. "After Belfast....what they lived through was not the old state of siege, but an aimless and *permanent state of emergency*. To survive in the city one had to stay informed daily, by radio, about the strategic situation of one's own neighborhood; everyone transformed his car into an assault vehicle..." (Virilio 1986:120). Keeping informed about snipers on the rooftops and bomb blasts in the neighborhood means collecting information from every available source. A commanding officer of the Royal Green Jackets described how "on riotous evenings in the Upper Falls I found it necessary to move around the area with a small transistor radio tuned to the BBC in my flak-jacket pocket, so that I would have some warning of a change of mood in the crowds that might result from any particular news item" (cited in Arthur 1987:214).

BBC reports on real-time events filter out the static of the local emergency. TV now sometimes performs the role previously reserved for tactical military intelligence. If the complexity and quantity of raw information threatens to overwhelm even the BBC's best reporters, imagine what it does to soldiers on-the-ground. No intelligence system is sophisticated enough to catch and process all relevant information, but modern C^3I systems often provide more information than commanders can actually process, necessitating "information discipline" (Newell 1991:130). Filtering information has become increasingly complicated as warfighting has shifted from 'real' territorial geo-spatial coordinates to immaterial perceptual fields (Der Derian 1993). Although postmodern war is often represented as a simulated reality of 'real-world scenario' war-games fought by cyborgs, computer networks and cybernetic devices, *informational* characteristics predominate (Arquilla and Ronfeldt 1993). The military conditions on the streets of Belfast and the fields of South Armagh suggest that a postmodern 'infowar' is occurring, albeit at a very low level. Many of the evanescent, temporal and representational aspects of cyberwar are present in low-intensity conflicts (LIC),[153] such as Northern Ireland.[154]

Superficially, LIC has little in common with cyberwar. Whether cyberwar can only occur in high intensity conditions, as in the Gulf War, or whether low-intensity conflict is governed by entirely different operational principles, remains unanswered. The progressive expansion of global war space — geographical locations in which weapons may be feasibly deployed — through the development of satellite technologies, electronic communication networks, and tactical weapons capable of functioning in Arctic, jungle, desert, sub-aquatic, and non-terrestrial environments, has made humans rather inconsequential, their unmediated methods of data collection and analysis obsolete. Autonomous weapons systems are radically changing human participation in warfighting. The 'smart' weapons used during the proto-cyberwar in Vietnam, and during the full-blown hyperwar of Iraq, were still guided by pilots through an electronic data up-link.

Future wars may be fought by 'brilliant' cruise missiles requiring no human guidance at all post-launch (see Arnett 1992). Despite its pathetic test-demonstration in the Gulf, the US Navy's Tomahawk may foreshadow this trend. Already, computerized defence networks such as the US Air Force's Airborne Warning and Control System (AWACS), and the Joint Surveillance and Target-Attack Radar System (JSTARS), and the US Navy's Aegis system are protecting human operators from information overload. Sun Tzu wrote that 'speed is the essence of war', but when computers set the pace, the velocity of war may exceed the capacity of human participants to filter the information, or to make tactical combat decisions.

Cyberwar, and the cyborgs necessary to fight it, are the latest evolutionary phase of the rationalization of slaughter (Pick 1993). Mechanization of war — the use of machines for killing— long predates computers, the introduction of which merely accelerated the total 'metalization' of the human body, and increased the feasibility of remote killing. According to Manuel DeLanda in War in the Age of Intelligent Machines (1991), armies not only utilize machines to distribute death, but are themselves predatory machines. Armies, as predatory machines, are evolving within the machinic phylum towards autonomy from human interference. This internal dynamic of war systems is manifest at the macro level of the military-industrial complex as a self-sustained feedback loop in the arms race. The war economy, capable of initiating the "onset of processes of self-organization" (1991:49), is autonomous of political controls (Virilio and Lotringer 1983). At a micro-level within war-machines, human beings are being displaced by intelligent machines (Bracken 1985).

Artificial intelligence establishes the preconditions for war fought without human intervention. With the introduction of artificial intelligence to military computer networks, "intelligence 'migrated' from human bodies to become incarnated in physical contraptions..." (DeLanda 1991:10). The penultimate expression of autonomy of the predatory machines may be the absence of human beings in combat, which will become increasingly possible as artificial intelligence components of the defence networks develop the capacity for understanding natural language.

Meanwhile, despite the fact that enormous satellites are humming in the deep black, the vast majority of recent wars have been fought against enemies who do not have computers. Since 1945, seventy-five percent of all wars have been waged by political entities that are not states, with armed forces which were not conventional armies (Van Creveld 1992:62). Britain, with its vast imperial experience, assessed the threat accurately long before most other Western states. Brigadier Frank Kitson predicted "a further swing towards the lower end of the operational spectrum" (1971:199). The era of decolonization, and low-intensity conflict, has been characterized by the use of low-level technology, and guerrilla tactics against internal or insurgent enemies: what the SAS call "wet operations." The British Army is the internationally acknowledged experts in the counter-insurgency, and counter-terrorist, industry. While conventional land warfare has been the modus operandi for most Western armies, the British Army has been slogging around in muddy colonies for centuries (Mockaitis 1990:173, 147).

While NATO war games are removing human beings from both actual combat and combat decision-making, counter-insurgency operations are increasingly relying on the input of human-beings, especially for intelligence needs:[155]

> covert Intelligence from agents still has its uses, particularly in regional conflicts where the level of technology involved is relatively low such as in Afghanistan, Central America and the Near East, but between nuclear alliances such as NATO and the Warsaw Pact high technology methods produce better and more reliable results (Gudgin 1989:119)

Global positioning systems and cruise missiles won't pay for your ammunition in Kurdistan. Low-intensity conflict requires human-generated intelligence, local knowledge, and mission oriented tactics (Vetschera 1993). Atavistic modes of intelligence collection — espionage, infiltration — take precedence over more sophisticated techniques in these conditions. Thus, an interesting inversion occurs: as the technological sophistication of the enemy declines, reliance on intelligence derived from human sources (HUMINT) increases.[156]

Warfighting based on long-range missiles derives only marginal usage from human generated intelligence, relying much more on panoptic satellites in the 'deep black' (Campbell 1993). Technostrategic panopticism, which takes the form of electronic intelligence (ELINT), radar intelligence (RADINT), and other forms of technical intelligence (TECHINT), constitutes a new crypto-military surveillance regime, structuring and permeating international relations between states in the postmodern era (Der Derian 1993:31). Technological, remote information sources include the interception of communications (COMINT), intelligence derived from signals (SIGINT), interception of images via camera (IMINT), and telemetry intelligence (TELINT).

HUMINT, lacking the ubiquity and panoptic surveillance power of technical intelligence systems, is easily dismissed as an unsophisticated informational grout, filling the gaps between the chunks of information provided by technical apparatus.[157] Admiral Stansfield Turner, former director of the CIA, said:

> What espionage people have not accepted is that human espionage has become a complement to technical systems. Espionage either reaches out into voids where technical systems cannot probe or double-checks the results of technical collection. In short, human intelligence today is employed to do what technical systems cannot do (cited in Burrows 1986:v).

When low-intensity conflict reverts to prior forms, most cybertech systems go blind. War in the post-modern era is not necessarily post-modern war: the war of the space age resembles, in fact, the war of antiquity (Mockaitis 1990:9). The methods of the guerrillas are almost impervious to the technical intelligence apparatus of the state and, as a result, the less sophisticated the army, the better equipped it is to defeat 'insurgency' (Beckett and Pimlott 1985:10). The British Army, predisposed to ad hoc, improvised operations, has exploited a natural advantage in the counter-insurgency game. Use of local, human sources is the British Army's forte;[158] Dyak headhunters guided patrols through the humid, green jungles of Malaya, Bedouin chiefs showed TE Lawrence how to ride a camel. British military intelligence[159] has traditionally been oriented toward immediate tactical needs,[160] decentralized administration,[161] and the exploitation of local knowledge in the field, making it very well suited to the predominant war tasks over the past fifty years.

LIC, because it is about what human beings *think*, is impervious to counter-measures based on mere firepower. Tactical HUMINT is central to LIC: "technology is a significant contributor in finding guerrillas. However, by the very nature of LIC, where the struggle is between the government and guerrilla for the loyalty of the people, HUMINT, particularly low-level tactical HUMINT, is potentially the most important intelligence weapon in our arsenal" (Stewart 1988:22). The war-machine cannot produce a military meta-theory to cope with low-intensity conflicts: hearts and minds and bellies don't respond to firepower. LIC "because it is unpredictable" is therefore extremely resistant to operational modeling (unattributed informant statement 1994; the opacity of LIC to operational analysis is quickly changing, see for example Dockery and Woodcock 1993), and to the application of certain kinds of technological counter-measures. "We have the means to develop intelligence-electronic warfare (IEW) systems for the conventional nuclear battlefield, but USIEW systems generally lack the flexibility required by LIC" (Stewart 1988:22). Electronic instruments for identifying, tracking and striking targets effective in uncluttered environments are vulnerable to jamming, spoofing, jarking and overloading in the urban battlespace (Van Creveld 1992).

Modern, high-technology weapons designed to fight each other are useless against rebels with machetes. An adage which all competent soldiers should memorize is: the very best weapon for fighting a guerrilla is a knife, the worst is a bomber; the second best is a rifle, the second worse is artillery (Hilsman 1967:433). Failure to apply this very simple formula for the use of appropriate technology was a central factor in the US Army's defeat in Vietnam (Cinncinatus 1981; Gibson 1986). Despite the US Army's 'swamp fox' tradition of guerrilla warfare, the Army persisted with its inappropriate "technowar" paradigm (Gibson 1986, Shafer 1988), refusing to adopt the "primitive" tactics of the Vietcong, or to 'know the enemy' (see Chapter Two).[162]

mastery of the neo-nomadic

The cyberwar paradigm, and the astounding technical capabilities of weapons systems, seems to bear little relation to current conditions. The primitive aspects of low-intensity conflict demand appropriately raw, anthropocentric, counter-measures; a very different tableau from the weapons systems deployed over Baghdad. While a comparison might seem far-fetched, low-intensity conflict and cyberwar, despite their formal differences, both revolve around information and speed. Arquilla and Ronfeldt (1993) argue that success in "warfare is no longer primarily a function of who puts the most capital, labor, and technology on the battlefield, but of who has the best information about the battlefield" (141). This proposition regarding the centrality of information holds true under all warfighting conditions.

The reorganization and expansion of instantaneous, global information systems (for example, INTERNET) have made information a strategic resource, and imply that all future wars will be about *knowledge* rather than firepower or position. Arquilla and Ronfeldt see two likely types of future war: cyberwar and netwar. Netwars target information and communications systems at a social and electronic level, seeking to damage, disrupt or modify what a target population knows, or thinks it knows, about its own position, capability and motivations.[163] While psychological operations have always been an aspect of war, netwars interfere with military epistemologies, not just with hardware, but with *knowledge systems*. Neither the armed force component of netwar, nor the participation of formal, hierarchical, military groups, are prerequisites for netwar; rather, netwar may occur between governments and non-state actors (guerrillas, terrorists, bandits, drug cartels, black market weapons dealers, advocacy groups, etc.), organized into transnational networks and coalitions.[164]

Cyberwar refers to explicit military operations conducted according to information-related principles, which may occur at any intensity. Disrupting, blinding, jamming, overloading, infecting, and destroying information and communications systems turns the balance of information and knowledge in one's favor. Knowledge systems also include military culture, "on which an adversary relies in order to know itself: who it is, where it is, what it can do when, why it is fighting..." (1993:146). Although cyberwar may involve complex technologies for C^3I, intelligence, 'smart weapons', and identification friend-or-foe, organizational and psychological factors are equally as important as the technical. Contrary to the view of cyberwar as a systems

236

of automated battlefields, autonomous computers and 'brilliant weapons' (Arnett 1992, also the US Navy's concept of 'hyperwar'), Arquilla and Ronfeldt maintain that cyberwar entails an integration and adaptation of man-machine systems, which *as a form* may be waged not only in low-intensity conditions, but with low-technology as well. By emphasizing speed, information, and organization, rather than technology or combat intensity, as the shared features of cyberwar and netwar, their model provides one possible resolution of the conundrum of how capabilities can be brought into relationship with conditions.

The formal, organizational mode which the war assumes is central: cyberwar entails the re-organization of military systems into interconnected, decentralized networks rather than hierarchies. Mongol military praxis is an early ancestor of hetrarchical networks being explored for cyberwar. During the twelfth and thirteenth centuries, the Mongols defeated the armies of the empire of Khwarizm, and the Polish-Prussian forces at Liegnitz, and almost every other army in Christendom. Although drastically inferior in numbers, their strategy of knowing the exact position of the enemy, while keeping their own location secret gave them an information advantage. They disrupted enemy communications by waylaying messengers, provided real-time intelligence through horse-mounted scouts ('arrow riders'), invented a decentralized C^3I structure utilizing clan and lineage structures as battlefield organizational maps, and conducted proto-psychological operations (PSYOPS) by cultivating the enemies belief that they were the dark forces of Gog and Magog, whose presence heralded the 'end of time'. They renamed themselves 'Tartars', after the biblical nether world.[165]

In exploring the Mongols, Arquilla and Ronfeldt are not simply developing elegant, pretty theories of warfighting, they are advocating the incorporation and adoption of the tactics, devices, equipment, and forms of the "wide-spread multi-organizational networks" of non-state "proliferators" as a counter-measure against a future-threat (1993:158). According to Brigadier Richard Simpkin, "...established armed forces need to do more than just master high-intensity manoeuvre warfare between large forces with baroque equipment. They have to go one step further and structure, equip and train themselves to employ the techniques of revolutionary warfare — to beat the opposition at their own game on their own ground" (1985:320). The armies of the state are obliged to adopt, adapt, and incorporate the information and network system of the nomad war machine: "success in warfare is gained by carefully accommodating ourselves to the enemy's purpose" (Sun Tzu 1983:71).[166]

It is not just military theorists who are advocating the incorporation of the modus operandi of the enemy, or using the Mongols as a shibboleth of post-modern war. Delueze and Guattari in Treatise on Nomadology: The War Machine (1992) advance the thesis that nomad techniques and technologies have always been co-opted into the sedentary war machine (355; also Virilio 1986). Indeed, symbiotic transfer between states and nomads is an aspect of the evolution of the war-machine. Let us momentarily digress (although it is not really a digression), and zoom in on the concept of the nomad war-machine, and its anteriority to the state apparatus, which was touched on in Chapter One.

Although the rise and reproduction of the states is generally seen as either a function or by-product of their war-making capacity (for example, Tilly 1990), Delueze and Guattari, elaborating on the theories of Pierre Clastres (1977), identify "war in primitives societies as the surest mechanism directed against the formation of the State: war maintains the dispersal and segmentarity of groups" (1992:357). The State is against war, and war is against the state. Furthermore, sovereignty of states is limited to that which is local, and internal (360): the "outside of states cannot be reduced to 'foreign policy,' that is, to a set of relations among States." Rather, multi-national organizations, ecumenical formations such as Islam, industrial complexes, terrorist networks, and tribal Chiefdoms are irreducible to the state, and conduct a very different type of warfare from that of the State's military structure — diffuse, polymorphous, and non-hierarchical.

If war has no functional, organic dependence on the state, from where does it arise? The war machine is the invention of the nomad. Certain aspects of pastoral nomadism facilitated warfare. Reverence of fighting skill in nomadic cultures, mobility of the pastoral economy, and the organization and movement of people and animals all "gave the steppe nomads an intuitive appreciation for logistics and maneuver that was almost entirely absent in contemporary sedentary militaries" (Barfield 1994:162). Horse riding allowed rapid movement across vast distances and, combined with the compound bow, gave flexibility and strength in battle (Barfield 1994:163). "If the nomads formed the war machine, it was by inventing absolute speed" (Deleuze and Guattari 1992:186), penetrating the smooth, open spaces of the desert as a mounted force-trajectory. The sedentary state, on the other hand, has always sought to regulate speed, to enclose and striate the open spaces of the nomad with gates, levies, roads, fortresses. The weapons-systems of the nomads relies on the speed-vector of the horse, the ballistic projectile as weapon (1992:394-403, see also Virilio 1986), and metallurgical appropriation.

War is not the object of the nomad war-machine. Raiding was more common than outright war, and such raids were controlled by rules defining the appropriate limit of stock theft and prohibiting wanton killing (Barfield 1994:163). Wars were often conducted with the goal of political unification, and "were designed to bring rivals under the political hegemony of a dominant tribe, not to dispossess an enemy people of their lands or animals" (Barfield 1994:164). Annihilation and capitulation result not from the interaction of nomadic groups with one another, but through the collision of nomads with States and cities: "from then on, the war machine has as its enemy the State, the city,...and adopts as its objective their annihilation. It is at this point that the war machine becomes war: annihilate the forces of the State, destroy the State-form. The Attila, or Genghis Khan, adventure clearly demonstrates this progression..." (Deleuze and Guattari 1992:417).

Nomads are not self-sufficient, but must interact with the outside world, whether this interaction is a symbiotic exchange of camels and medjool dates between nomadic Bedouin and sedentary Berber, or the "predatory warfare" of Malay pirates in the Andaman Sea. The structure of the interactions between sedentary and nomadic groups determines political development. "It is this problem of organizing relations with the outside world, and not internal affairs, that drove the development of political organizations capable of organizing and sustain a war machine among steppe nomads" (Barfield 1994:166).[167] In other words, hierarchical political institutions within nomadic societies do not develop as the result of internal social dynamic, but only as a result of interaction with external state societies (Irons, cited in Barfield 1994:166). This would seem to confirm the general thesis of this project that insurgents/paramilitaries and states nurture each other's political evolution and military prowess through interaction.

The evolution of war machines depends both on innovation and incorporation of enemy tactics and technologies. The Mongol army incorporated captured Chinese and Muslim military engineers, who provide the Mongols with siege engines such as catapults for capturing fortified cities (Barfield 1994:174). States appropriate and co-opt the nomad paradigm insofar as they harness speed: this is the basis of the evolution of the machinic phylum. DeLanda in War in the Age of Mechanical Reproduction (1991) elaborates on this military evolution:

> Thus the modern army which began by structuring the battlefield in a form directly opposed to the nomad paradigm, was later forced to adopt the methods of its rival under the pressure of both colonial and machine warfare (1991:13).

These principles, though based in antiquity, do not disappear. Neo-nomadic groups — terrorists, guerrillas, networks — seek to plunder the state. The speed of trajectories launched in smooth space continues to characterize modern weapons systems. "And each time there is an operation against the State — subordination, rioting, guerrilla warfare or revolution as act — it can be said that a war machine has revived, that a new nomadic potential has appeared" (Deleuze and Guattari 1992:386). The State's war machine, having appropriated, territorialized, and subordinated the nomad war machine to political aims, deploys this altered form against other states. But what happens when States use their armies against nomads? "Turning the war machine back against the nomads may constitute for the State a danger as great as that presented by nomads directing the war machine against States" (1992:419). In a literal sense, indigenous tribal peoples have been decimated recently by government troops in Bangladesh, Papua New Guinea and East Timor; over 50 million tribal people are estimated to have died in the hundred years before 1920 (Bodley 1992:37). In a more metaphoric sense, Simpkin's imperative that the British Army should "beat them at their own game" by using small groups of highly mobile special forces units shows the degree to which states continually appropriate the forms, weapons, and structures of the nomad war machine and use them against their originators.

The network is the form of the nomad, which the state imitates to survive. The Mongols, according to Arquilla and Ronfeldt, were organized as a network rather than a hierarchical institution (1993:152). The Vietcong, like the Mongols, easily defeated opponents "whose forces were designed to fight set-piece attritional battles" (152). In Northern Ireland, the genealogical networks of Republican paramilitary organizations elude the states' attempts to penetrate them.

> Perhaps a reason that military (and police) institutions have difficulty engaging in low-intensity conflicts is because they are not

meant to be fought by institutions. The lesson: institutions can be defeated by networks, and it may take networks to counter networks. The future may belong to whoever masters the network form (Arquilla and Ronfeldt 1993:152).

Defeat can only be coaxed by the imitation of the form of the guerrilla network. Arquilla and Ronfeldt's revelation that military institutions fail in low-intensity conflicts because only networks can counter networks (1993:152) not only demonstrates the degree to which the State must engage in a mimetic relationship with its opponents in order to survive, but that *victory depends on formal mastery in a Taoist sense.*

It is in low-intensity conflict that the cellular, internecine, genealogical structure of the nomad war machine emerges most clearly. In low-intensity conflict, we see the highest expression of the state's appropriation of the nomad techniques of speed, autonomy, self-organization, exchange, and anthropocentric information gathering. The State's war machine always retains an element of the nomadic, "something of 'the tribe in the desert'" (Pick 1993:259). The re-deployment of the war machine against the nomads constitutes a mimetic exchange relationship, which, in some sense, functions *against the state.* As will become clear, the self-organizing factions at the fringe of the state war-machine conspire against the state, desubordinating themselves from political control, becoming indistinguishable from the neo-nomadic guerrillas whom they nominally oppose, and expressing their autonomy from the state, taking war as their 'analytic' object, and seeking to promote the continuation of the conditions of war, the oxygen in which special forces of the military nether world thrive.[168]

The principles of the nomad war paradigm are speed, maneuverability, and acceleration. The essence of nomad war is "dromocratic" (Virilio 1986): kinetic, circulatory, and swift. As Colonel Delair said, "stasis is death" (cited in Virilio 1986:13). Genghis Khan, pounding on horseback through the steppe, with his Mongol hordes organized by clan and lineage for war, is the first example of dromocratic, nomadic warfare (Deleuze and Guattari 1992:392, DeLanda 1991), which the state will adopt, exploit and transform. The symbiosis or incorporation of nomadic principles into state war machines constitutes the *dromocratic revolution*. The introduction of speed as dromocratic technique alters the nature of war from movement of mass in space to movement of energy through time.

> Space is no longer in geography — it's in electronics. Unity is in the terminals. It's in the instantaneous time of command posts, multi-national headquarters, control towers, etc....There is a movement from geo- to chrono-politics: the distribution of territory becomes the distribution of time. The distribution of territory is outmoded, minimal (Virilio 1983:115).

Sun Tzu was the first to point out that "speed is the essence of war," and this acceleration (of which only nomadic armies were capable) was the genesis of dromologic warfighting. Speed is still central to operational doctrine. Critics of post-modern militarism use the example of the Mongols, and the truncated adage of Sun Tzu, to conceptualize speed as an element of cyberwar (Virilio 1986; Der Derian 1993), or in order to develop a concept of military evolution (Deleuze and Guattari 1992).

Military theorists, on the other hand, such as Duffy (1981) and Bellamy (1983), apparently stumbled on 'dromology' (although they don't call it that) and the Mongols from a completely different route. In addition to critical or theoretical relevance, their work seeks to trace an intellectual genealogy for 'maneuver theory'. Maneuver, attrition and revolutionary war are the three key types of land warfare conceptualized in Western military theory, fundamental to understanding how British, German, NATO, post-Soviet and US armed forces actually conduct and conceptualize war. Simpkin in <u>Race to the Swift</u> cogently reviews and contrasts the evolution of maneuver, attrition and revolutionary styles of land warfare. An adherent of attrition theory (also known as 'position theory'), "... simply seeks to achieve a shift of relative strengths in his favour by imposing on the enemy a higher casualty rate, or more broadly "attrition rate", than he himself suffers" (1985:20). This model is essentially static, two-dimensional and based on the idea that holding ground confers a tactical advantage, and that body counts can be equated with victory. Attrition warfare, which lost the US Army the Vietnam war, resembles the sedentary state war machine of Deleuze and Guattari,.

Maneuver theory, on the other hand, "regards fighting as only one way of applying military force to the attainment of a politico-economic aim — and a rather inelegant last resort at that" (1985:22), and may actually seek to forestall combat. Ground *per se* is unimportant; maneuver theory relies more on pre-emptive targeting ("decision by initial surprise") of enemy resources, the application of calculated risk, and the build-up of speed and momentum. Simpkin's articulation of 'maneuver theory' sounds very much like the dromological nomad war machine; which of course, it is. Maneuver theory represents the adaptive response of the state war-machine to a nomad paradigm. Attrition warfare, on the other hand, represents the static mass of the state's war-machine, who's highest and noblest expression is *siege*. Although Simpkin sets up an opposition between these two forms, this is for purely pedagogic purposes; when actual warfighting begins, the theories are complementary: "Manoeuvre theory represents an added dimension superimposed on attrition theory" (1985:23). Attrition, large-scale maneuver, small-scale maneuver, and revolutionary warfare share a 'continuum of application' of the same devices and principles of battle.

Do these concepts of speed, mobility and decentralization (*Auftragstaktik*) in maneuver have an intellectual biography? According to Bellamy, Sun Tzu's concepts of 'maneuver' and 'turning' in battle,[169] were probably absorbed by the Mongols under Genghis Khan: "the Mongols may have encountered intellectualized theories of war among captive Chinese, but the scope and perspicacity of their operations and strategic vision must be ascribed primarily to instinct" (Bellamy 1983:53). The trajectory of Sun Tzu's kinetic theory extends to the Russian Army, who had access to the numerous Russian translations of The Art of War, and adopted and developed what became known as 'maneuver theory' for land warfare during the Russian Civil and Russo-Polish wars (Bellamy 1990:81).

The two most significant forms of modern land warfare, manoeuvre warfare and revolutionary warfare, both derived from Sun Tzu. The one apparently passed to the Russians via Genghis Khan, to be re-expressed by Trindafillov and Tukhachevskii. And it was on Sun Tzu, Lenin and Tukhachevskii that Mao Tse Tung drew for his doctrine of revolutionary warfare (Simpkin 1985:311).

In other words, Sun Tzu produced three different branches of dromological war theory. The first was refined by the Soviets, and came to fruition in the principle of *simultaneity* in Tukhachevsii's 'deep-operation' theory of maneuver warfare (see Simpkin 1985:37, and more generally Deep Battle). The second dromological quickening was by the Chinese communists in the theory of revolutionary war.

A separate evolutionary branch carried the theory into Western European land warfare praxis. Liddel Hart (1927), fascinated by the Mongols, believed that they were an early example of blitzkrieg. Liddell Hart's concept of "deep strategic penetration" in armored maneuver warfare was influenced by the thirteenth century Mongol campaigns, and by Sherman and Forrest in the American Civil War (see Bond 1977:48 fn.). The concepts of 'direct' and 'indirect' approaches to warfare, elaborated by Liddel Hart and Fuller (1942), derive from the thought of Sun Tzu (see the discussion in Bellamy 1990:190). Furthermore, the British military intelligentsia derives strategic lessons not only from members of their own camp, such as Fuller and Liddel Hart, but also studies Mao's interpretation of Sun Tzu in Yu Chi Chan (Guerrilla Warfare), General Giap's utilization of Mao, and Soviet utilization of Sun Tzu and Mao. All of the theories are refined, incorporated, and utilized in military praxis. Needless to say, the theories also interact (see for example Bellamy's discussion of Soviet OMG and NATO's FOFA doctrine 1990:121-123).

Sun Tzu's Taoist approach to war, derived from Lao Tse and the I Ching (if we really have to probe history), produced three different strains of conceptualizing warfare as a function of speed ('tempo'), initiative, and application of force. Soviet 'deep battle', Maoist revolutionary war, and Western 'maneuver theory' all express the same principles. Maneuver warfare can be considered as the deepest integration of nomad technique by Western state armies. Revolutionary war, which Mao believed could only be undertaken by guerrillas, is an elegant distillation and formalization of nomad military praxis, and can actually be undertaken by anyone with the right training and equipment. The US Army's technowar in Vietnam exemplifies the misapplication of a strategy of attrition to a revolutionary, guerrilla war scenario.

In Northern Ireland, the British Army conducts low-intensity counter-terrorist operations according to the principles of maneuver warfare, filtered down from Sun Tzu. PIRA, meanwhile, conducts a low-intensity guerrilla war against the British Army according to doctrines derived from *identical sources*. Military intellectuals in the Provisional IRA, quite familiar with Mao's adage that 'political power comes out of the barrel of a gun', have their own tradition of studying military history, especially EOKA, the Vietcong, and the ANC. When PIRA snipes a soldier, or the SAS conducts an ambush, not only do their operations bring to bear a high degree of shared knowledge about one another's habits, tactics, potentials, and dispositions, but the theoretical ammunition suits an armalite as well as an SA-80. This intellectual genealogy demonstrates yet another way in which the counter-insurgency system functions as an exchange loop for strategic software. The interactions, both concrete and theoretical, between PIRA and the British Army also display counterinsurgency praxis as the most concise example of how the state adopts the techniques of the nomad war machine.

French post-modernists, such as Deleuze and Guattari, incorporated Virilio's theory of 'speed' in war ('dromology') to develop an anthropology of the nomad war machine (1992). James Der Derian, writing post-structuralist international relations theory, exploited Virilio's concept of 'speed' to critique the cyberwar of the Gulf in his excellent book Antidiplomacy (1992). Manuel DeLanda, drawing on Deleuze and Guattari as well as Virilio, examined the application of AI to weapons systems, and the machine phylum evolution of armies into autonomous machines (1991). Military professionals such as Simpkin (1985), and military historians such as Bellamy (1990), in promoting the co-optation of guerrilla technique, or stressing the continuity between revolutionary and maneuver forms of combat, remain unaware of the ideas connecting them to French post-modernists (and no doubt, of the loop of ideas which connects them directly to PIRA). Arquilla and Ronfeld's "Cyberwar is Coming!", written from within the RAND corporation, represent the first synthesis of post-structural and sub-contracted military theory (1993). The question which emerges is, how are these techniques and concepts manifesting on-the-ground in Northern Ireland as a war-exchange system?

no one is innocent in West Belfast

The myth of cyberwar, "a technologically generated, televisually linked, and strategically gamed form of violence that dominated the formulation as well as the representation of US policy in the Gulf" (Der Derian 1992:175), displayed on glossy recruiting posters and televised on the BBC motivated enlistments and re-popularized the Army in Britain. Infantry soldiers trained for high-intensity military operations in the Gulf and the Falklands, found internal security duties in Northern Ireland onerous.

> I'd sooner go back to the Falklands tomorrow morning than Northern Ireland. I suppose I was more likely to be killed in the Falklands because of the amount of ammunition fired, but I feel safer in a conventional war than walking about the streets of Belfast like a figure eleven target (Sergeant, 3 Para cited in Arthur 1987:250).

The bullets and bombs in Tyrone and [London]Derry did not correspond with the soldier's military imaginings of heat-seeking missiles gliding towards their targets in the empty desert sky above Baghdad. Baudrillard wrote that "If military power once again finds a theater for war, a restricted space...for war...it will again be possible to exchange war" (1990:15). Local, low-intensity conflicts are the only scenarios that provide, according to military analysts, a limited space for war (Newell 1991:149). In Ulster, violence is compressed into a very compact theater, limited by geography, technology and law. Within those confines, the war is total, deadly, and omnipresent. According to a restricted British Army training manual, "There is no safe place in Belfast. Wherever they are and whatever they are doing, all rank must be alert to the possibility of sudden attack..." (MOD

1972:13). Northern Ireland is simultaneously a low-intensity conflict, and a total war.[170]

In 1992, 10,500 regular British Army, 6,000 Ulster Defence Regiment (UDR), 1,000 Royal Air Force (RAF) and 250 Royal Navy were deployed in Northern Ireland, supported by the Royal Ulster Constabulary (RUC). British troops have been on deployed on roulemont, 'cara cara', and 'spearhead' tours in Ulster for more than twenty years; in some sense, internal security operations in Northern Ireland are the military norm, and conventional warfare the exception. Why the conflict continues, the role of the Army, and possible solutions are conundrums beyond the scope of this chapter (see Whyte 1991). Suffice to say, the British Army typically is represented in popular discourse as a restraining influence on a situation which would otherwise become a chaotic bloodbath. While this may or may not be true (Rolston 1991), the containment and compression of the war into the geostrategic confines of Ulster has been achieved not only by the presence of the British Army, but through mechanisms internal to the state.

Permanent, entrenched emergency legislation contributes to the manageability of the state of emergency, as we saw in the previous Chapter. The war is prevented from reaching mainland Britain by the power of exclusion derived from the Prevention of Terrorism Act Section 5(1), which has turned Northern Ireland into a "dumping ground for terrorists" (Evelegh 1978:88). According to a British Army major, confinement of the war to one corner of Britain results in "displacement [of terrorism] rather than prevention..." (Evelegh 1978:88). Boundaries between communities within Northern Ireland also help contain the violence. After 1969, boundaries between Republican and Loyalist enclaves were progressively rigidified in the form of 'peacelines', while political parades symbolically penetrated 'sanctuary' space, providing communication and exchange between Loyalists and Republicans (Feldman 1991:29-31). Cordons around Belfast City Center and permanent barricades push the war out of the center, up the Falls Road, and into the no-man's land of Ballymurphy and Whiterock.

Barricades are a major part of life in war zones. According to a restricted 1973 Army Training Manual: "recent events in Northern Ireland have emphasised the importance of both field defences and obstacles in urban warfare" (1973:9). Barricade types may include palisade barriers with metal pickets, and snowflake barriers using dannert coils, which "may be erected either temporarily or as a permanent part of the protection around security force bases. The traditional obstacles of barbed wire and knife rests are used in these situations" (Army Training Manual 1973:10). Barricades, obstacles, field defences, and legal restrictions provide rigid spatial boundaries to contain war.

Despite the efforts of PIRA to bring the war 'out of bounds', bomb blasts, mortar attacks, and trip wires are concentrated in North and West Belfast, South Armagh and Tyrone. From 1969-1989, almost 40% of the total casualties have been concentrated in North and West Belfast (see Table 1 below). For residents of North Down, violence scarcely penetrates everyday life, except through media images. These anomalous patterns led the Northern Ireland tourist board to accurately claim in 1993 the province had a crime rate one-fourth that of Sweden, and a murder rate one-fifteenth that of Washington, DC.

When PIRA takes the war out of its territorial boundaries through operations in Germany, Spain and the US, they violate the informal, non-articulated policy of keeping the war at home. Assassination of unarmed PIRA members (e.g., by the SAS in Gibraltar in 1989) is hardly a pragmatic approach for British security forces to take, especially when paramilitaries could be more easily arrested, and when their deaths create public scandals. A more realistic interpretation of such illogical operations is that these assassinations function as a stern symbolic rebuke to PIRA for violating the unspoken containment policy.

Constituency	1983 electorate	% of NI elec.	Fatalities	%NI fatal-ities
North Down	61,574	5.86	8	0.29
Strangford	60,232	5.73	23	0.83
South Down	66,987	6.37	81	2.93
Lagan Valley	60,009	5.71	75	2.71
Upper Bann	60,797	5.78	103	3.73
Newry & Armagh	62,837	5.98	352	12.74
Fermanagh/S. Tyrone	67,880	6.46	228	8.25
Mid Ulster	63,899	6.08	119	4.31
Foyle	67,432	6.41	275	9.95
E. Londonderry	67,362	6.40	59	2.14
North Antrim	63,254	6.02	27	0.98
East Antrim	58,863	5.60	24	0.87
South Antrim	59,321	5.64	53	1.92
Belfast North	61,128	5.81	544	19.69
Belfast West	59,750	5.68	544	19.69
Belfast South	53,694	5.11	137	4.96
Belfast East	55,581	5.28	111	4.02

Source: McKeown 1989:51.

Containing the conflict within demarcated geo-spatial boundaries means controlling the border with the Republic of Ireland, a notoriously difficult task for the Army. The border has no natural contours, penetrates fields and farms, and is often overgrown. PIRA skips to and fro between Great Britain and the Republic of Ireland. International legal jurisdictions prevent British forces from operating in the Republic, effectively ruling out hot pursuit across the border. PIRA can thus engage the Army in a 'catch me if you can' game back and forth across unmarked, green pastures. An unattributed dossier from 1972, Incursions of the I.R.A. from Bases in the Irish Republic, points out the extent of the problem. The cemetery at Dundalk was used as a training base, and remote fields in County Donegal have also been used for weapons training (unattributed informant statement 1991). Abandoned farmhouses provide ideal bases for raids into Northern Ireland, with run-backs along disused, unapproved roads. That nationalist areas such as Buncrana and Monaghan provide safe haven, and that PIRA's permanent HQ is based in Dublin (unattributed informant statement 1991) seems to confirm that conflict does have an international aspect. These raids across the frontier resemble border skirmishes during the Indian wars in North America, and in certain ways, parts of Northern Ireland resemble an Indian reservation.[171]

Northern Ireland is coming to approximate a closed system in a limited space. As the space for war contracts, the category of enemy and the information requirements of the war system expand. For a British soldier on patrol in Divis Flats, ten year old girls in their bedroom slippers holding milk bottles are the enemy; combatant and non-combatant distinctions no longer exist. Not only are combatants indistinguishable from the general population, but war as a permanent, social condition, makes everyone into a combatant. "[W]hole societies were at war, not just their armed forces" (Dandaker 1990:104; see also Howard 1983:7-35; Giddens 1987:103-15). Indeed, as Gerry Adams (paraphrasing Bobby Sands) admonishes us, everyone has their part to play in the struggle for freedom (1986). In a small space, war can be total: everyone can participate in the local defence network, regardless of whether they own a PC.

Regardless of technology, the operational principles of cyberwar/netwar — decentralization of command and control, autonomous decision-making, and exploitation of chaos — are, as Arquilla and Ronfeldt suggest, central aspects of low-intensity conflicts. Because the human factor characterizes limited, low-intensity conflict, cyberwar principles materialize at the level of human conduct, and effect the way in which human organize themselves for war. Before assessing the levels and modes in which cyberwar principles manifest, a rough overview of the operations of computerized defence networks in a high-intensity cyberwar is in order.

Computers have wide-spread application for combat operations in large-scale, technologically sophisticated wars. The 'electronic battlefield' depends on computerized information processing to maintain "an extensive network of near instantaneous communications between central authorities and lower echelons as well as between different sub-units within military organizations" (Dandaker 1990:92). Computers allow for decentralization of information processing in combat situations, obviating the need for any central headquarters:

> As information technology and communications improve, and as headquarters become more automated and more vulnerable, it is possible to foresee the collation of Intelligence and the filtration of information being done elsewhere and fed in finished form to a smaller and dispersed headquarters in the field. The screen of the computer's visual display unit, however hypnotic the effect on the viewer, merely displays one page out of what is, in effect, an electronic filing cabinet... (Gudgin 1989:134).

Computer networks used in electronic battlefields rely on a dispersed pattern of decision-making. However, remote decision making by local commanders, whether computers or humans, also has a place in the history of traditional warfighting. Military command, control, communication, and information systems, C^3I,[172] act to control the conduct of war and dissipate the chaos of battle ('friction') at an operational level. Computers merely multiply a capacity or predisposition already present within the system.

The most common means of organizing and controlling warfare is through advance planning, routinization, stabilization, centralization, and direct orders. The Prussian Army, the North Korean Army, and Tamerlane provide examples of classic use of *Befehlstaktik* within a C^3I system (see Bellamy 1990:201). Floppy, decentralized command and control structures can also be used to control the conduct and chaos of military operations, although this form is less common. Decentralization in C^{3}I lowers decision-making thresholds by granting local responsibility to field commanders. Uncertainty (or 'friction') is diffused throughout the hierarchy, so that even if the entire general staff is blown up, the war can continue. Decentralization, although it increases adaptability, may result in organizational fragmentation ("I thought you were in charge!"). *Auftragstaktik* ('directive control', 'mission tactics' or 'mission-type control') are tactical devices in command and control systems designed to cope with fragmentation. Commanders establish the objectives, while subordinates have discretion to select the means of implementation. The intention of orders takes primacy over exact specifications. *Auftragstaktik* effectiveness depends on inter-organizational trust, and the existence of a common organizational culture (Offerdal and Jacobsen 1993:211-212). Chaos is not eliminated from the conduct of war, but harnessed.[173] Although decentralized schemes are often more successful during actual battle than rigid, authoritarian command structures, they are uncommon because they provoke anxiety in top brass. Although local initiative is increased, the overall flow of information is decreased. From the top of the hierarchy, decentralized schemes appear chaotic (Newell 1991:19, 131).

Chaotic hetrarchies and decentralization may be revolutionary approaches for high-intensity computerized cyber war, but *Auftragstaktik* has always been a principle of low-intensity conflicts, especially internal security and colonial policing. Air Vice-Marshal R.E. Peirse, commander in Palestine until 1936, recognized that "the only policy that can lead to success in this kind of warfare is decentralisation" (cited in Mockaitis 1990:158). In combat situations, local junior commanders were given substantial responsibility and empowered to cope with local situations; they had a much better idea of who was tossing the bricks. Control was relinquished at the center and dispersed to the bottom of the officer corps; this dispersal of command encouraged autonomy. Intelligence information on enemy tactics and movements flowed both up and down the chain of command (Mockaitis 1990:158, 166). Decentralization, informality and non-doctrinal chains of persuasion, characteristic of the British Army, have allowed it to function as an 'untidy' yet harmonious coalition of goods and services (Charters 1989:178).

Decentralization and *Auftragstaktik* in C^{3}I schemes are not the evolutionary apex for predatory machines. The next developmental stage in the evolutionary sequence is autonomous self-organization of the network, such that the chaos of combat actually generates order in the computerized defence networks. The application of chaos theory to war modeling (Newell 1991:7) has enriched the complexity of theoretical models of operational and tactical warfighting (Simpkin 1985; Newell 1991; DeLanda 1992; Dockery and Woodcock 1993). Computer networks for use in the electronic battlefield during high-intensity conflict, ARPAnet and MILnet, were designed by the RAND Corporation on this principle (the UK has its own system called 'Wavell'). According to DeLanda, "the development of a network capable of withstanding the pressures of war involved a scheme of control that would allow the network to self-organize" (1991:120). Non-sequential approaches to problem-solving allow "predatory machines to operate under increasingly complex circumstances" (1991:166) in an electronic battlespace. The dispersal of the control mechanism in the "thinking" apparatus of predatory machines constitutes a new variety of command and control system: what DeLanda called "pandemonium". There is no hierarchical structure of decision-making in this type of dispersed network of control — disorder is 'planned' into the system. Because a hetrarchy of self-organizing nodes (or 'demons') capture control within the program when invoked into action, the data base (or patterns in it) controls the flow of computation (1991:164). "The migration of control from programs to data permits external events to trigger internal processes" (1991:156). Chaos allows the predatory machines to self-actualize as autonomous military systems, devoid of human intervention or control.[174]

A hetrarchical network like ARPAnet has only a limited application in low-intensity conflict, although computers such as the RUC's *Vengeful* are used by patrols and PVCP's to cross-reference registration, ownership and make-model-serial numbers of vehicles at checkpoints. Computers used operationally in LIC's often seem most useful for mundane counter-terrorist tasks. Decentralization of command and control, autonomous decision-making, and exploitation of chaos are nevertheless important aspects of low-intensity conflicts. But where do they appear? Self-organizing chaos manifests not in computer systems, but in military intelligence (MI) systems. MI in LIC perfectly mimics the operations of computerized defence networks in a high-intensity cyberwar, although it does not duplicate their function. Thus, the chaotic decentralization forms associated with cyberwar manifest in microcosm during limited, low-intensity conflict in the realm of intelligence operations conducted primarily by special forces.

Kitson (1971) described the basic intelligence needs during a counter-insurgency as the transformation of background information (which is strategic and general) to contact information (which is tactical and specific). Information may be collected in an overt fashion through public sources such as Republican newspapers such as AP/RN, and cordon and search operations; confidentially through informers and telephone free-phone solicitation; and in a clandestine fashion through running agents and spies (Jeffery 1987:129). Background or strategic intelligence, the center of operational planning, is typically based on open sources. These sources are neither 'hot' nor immediate, but emphasize the complexity of the issues, advocate operational caution and contain mostly political analysis. Strategic intelligence is normally derived from civil authorities, social scientists, or good political journalists. Briefings are provided to commanders, and also to men on the ground (Jeffery 1987:141). In contrast to specific contact information, strategic intelligence "tends to hedge its bets and speak in generalities. (This is presumably why many academics are attracted to intelligence work.)" (Jeffery 1987:142).

Military intelligence is designed to "emphasize the dynamics between obtaining information on an enemy and acting on it," which may be termed the "intelligence-operations interface" (Stewart 1988:20). Covert operations during low-intensity conflict, ranging from subtle political manipulation to outright paramilitary interference,[175] often fall under the control of intelligence agencies: "[t]he information and technical requirements for covert action will to a large extent derive from the intelligence system anyway, and given this close relationship between intelligence and covert action, it seems appropriate that both functions should be performed by the same agency as far as possible" (Bloch 1983:19).[176] The absolute need of the system for very detailed knowledge of the enemy is especially evident: "MI's strong suit is all-source analysis or the fusion of disparate pieces of information to form a kind of mosaic of the enemy situation to determine capabilities, vulnerabilities and intentions" (Stewart 1988:20). During low-intensity operations military ethnography and military praxis are conflated.

In limited, low-intensity situations, "a particular coordination problem arises with the existence of forces which generally have both an intelligence gathering and an operational function" (Jeffery 1987:127). This control problem results from the inherent structure of any covert system: "LIC MI operations are centrally controlled and decentrally executed. Unlike in mid- to high-intensity situation where MI units are organic to or directly supporting committed combat elements, the LIC environment requires centralized control of disparate intelligence operations that may be executed from various locations" (Stewart 1988: 20). In other words, infrastructural factors in LIC conditions encourage military intelligence agencies to self-organize. War as a system, no matter how small, tends to self-organize, to evolve towards autonomy regardless of scale.

This tendency is quite normal: in the absence of external political control, security services — especially of the secret variety — will establish their own mandate. The "culture of secrecy" (Ponting 1990:1) in Britain, however, encourages a high degree of autonomy in the realm of military intelligence and covert action. Although most Western democracies protect information in the interests of national security, Britain has an extensive system to control the flow of official information . Unofficial pressures, and a broad range of legal statutes,[177] prohibit disclosure of official information, and preserve government secrecy.[178] According to Weber, "the concept of the 'official secret' is the specific invention of bureaucracy" (1984:48). According to Sir Martin Furnival-Jones, ex-head of MI5, "[i]t is an official secret if it is in an official file" (cited in Bunyan 1976:16). Both the nature of state secrets, and the process whereby such secrets become secret are obscure: "...in a secretive country the extent of secrecy is itself a well-kept secret" (Ponting 1990:67).[179]

Britain has no form of external control over secret security services (Ponting 1990:27). Lacking judicial, public and parliamentary reviews, intelligence organizations function independently of political control. Informal control processes, flexible structure of ministerial control of intelligence agencies,[180] deferring decision-making by 'referring upward', operational authorizations given by committee,[181] lack of direct supervision by the Joint Intelligence Committee (JIC),[182] and secret organizations' tendency to increase secrecy, allowed intelligence to run amok. "The maze of committees and liaison procedures has the effect of obscuring the way in which intelligence and covert action policies are formulated" (Bloch 1983: 53). Austen Chamberlain addressing the House of Commons as Foreign Secretary in 1924 said,

> It is of the essence of a Secret Service that it must be secret, and if you once begin disclosure it is perfectly obvious to me as to honorable members opposite that there is no longer any Secret Service and that you must do without it (cited in Andrew 1985:500-1).

Decisions on operations and intelligence collection matters are thus largely discretionary (Rule 1973:47). Non-accountability of intelligence services allows agencies to function independently, to 'self-organize'. Each operative in the system makes independent decisions, capturing control within the system when possible. Lacking a central source of command, each agent operates like a 'demon' in the ARPAnet. During counter-insurgencies there is an "inevitable tendency for special forces to become private armies, to drift away from the normal channels of command, and thus become a law unto themselves" (Charters 1979:57-8).

> it is one of the contradictions of the intelligence profession...that the views of its substantive experts — its analysts — do not carry much weight with the clandestine operators engaged in covert action. The operators usually decide which operations to undertake without consulting the analysts (Marks and Marchetti 1974:39).

Command and control structures which would normally govern military operations are inoperative when such operations are conducted covertly. During counterinsurgency, the war machine's tendency towards operational autonomy is amplified. Accountability is sacrificed in the interest of secrecy, upon which operational effectiveness depends:

> Be clear on one point above all else. The intelligence world is not answerable to Secretaries of State. It is accountable to nobody — not the Prime Minister, not

Parliament, not the courts. An intelligence department decides what information politicians should be given and they're rarely, if ever, given the full facts (Colin Wallace cited in Irish Times 24 April 1984, cited in Bloch and Fitzgerald 1983:54).

During the early phases, the conflict in Northern Ireland was a "corporal's war," because, in contrast to conventional wars, the balance of responsibility rested with NCO's rather than officers (Dodd 1976:67). Military strategy was an unsophisticated pastiche; operations circa 1972 deriving their methodology from colonial counter-insurgency. Since the mid-1970's, however, the conflict in Northern Ireland has been an intelligence war fought by self-organizing, secret, autonomous special forces, including MI5, MI6, the SAS, and other less well known units such as 14 Intelligence Company. According to a British Army sergeant, "[f]rankly, it's an Intelligence war now. I think that's the nature of the conflict" (45 Commando, cited in Arthur 1987:245). The covert operations of secret organizations have become a self-organized nether world:

> If you are in the Int world — especially army intelligence, you are in a totally unreal world. You are always at war, even in peace-time. You don't exist, you have no legal standing. You see conspiracies everywhere. Ordinary rules don't apply. You feel it is unfair when you're asked to abide by the normal code of society (Former Army Press Officer Colin Wallace

cited by McKittrick, Irish Times 21 March 1981:5).

Invisible factions of the British military run the war-machine in Ulster. Under normal conditions, MI5 controls internal intelligence in Britain, while MI6 is responsibility for external operations. During the 1970's most MI6[183] resources were deployed inside the UK: "through most of the 1970's a colossal distortion was permitted in MI6's work...in these years, MI6 became preponderantly an internal security arm" (Hugo Young, Sunday Times 14 Feb 1980, cited in Bloch and Fitzgerald 1983:211). Operations became extremely convoluted in Northern Ireland, where the changes in constitutional status of the province disrupted the internal and external boundaries of the state. With the RUC's intelligence files in disarray, MI5 and MI6 both established their own intelligence networks.[184] By 1973-4, MI5 and MI6 were engaging in a serious scuffle for territorial supremacy in Northern Ireland.[185]

Internecine infighting over intelligence turf occurred without any external accountability, and MI5 in particular seems to have been fulfilling its own political agenda, which did not correspond with that of the Labour Government in power. MI5's quest for superiority led to the removal of many of MI6's field officers in Ulster, and of senior MI6 men in Lisburn and Belfast, who were replaced by MI5 (Bloch and Fitzgerald 1983:225; Geraghty 1980:195-6).[186] Extreme right-wing 'ultras' in MI5 may have used Northern Ireland to launch an attack against the elected government they were meant to be serving (Foot 1990:79; Wright 1987), and maneuvered to keep the Army deployed in Northern Ireland. For this 'ultra' faction, the left-wing government of Prime Minister Harold Wilson threatened the military-intelligence complex more than PIRA. Anxiety about communist subversion inside the elected government corresponded with the emergence of a discourse of PIRA as 'communist agitators,' which influenced security policy in the early 1970's (see Chapter Two). The autonomous network of the military intelligence nether world had thus organized itself *against the state*.

The fratricidal intelligence war in Northern Ireland caused collateral damage across the board: recalcitrant agents who resisted the new intelligence order, or sought to expose the 'black' psychological operations of MI5, were removed, discredited or imprisoned (Holroyd 1989; Cavendish 1990; Foot 1990). Publication of any allegations against the 'state within the state', for example by ex-MI5 agent Peter Wright in Spycatcher (1987), was blocked with the Official Secrets Act in Britain. Although a substantial number of corroborating accounts of the intelligence war in Northern Ireland in the 1970's have now been published (see Dorrill 1991, 1993), bits and pieces of the story have yet to be unveiled.

point blank

Military snafus and the negative publicity surrounding atrocities such as Bloody Sunday, led to a natural preference for intelligence and covert operations over firepower. The eventual stabilization of the military intelligence food chain, and the current unchallenged ascendancy of MI5 (now DI5) in Northern Ireland, dampened the display of military paraphernalia. Saracens no longer cruise past Woolworth's in Belfast City Center — the war has gone underground. By the mid-1970's shoot-to-kill tactics by infantry soldiers, even during ambushes, were no longer tolerated.[187] Uniformed soldiers manning PVCP's or patrolling in hardline Republican areas now rarely have 'contact' with PIRA. Casualty figures demonstrate that the war has shifted from overt combat, based on firepower, to a covert, permanent war involving very few, but very heavy, players. Between 1976 and 1987 only nine armed PIRA members were shot by uniformed soldiers, while 25 were killed in ambushes, most often by the Special Air Service (SAS).

The military application of lethal force is restrained above ground for infantry soldiers, but in the subterranean world of covert counter-terrorist operations, the SAS play by 'big boy's rules' (Urban 1992). Big boys' rules constitute a normative code for the conduct of war, operating between military and paramilitary groups, independent of political controls. The emergence of informal codes between combatants exemplifies cooperation in conflict which is not based on collaboration, and which serves to increase the manageability and sustainability of violence, allowing violence to be exchanged in a controlled fashion. Anthropologists such as Feldman (1991) and Burton (1978), and investigative journalists such as Urban (1993), have recognized that paramilitary combatants in Northern Ireland "develop performance repertoires that permit the exchange of ideological codes within a closed circuit" (Feldman 1991:85). The violence has a certain internal logic.

Gift exchange in anthropological theory illuminates how normative codes governing violence may develop. All exchange systems are based on repetition, reciprocity, and the expectation of continuity. Repetition of transfer may lead to the eventual development of normative codes which govern the transactions. Defection from the inherent system contract will be greeted with condemnation on moral grounds, since defection constitutes a breach of an implicit obligation which may very well be economic, but is phrased as a moral relationship. Gift exchange systems tend to escalate the intensity and level of transfers when such transfers are connected with status. Potlatch exemplifies this principle. If low-intensity conflict is interpreted as an exchange system, the tendency to escalate the exchange may lead not to stability, but to an oscillating cycle of extreme violence and relative calm. Limits, however, prevent escalation. Law is one such limiting device, codes of cooperation between paramilitary combatants are another. An example of cooperation between groups who are otherwise enemies is the apparent execution of Loyalist paramilitary Lenny Murphy by PIRA with the full cooperation of the UDA (see Dillon 1989).

Because covert, military intelligence organizations 'self-organize', guard and promote their autonomy from external, political controls, and structure their internal organization as *networks rather than hierarchies* , external political orders do not necessarily exert any control over these agencies: "talking to people who have served in senior positions at Stormont it becomes apparent that...they did not consider themselves to be in real control either of the RUC's or of the Army's special operations" (Urban 1993:167). Following the collapse of the 1972 cease-fire, PIRA also escaped Sinn Fein's control (Smith 1991:237). Although most cursory journalistic reports lump PIRA and PSF together, tension between physical force strategy and political strategy underscore the relationship between the military and political factions of the Republican movement. "Within the Official IRA, the Provisional IRA, and the INLA there have been repeated conflicts over whether the military council or the political wing of the organization should define and direct the overall revolutionary strategy of the organization" (Feldman 1991:105; Coogan 1980:14-15). While it should now be clear that covert military organizations often exert their autonomy against the state, covert *para*military organizations are just as likely to exceed political control (as for example in the hungerstrikes; see Chapter Six).[188]

When military and paramilitary organizations break free of political control simultaneously in a confined theater, they have far more in common with one another structurally and professionally than with their respective parent political organizations. A reciprocal discourse of 'professional' violence expresses not only sympathy for one's opponent, but an essential similarity of function. Special Branch and Army intelligence operators establish long-term relationships and "close working bonds with people whom they suspected or even knew to have been responsible for killing their colleagues" (Urban 1991:106). These organizations not only accommodate each other's form, but participate in a reciprocal exchange relationship. The object of transfer or exchange for military organizations is death. This structure is suggested by the concept of the 'clean kill'.

Even when PIRA suspects can be safely disarmed and arrested, the SAS prefer to 'take' the target. For both Republicans and British military, shooting an armed suspect is perfectly acceptable, and is considered to be a 'clean kill'. "Shoot-to-kill stories are also disgruntled recognitions of the formation of a shared material language and political culture between state and paramilitary" (Feldman 1991:104). 'Clean kills', rarely reported on the news outside of the UK, are not atrocities, but simply programmatic, banal events in a long, long war.

> To be caught with weapons by counterinsurgency forces is considered by paramilitaries an automatic death sentence in the field. To possess a weapon establishes the paramilitary as a legitimate target for a 'shoot-to-kill' whether he is offering resistance to the arresting authorities or not (Feldman 1991:102).

Similarly, Urban points out that "the knack is to get IRA terrorists, armed and carrying out an operation, to walk into a trap" (1993:164). Brigadier Peter Morton, commander of the Parachute Regiment, wrote in Emergency Tour: 3 Para in South Armagh that "it was certainly a pity that the first occasion on which a terrorist was killed by the SAS was not more clear-cut; the ideal would have been to shoot an armed terrorist" (cited in Urban 1993:164). Thus, 'big boy's rules' depend to a large degree on public perceptions of legitimate use of force: "If it is not perceived to be an immaculately clean kill, it is automatically assumed to be wrong" (member of the security forces cited in Urban 1993:165). Such codes surrounding shoot-to-kill practice also reflect the restrictions imposed by the law, which security forces must accommodate and work around.

Brigadier Frank Kitson's introduction of gangs and counter-gangs into Northern Ireland exemplifies how special operations adopt not only the technique of the nomad war machine, but actually utilize 'tame' paramilitaries. While in command of 39 Brigade in Belfast, Kitson initiated various surveillance operations by plain clothes soldiers, including the Mobile Reconnaissance Force (MRF). Kitson recruited 'turned' PIRA members called 'Freds' who were barracked in Palace Barracks in Holywood, outside of Belfast. The MRF, accompanied by 'Freds', drove through hardline areas, identifying suspicious persons. The MRF used any pretext to penetrate Republican neighborhoods: door-to-door cosmetic sales, a massage parlor, and an infamous operation known as the Four Square Laundry. Laundries and haberdasheries had been part of the Cold War era spy-trade as covers for intelligence gathering (see Cavendish 1990). In this case, soldiers conducted recon by driving around in a conspicuous laundry van, picking up and delivering nappies, and inspecting the literal 'dirty laundry' of PIRA suspects for nitroglycerine traces. This operation ended in disaster for the MRF when a 'Fred' was 'turned' by PIRA, and blew the MRF cover, leading to the ambush of the Four Square van in Whiterock, and the death of one soldier.

'Clean kills', 'big boy's rules' and ambush operations conducted by special forces governed by an implicit shoot-to-kill policy emphasize the human factor in low-intensity conflict. Death, in these scenarios, is distributed at point blank range. 'Point blank' derives from the French point-blanc meaning white mark, or bull's eye. "Technically, 'point blank range' means close enough to aim directly at the mark and not to have to compensate for fall of shot" (Bellamy 1990:280fn.). Although long-range projectile weapons have increased the possible distance between eye and target, in Northern Ireland, death is still exchanged at very close range (we return to problem of visual targeting in the Conclusion).

In Northern Ireland, the other element of point blank exchange is information. Human sources are the predominant mode of gathering tactical intelligence. HUMINT derives from long-term informers and interrogation of suspects by security forces. The gathering, processing and continued availability of human source intelligence to security forces exemplifies another way in which counterinsurgency functions as an exchange system. On one hand, interrogation techniques have been circulated through the counterinsurgency loop, and on the other hand, the practice of informing establishes a reciprocal relationship between paramilitaries and security forces. According to Foucault, "power never ceases its interrogation, its inquisition, its registration of truth" (1984:203). Interrogation, inquisition and video-taped interviews in Castlereagh detention center produced truth.

During any low-intensity conflict, interrogation is the cheapest and easiest source of information available (Gazit and Handel 1980). Interrogation techniques used in Northern Ireland were adapted from prior experiences of the British Army during wars of decolonization:

> the techniques developed by the joint Services Intelligence School at Maresfield in Sussex and based on experience in previous counter-insurgency campaigns, as well as the Korean War, were applied in Northern Ireland. This involved the so-called 'Five Techniques' of sensory deprivation....These methods were used by the Royal Ulster Constabulary (RUC) Special Branch officers (under the guidance of army intelligence personnel) so that the maximum

intelligence results could be secured from the implementation of internment (Jeffery 1987:135).

The Compton Report justified the use of the 'Five Techniques' on the grounds that "interrogation of a small number of persons arrested in Northern Ireland who were believed to possess information of a kind which it was operationally necessary to obtain as rapidly as possible in the interest of saving lives" (Compton Report 1971:13). Only fourteen men were subjected to the Five Techniques, but over 3,000 suspects were interrogated by the RUC during 1971. According to the Compton Report, interrogation sweeps in 1971 produced a huge volume of information: over seven hundred IRA members were identified, and responsibility was determined for eighty-five incidents (see also McKinley 1987).

The 'Five Techniques' consisted of "hooding the suspect, subjecting him to a high-pitched noise, making him stand against the wall, and depriving him of sleep and proper diet — all classic techniques of sensory deprivation pioneered by the KGB" (Taylor 1980:20). Interrogation techniques can be considered as a form of subaltern knowledge, which has a very restricted circulation limited to practitioners and victims. One encounters these techniques either by being trained to use them, or by being subjected to them: the Commander of Land Forces (CLF), Sir Anthony Farrar-Hockley, instrumental in the introduction of the techniques to Northern Ireland, had experienced them first-hand during the Korean War at the hand of Chinese Communists.[189]

The application of these techniques to paramilitary suspects has led to the development of anti-interrogation training by PIRA. Deciphering and naming the "apparently random and unpredictable assaults... became a powerful countertechnique" (Feldman 1991:142): "The Special Branch would be questioning Ronnie [Bunting] and he would have said, 'Oh I see, you're trying such and such a method.' So they'd try something else, and he'd name that one too. It fucked the Branch up completely because some of them didn't even know the names for what they were doing and he did" (INLA member, cited in Feldman 1991:142).

Interrogation and the five techniques used during internment resulted in public scandal, an Amnesty International Report (1975), an International Committee of the Red Cross investigation, and a legal case brought against the UK by the Republic of Ireland to the European Court of Human Rights at Strasbourg.[190] Spontaneous, overt verbal examination, reliant on force or threat of force to elicit information, obviously does not reflect well on the state using it. But, as the Senior Psychologist for POW Intelligence at the Ministry of Defence reminds us, "torture is not a method of interrogation *per se*" (Cunningham 1972:31). Like other war technologies in Northern Ireland, interrogation was developed and refined into a set of scientific procedures for extracting information from suspects. Indirect and clandestine methods properly applied circumvent the need for direct examination. Organizing sources, influencing events, manipulating the social structure of the subjects to stimulate confession (also known as 'group dynamics'), and creating an "intelligence environment" have had a far wider application in Northern Ireland as intelligence methods than the scandalous 'Five Techniques'.

Resorting to the use 'interrogation-in-depth', in fact, signified a breakdown in the Northern Ireland intelligence environment in the early 1970's. When it became apparent that the intelligence files of local agencies were outdated, spontaneous overt verbal examination was used by intelligence agencies as a last resort. Extraction of information and the elicitation of tactical intelligence was not necessarily the goal of internment-era interrogation. Military interrogation during the Korean war was a minor part of the Chinese programme (Cunningham 1972:32), but British and American POW's were often hanged, flogged, or tortured for failing to provide autobiographical details.

> Since ancient times autobiographical information has been a priority target of almost all types of interrogation because it leads to the selection, induction and control of informers or stool pigeons. The selection, control and operation of secret agents and informers has always been one of the most important tasks of every interrogation agency. So too has the interrogation of innocent victims to provide a traffic to cover these operations (Cunningham 1972:32).

The fratricidal intelligence scuffle, and the eventual ascendancy of MI5, alluded to above, can be interpreted as an attempt fill the intelligence gap. Interrogation during internment was a means of establishing network of informers, contacts and 'players', necessary in order to reconstruct a new intelligence web in Northern Ireland.

Interrogation does not necessarily aim to produce information about operations, but information about genealogies. Autobiographical detail and explication of genealogical structure leads to the control of informers. Genealogies were considered to be the underlying organizational structure of paramilitary groups during the period when counterinsurgency was organized vis-a-vis a peace-keeping paradigm focused on tribalism. When the state subjects the body to torture, the body becomes the site through which information about the tribe can be extracted. Security forces in Northern Ireland are fully cognizant of the genealogical networks within PIRA: interrogation is one method of gaining organizational details of tight-knit families. The dynamic relationship between information, bodies, tribes and torture shows up in the anthropology of torture. Initiation rites enacted on the body, such as scarification and cliterectomy, subject the body to torture. The intensity of suffering during the ritual invests the body with the ethos of the tribe; knowledge is thereby inscribed on the body (Clastres 1977:151-153). For Republican paramilitaries, the violence of interrogation served as a political radicalization mapped onto the body: "either the body was extended by the technology of weaponry in the military operation or it was extended into a weapon in the interrogation center and the prison" (Feldman 1991:143).

Arrest and interrogation are one of the main contact points between paramilitaries and security forces, and informers are very often recruited in police stations or Army bases following arrest through offers of clemency in exchange for information (Urban 1992:104).[191] Interrogation and/or threats of physical violence are often unnecessary, although the other techniques employed, such as blackmail, may be unscrupulous. Both the Parker Committee Report and the Diplock Report emphasized the tactical value of HUMINT in Northern Ireland as part of an exchange system. According to the Diplock findings, the success of the intelligence war depends on a reciprocal exchange of information between intelligence agents and their informants: "if there were any weakening of the implicit trust that this understanding would never be broken by the security authorities these sources of information would dry up. The intelligence which they provide is operationally essential to the army's role in protecting life..." (1972:10). Informers, once established, are carefully nurtured by intelligence agencies so that the loop can be maintained. Informers are utterly despised by Republicans because they signify the states penetration into the cultural network of genealogies.

Paramilitary disgust towards informers results from the informers' literal violation of the sanctity of the tribal net. Informers not only open up the military netherworld to penetration, they commit cultural treason by doing so. Informers participate in a black-market economy of information, whose currency of exchange is actual money. The idea that information, and by extension loyalty, can be *bought* violates the principle of tribal, cultural loyalty which (para)militaries proclaim. Executing informers signifies an absolute reaffirmation of control over the flow of information. Weapons' "jarking" exemplifies how informers can have a central role in the system of information exchange on the ground in Northern Ireland. For many years, security forces were perplexed by the discovery of arms caches. Rather than ambushing and killing outright whoever eventually collected the weapons, security forces began tracking the weapons to get information about PIRAs military command structure. Intelligence specialists placed mini transmitters into the weapons found in arms dumps, the locations of which had usually been pinpointed by informers from within the Provisionals. The transmitters would be activated when weapons were moved. Later, more sophisticated devices, including voice-activated microphones, were installed which allowed security forces to track the weapons, and either carry out arrests, identify 'players', or shoot PIRA suspects red-handed. PIRA, after a while, eventually discovered that weapons were being tampered with. Having identified which of their volunteers had information regarding the location of weapons caches, PIRA could easily deduce who had informed to the security forces. These informers could then be executed; the shootings of James 'Jas' Young in 1984 and Gerard and Catherine Mahon in 1985 being cases in point. "After years of success, jarking became a mixed blessing for the intelligence operators, because it provided the IRA with a method of uncovering informers" (Urban 1992:120). The continuity of the practice reveals that intelligence information is at the heart of the military exchange circuit.

the tribes at war

Wars are no longer being conducted by states, rather "from Sri Lanka to Northern Ireland almost all are now fought by organizations that do not have a clear territorial base and cannot be targeted by nuclear weapons even when they are available" (Van Creveld 1992:63). Although states can no longer exchange war in a classical form, war is still being exchanged. The antagonists are often "small, dispersed and concealed groups..." slinging SAM's at Goliaths with increasing accuracy. Large, cumbersome state armies and small, agile nomadic ones, fighting each other is nothing new under the sun. But this atavistic, asymmetrical relationship, dating from antiquity, has become the predominant mode.

Revolutionary warfare is "both the most probably form by far of future armed conflict in the first and second worlds, and the most effective means of first and second world intervention in the third....special forces should be at the mass center of future armies..." (Simpkin 1985:317). Simpkin was not the first to advocate this future role for special forces; Sir Robert Thompson in <u>Defeating Communist Insurgency</u> recommended the use of special forces to make contact with "aboriginal tribes, living in the jungles and mountains" and "to recruit and organize them into units for their own defense on the side of the government" (1966:108). The adaptation of large, sedentary state armies to the nomadic paradigm of the guerrillas, though nothing new, is perhaps increasingly pronounced and operationally necessary (Arquilla and Ronfeldt 1993).

Despite the fact that nomadic paramilitary groups and state special forces are apparently mismatched, they are quite capable of adapting to each other. In LIC, these forces can co-exist in an orderly, predictable fashion. There is a pattern in the muddy chaos of low-intensity conflict — and surprise, surprise — it's an oscillating equilibrium between military forces. In anthropological circles, the equilibrium model of structural-functionalism (in which conflict is the normal condition between segmentary tribal groups) is unequivocally condemned. However, in military operational research, oscillating equilibrium models are now being used to describe and predict how special forces and insurgents interact as part of a system in which resources, manpower, and death circulate. In their article "Recruitment, Disaffection, and the Tactical Control of Insurgents" (1993), Woodcock and Dockery use predator-prey models from population dynamics to evaluate how state military and insurgent forces interact. Overall, their project concerns the operational modeling of combat, as a dynamic system, using chaos mathematics and catastrophe theory. In this particular essay, they "address the particularly vexing problem of modeling the command and control response of military forces to low intensity conflict as exemplified in insurgent movements" (1993:115).

Low-intensity conflict poses difficulty for operational modeling because it "sits at the interface between military combat and wider societal processes involving the development, allocation, and control of economic, political, and other resources" (115). In other words, because it has certain cultural characteristics, LIC is more difficult to predict than mechanized tank warfare. While anthropologists have been fussing over the complexities inherent in the cultural aspects of war for decades, Dockery and Woodcock seem slightly surprised that LIC, when subjected to predator-prey analysis of biological systems, produces "counter-intuitive results" which appear to be the outcome of *cultural* factors.

According to Dockery and Woodcock, state and insurgent forces show a tendency to oscillate in terms of military power. While a high level of resources helps to sustain oscillation between forces, the introduction of limits (density dependent levels of recruitment, for example) into the equation, brings the oscillating cycles to equilibrium. A limit cycle may result from these sustained oscillations, and contribute to a stabilization of conditions which support force co-existence. An increase in the level of resources or manpower allotted to the state military will destabilize the system, and may result in the "paradoxical enrichment" of the insurgents (135). "Providing a relatively low level of resources to a national military or security force can lead to a situation where the level of these forces (Force 1) and the insurgents (Force 2) reach a condition of force coexistence..." (137). So while one might think (especially if one is wearing a military cap) that the application of massive amounts of firepower could decimate the insurgents, once a system of force-coexistence is established and running smoothly, the introduction of more guns or butter into the loop only destabilizes the system.

Limiting technological and man-power resources available to state militaries results in a stable military equilibrium between state and insurgent forces: force co-existence. While this may prolong a low-level conflict, the alternative is to risk the insurgents gaining the military upper hand. This supports the hypothesis that the lower the intensity of a war, the more intractable it becomes. What then is the solution? By conceiving of force levels as oscillating cycles, "Force destruction can take place as the size of a military or insurgent force reaches a very low level in an oscillatory cycle" (136). If you want to destroy the system, the trick is to predict the cycles and strike with precision. Otherwise, leave it alone. Allowing the forces to co-exist (e.g., by restricting the British Army's access to resources, and their recourse to force in dealing with Republican paramilitaries) seems to be the preferred means of dealing with LIC in

Northern Ireland. Such low-level force co-existence frustrates soldiers.

A central argument of this chapter was that the cyber capabilities of the postmodern era do not match the rough and ready conditions of low-intensity warfare. LIC shows a marked tendency to revert to prior forms, making certain kinds of technology not only useless, but counterproductive. The notion that war can 'revert to prior forms' assumes that war is in some sense undergoing a process of evolution. While the notion that societies 'evolve' may be outmoded in anthropological theory, military historians cling to this idea. Bellamy, for example, advances the pseudo-anthropological thesis that warfare was a product of agricultural or "nomadic herding societies" and has been made increasingly obsolete as a rational political tool by the exponential lethality of industrial weapons. War, according to Bellamy, is now being deployed internally, and receding to the less-developed areas of the globe (1990:240-242; see also Virilio and Lotringer 1983). War, however, is hardly outmoded as a rational tool of politics; it is now simply neo-nomadic and must be conducted appropriately. War can no longer be conducted by sedentary state armies as a series of large-scale, classical attrition battles, but will be diffuse, polymorphous, human-focused, and non-hierarchical. Special forces will co-exist with paramilitaries in stable, low-intensity force/counter-force systems.

As Bellamy suggests, "all operations will become special operations" (1990:244). And what will special operations become? Special forces will increasingly imitate their opponents, establishing mimetic, antagonistic relationships with neo-nomadic guerrilla groups. Special forces will become nomads, outside the boundaries of the state which deploys them, but inside the boundaries of their own tribe. "Everyone's got their own tribe, their own group identity, and that's the strongest thing we all cling to at times of stress" (Sergeant, 2 Royal Green Jackets, cited in Arthur 1987:50). The tribe, rather than the state, will control the application and expression of violence. According to a former member of the Parachute Regiment and Territorial SAS:

> Everyone is a killer to a greater or lesser extent. The Paras just took that capability and articulated it. They dressed it up in a maroon-red beret and gave it an identity. They replaced its primitive bone-club with a high-velocity rifle, so that it could reach out and kill its enemies from afar. You'd spend eighteen years learning to be 'civilized', to contain the violence which boiled in your brain, and now the Paras opened the floodgates and told you to let rip. Only now they controlled the floodgates (Asher 1991:32)

This statement illuminates a vital component of how military systems control and manage violence: it is not the state which controls the 'floodgates' but the Paras, the Regiment, the Tribe who give the order to fire, and whom the soldier obeys and owes allegiance. The State is very largely superfluous in the equation. Dockery and Woodcock suggest that the absence of states in war between the tribes does not signify a breakdown, but an emergence of a new military order. While the traditional military logic of "more, bigger, faster" invites catastrophe, states can keep a cap on the tendency of LIC force/ counter-force structures towards potlatch by restricting military access to resources.

Although states may harness the nomad war machine, network forms can never fully be imitated because states do not work that way. Attempting to adapt and master the form produces a mutant variety of special forces, who as we have seen, establish dangerously mimetic and interpersonal relationships with their opponents, who have a natural tendency to exceed political control, and who will seek to organize against the state. The tribe can be brought in from the desert, but it can never be made trustworthy unless you abide by its codes.

Chapter Six:

The Death of Culture: Hungerstrikes against Counter-Insurgency

"Death, as we may call that unreality, is the most terrible thing, and to keep and hold fast what is dead demands the greatest force of all"
—— Hegel, *Phenomenology of Spirit* (1977:79)

"...for in Northern Ireland memories are very long"
—— *Diplock Report, 1972, para 13-14*

"The monster...neutralizes me to the point of death. The monster keeps me naked. It feeds me...I know why it won't kill me. It wants me to bow before it, to admit defeat" (Sands cited in O'Malley 1990:52). On 5 May 1981, the Right Honorable Robert Sands, esquire, Member of Parliament for Fermanagh and South Tyrone and member of the Provisional Irish Republican Army, died on hunger strike in Long Kesh prison in Northern Ireland. After sixty six days, having refused food and medical intervention, his body had digested itself.[192] Sands' death was followed by the deaths of nine other Republican paramilitaries. Although only two prisoners had been convicted of violent crimes, all were serving ten years to life.[193] The fast was initiated against the British government in order to reestablish 'special status' for the prisoners as POWs. 'Special status' was composed of five negligible demands: to refrain from prison labor, to wear one's own clothes, to associate freely with one another, to have one letter or visit a week, and to have lost remission time restored. But as Kieran Doherty whispered on his deathbed, "oh, the demands, there's more to it than that" (cited in O'Malley 1990:7).

Indeed, there was more to it than that. The removal of 'special status' was the cornerstone of criminalization, and criminalization was at the center of counterinsurgency strategy. Counterinsurgency cannot be understood without examining its effect on the people subjected to it. The hungerstrikes were a direct response to the removal of special status, and a visceral rejection of British counterinsurgency. The military legitimacy crisis percolating in Northern Ireland between the Republican movement and the British Army reached a dénouement during the hungerstrikes. Just as the sum of the demands constituted, in effect, special status, so special status meant, in effect, being a prisoner-of-war. Acquiring 'special status' was a Republican double-bluff to force the British government to acknowledge PIRA's military legitimacy: if PIRA members were awarded prisoner-of-war status, these 'terrorists' would become, *ipso post facto*, an army on equal foot with the British. Compliance with the prisoner's demands would have redefined the conflict in Ireland as a war, conferred legitimacy on their 'terrorist' activities and jeopardized Britain's claim to sovereign political authority in Ireland. Capitulation was, clearly, out of the question. As Ulster Secretary Humphrey Atkins said, "I am afraid we will just have to let them die" (cited in Sunday Times, 1 March, 1981:3).

In the political culture of Northern Ireland, where violence structures and permeates everyday life, the legitimacy of 'violence' is a vital point of contention between the Provisional IRA and the British State. As David Apter points out in "A View From the Bogside," "violence creates its own ordering discourse" (1991:2), but the functioning of the discourse depends on how, and by whom, 'violence' is defined. The crisis of military legitimacy centered on this deployment of definitions.[194] Contrary to their opponents' claims, both Great Britain and PIRA behave according to a stable, pre-established, informal logic of conflict which governs violent interactions. "One thus has to acknowledge," writes

Bourdieu, "that practice has a logic which is not that of logic" (1977:109). In the logic of conflict, definitions of what constitutes war vs. terrorism are consistent between PIRA and the British State. The question at issue is: who is the terrorist here? The topsy turvy roles of victim and aggressor are constantly (re)negotiated, according to who can be classed as a 'criminal terrorist' and who can be classed as a soldier. Although both may use a gun, one man's terrorist is — as we all know — another man's freedom fighter.

Both Britain and the Republicans claim to be the victim of the other's political violence. Within the structure of the discourse surrounding political violence, the ur-source of violence must always be external, thus PIRA characterize attacks on British forces as a valid military response to a history of occupation (see Leeman 1991). Mairead Farrell, assassinated by the SAS in Gibraltar (Jack 1988; Williams 1989) said, "I've always believed we had a legitimate right to take up arms and defend our country and ourselves against the British occupation" (PBS 1988). Great Britain, in turn, claims that PIRA tactics are indiscriminate and 'unfair', and that the state is the victim of 'terrorism'. As Burton Mack points out in Violent Origins; Ritual Killing and Cultural Transformation, there is always a "curious metamorphosis of victims in myths--now monster, now saviour" (1987:17).

By inverting the roles of victim and aggressor, the hungerstrikes embodied the penultimate resistance to the counterinsurgency program. In a sense, the hungerstrikes play with the *modus operandi* of counter-insurgency. Because counter-insurgency conceals the destruction of the opposition behind a veil of secrecy, hungerstrikes spectacularized death at the hands of the state. Insofar as counter-insurgency used imported law to enforce a military regime, hungerstrikes called on antiquated, indigenous justice as the source of legitimacy. Where the state had eschewed its responsibility for fair political representation of members of the polis, the hungerstrikes affirmed a representative body politic beyond the pale of the state. Where silence was a sign of guilt in the legal code, the hungerstrikes used silence to denounce that legal code. Where counterinsurgency used food denial to destroy resistance, hungerstrikes used starvation as resistance. Where counterinsurgency created strategic hamlets of geographic control, hungerstrikes showed the innermost locus of state's political geography, the prison, to be out of control. In effect, where the British government suppressed opposition by altering its terms of reference (Ellman 1993:88), the hungerstrikes called a spade a spade.

The struggle for military legitimacy was fought on the battleground of the body, and hunger was the primary weapon. Hunger, as Maud Ellmann points out in The Hunger Artists, "exemplifies the fact that the body is determined by its culture, because the meanings of starvation differ so profoundly according to the social contexts in which it is endured" (1993:4). The capacity of the 1981 hungerstrikes to wound, inspire and disgust participants and witnesses derives from the socially constructed meanings of hunger, starvation, and imprisonment in Ireland. The hungerstrikes retrieved and reenacted particular popularly-constructed histories of Ireland. Hunger, as an "operator... in the social process" (Ortner 1989:131), symbolized a social relationship of domination and subordination between England and Ireland, of which hunger, itself, was the central metaphor. The hungerstrikes, as a ritual enactment of spectacular death, also 'told stories' about the continuity of struggle, sacrifice and death within the Republican movement.[195] Said points out that "nationalism...fastens on narratives for structuring, assimilating, or excluding one or another version of history" (1989:221; see also White 1987). These 'texts' were written on the starving bodies, both literal and metonymic, of the hungerstrikers. The hungerstrikes were a way to "control the cultural terms in which the world is ordered and, within it, power legitimized" (Comaroffs 1991:24).

tradition, famine, law

The hungerstrikes have been interpreted as a legal protest (Devlin 1993), as an ascetic technology of the body (Ellman 1993), as cult of self-sacrifice which confirmed a dogmatic theology of violence (Sweeney 1993), as a typical passive-aggressive ritual of victim-bonded Irish society (O'Malley 1990), and while these explanations are all compelling, the most convincing interpretation of the hungerstrikes is as a military strategy using prisoners bodies as weapons (Beresford 1987; Clarke 1987; Feldman 1991). As a specifically military project, the hungerstrikes had the potential to transform the conflict in Northern Ireland into an international war: "it was hoped...the Republic could be forced into severing diplomatic relations with Britain, bringing the country to the brink of civil and international war" (Clarke 1987:138).

Irish hungerstriking begins with St. Patrick's (the patron saint of Ireland) fast against God. While fasting or hungerstriking may derive from a Catholic tradition (see Bynum 1987), the religious practice of abstinence may actually have originated in the older civil practice of fasting for redress (Ellman 1993:12). Whether fasting as protest has a religious or litigious origin seems to be a question of the chicken or the egg variety: modern Irish hungerstriking exploits both civil and Catholic traditions.

The Anglo-Irish conflict and the Irish Civil War were punctuated by hunger strikes: in 1918, prisoners at Dublin's Mountjoy jail refused to eat after being brutally treated for refusing to wear prison clothes and do prison work.[196] Thomas Ashe, who led the strike, collapsed after forced feeding and later died. During his funeral, attended by forty thousand mourners, IRB (Irish Republican Brotherhood) leader, Michael Collins said, "The volley you have just heard fired is the only speech which it is proper to make over the grave of a dead Fenian." In 1920, Terence MacSwiney, Lord Mayor of Cork and the commanding officer of the IRA, died on hungerstrike in HMP Brixton. His protest that Britain had no jurisdiction in Ireland earned him a two year sentence for sedition. In 1940, Tony D'Arcy and John McNeela died in St. Bricin's Military Hospital. In 1946, IRA Chief of Staff, Sean McCaughey died in Portlaoise prison. In 1972 at Crumlin Road Jail in Belfast, Billy McKee and Pronsias McArt went on a thirty-seven day hungerstrike for political status. In 1974, the Price sisters and Hugh Feeney went on hungerstrike for repatriation of Irish prisoners in English jails, which was terminated by forced feeding. In 1974 and 1976 respectively, Michael Gaughan and Frank Stagg died on hungerstrike.

Prison hungerstrikes in Ireland have typically *not* been about prison conditions per se but rather about the status of prisoners. Britain has often made 'special status' concessions to convicted paramilitary prisoners. In 1972, most IRA prisoners had *de facto* 'special category status'.

Long Kesh prison, essentially a detention camp, was used for holding prisoners without trial for periods of up to seven years. Prisoners were housed in Nissen hut dormitories, segregated by paramilitary allegiance, organized by military command structures and allowed to drill with dummy guns.[5] Following the implementation of the joint security strategies of 'criminalization' and 'ulsterization' in 1975, detention without trial was phased out. By deploying police rather than the Army, the violence in Northern Ireland became a 'law and order issue' rather than a military one. Prisoners convicted of 'terrorist' offenses after March 1, 1976 were to be treated as ordinary criminals. The Nissen hut compound at Long Kesh was used for the remaining 'special status' prisoners, and a complex of blocks (known as the 'H-Blocks') was built for ordinary criminals.

The removal of special category status negated the former distinctions between paramilitary prisoners and 'common' criminals. Ciaran Nugent, the first person convicted under the new regime, refused to wear prison clothes: "you must be joking me," he said (Beresford 1987:16). Refusing prisoners' garb, Nugent wore the only thing available, the blanket in his cell. Prison uniforms criminalized Republican soldiers. Francie Brolly's "H-Block Song" was widely sung in the blocks:

> So I'll wear no convict's uniform
> Nor meekly serve my time
> That Britain might brand Ireland's fight
> Eight hundred years of crime

Paramilitaries identify themselves as first and foremost as soldiers; to wear a prison uniform is to wear the uniform of a foreign state. A woman named Mary McDermott visited Kieran Doherty until he died on hungerstrike. She described visiting Doherty after the end of the first hungerstrike, when the prison authorities had issued him with a suit of civilian clothes from Marks and Spenser:

> I said, 'Well good heavens, you won't want me to visit you any more, I says 'a fine good looking fellow like you dressed up like this won't want to be seeing me.' His face contorted and he said, 'this is prison uniform, I don't want it. This is prison uniform.' I said, 'well it looks lovely on you' but he didn't like that, he didn't like that at all. He was in gear as he called it, this was prison gear..." (cited in Clarke 1987:136).

By September of 1980, of the thousand men without 'special status', almost half were "on the blanket" (O'Malley 1990:21); confined to their cells twenty-four hours a day, without reading materials and allowed no visitors. The refusal to comply with prison authorities resulted in a series of beatings and prison riots and protests (see Coogan 1980). The 'no-wash' protest, which had begun when prisoners were denied towels to cover themselves while they washed, escalated into the 'dirty-protest.' Denied permission to use the toilets unless they wore a uniform, and denied buckets in their cells for excrement, they smeared it on the walls.[197] Excrement was considered to be less defiling than prison garments, which for Republican prisoners were structurally equivalent to wearing the uniform of the enemy. The dirty protest, which ran for over a year, failed to gain 'special-status' and was abandoned in favor of a hungerstrike.

The prisoners understood that a hungerstrike would be the eventual conclusion of the prison protest: "sooner or later prison protest arrives at the life and death issue of a hungerstrike" (McGeown cited on BBC 1993). Hungerstriking was within the cultural repertoire of Republicanism, and not at all 'unthinkable'. Because culture is historically predetermined and "continuity with the past constantly restricts freedom of choice" (de Saussure 1983:74; Sahlins 1985), present praxis is always informed by pre-existing narratives (Bourdieu 1990).[198] Republican historical traditions provided an implicit legitimation for the hungerstrike. Sands wrote: "It is my political ideology and principles that my captors wish to change. They have suppressed my body and attacked my dignity...When thinking of the men and women who sacrificed life itself, my suffering seems insignificant" (cited in O'Malley 1990:50). "By hungering, the protesters transform their bodies into the 'quotations' of forbearers and reinscribe the cause of Irish nationalism in the spectacle of the starving flesh" (Ellmann 1993:14). In an official statement, smuggled out of Long Kesh, the prisoners wrote

> We, the Republican Prisoners-of-War in the H-Blocks, Long Kesh demand, as of right, political recognition and that we be accorded the status of political prisoners....not wishing to break faith with *those from whom we have inherited our principles*, we now commit ourselves to a hunger-strike (PRO statement from H-Block Blanketmen, issued from Long Kesh Camp, 10 October 1980, published in AP/RN 18 October 1980).

"Unquestionable tenets exist in secular political ideologies," write Moore and Myerhoff, "which are as sacred in that sense as the tenets of any religion" (1977:3). The hungerstrikes replicated historical events, and invoked prior exemplary hungerstrikes as confirmation of present practice. Bobby Sands wrote in his journal: "I remember...how this monster took the life of Tom Ashe, Terence McSwiney, Michael Gaughan, Frank Stagg and Hugh Coney, and I wonder each night what the monster and his black devils will do to us tomorrow" (cited in O'Malley 1990:51). Vincent Buckley in Memory Ireland writes that, "The strikers' own statements used, and grew from, a highly traditional store of concepts and attitudes, representing more than a hundred years in which their paradigm was established. The modern strikers are coding their understanding of history by replicating exemplary events" (Buckley 1985:174; see also Ellman 1993:11).

Hungerstriking, taken to the death, is extreme. If hungerstrikes have become an element of Irish Republican political culture — if, in fact, they are some sort of quintessentially Irish form of protest, as certain authors have suggested — the question remains to be answered, why?

The starvation of the hungerstrikes bespeaks a political mythology of Republicanism, by invoking the actual physiological conditions of famine, which is a very powerful cultural image in Ireland. During the Irish famine of 1846-1848, an estimated two million people died from both hunger and disease, while another million and a half emigrated to Canada and the US. This fact is well remembered in Ireland, especially in Republican Ireland. In 1991, The Sinn Fein Education Department issued a pamphlet called the "History of

Republicanism 1798-1990", which states in a discussion of the famine: "Deliberate economic policy in relation to Ireland forced the colony into a position of *feeder to England's economic superiority*...Amidst the horror of countless thousands of deaths the British exported foodstuffs from Ireland, sufficient to feed the Irish people twice over" (Sinn Fein 1991). Reproducing a spectacle of starvation during a hungerstrike exposes the economic colonialism of Britain as based in the production and consumption of foodstuffs.

The exploitation of Ireland as an agricultural colony is well-documented (see, for example, Woodham-Smith 1990): in Ireland, food has unquestionably been used as a weapon. Edmund Spenser wrote that "Great force must be the instrument but famine must be the meane for till Ireland be famished it can not be subdued" (1949:244). Restricting access to food in order to control uncooperative populations later became a feature of British counter-insurgency strategy. In Malaya, food denial was the "key operational concept linking the hearts-and-minds campaign with military operations" (Mockaitis 1990:116). Restricting insurgent access to food was central to the comprehensive strategy of the Briggs Plan: forced resettlement of villagers into 'strategic hamlets' allowed the food supply to be strictly controlled through rationing, thereby flushing rebel sympathizers into the open.[199] "Counterinsurgents (like old wives) believe the surest way to the hearts and minds of peasants is through the stomach" (Shafer 1988:117). The main complaint of the blanketmen in the H-Blocks before the hungerstrike was hunger; allegations were made that prison officers were withholding food (H-Block X-Mas 1979).[200]

During the period of the hungerstrikes, graffiti in the Republican areas of West Belfast read, "Blessed are those who hunger for justice". The hungerstriker 'hungers' for justice, acting as a symbolic representative of the Irish Republicans, who are also 'famished'. As Sands wrote, "...human food can never keep a man alive forever" (1991:Day 5). That the hungerstrikers died slowly of starvation, rather than prematurely from toxic poisoning, brain damage, kidney failure, or any other ailment, shows the centrality of hunger as a symbol during the strikes. In order to emphasize the effects of hunger, *medically unfit men were removed from the hungerstrike.*

The hungerstrikes invoked images of the hungering bodies of famine victims, toying with the notion of food as a circulating currency in a system of economic exchange.[201] The famine was both the proof and the consequence of the inequitable relationship of production and consumption between the British Empire and Ireland. The idea of Britain as a predatory consumer was already present in popular culture. Swift's <u>A Modest Proposal</u> (1969) extended the metaphor of English 'consumption' of Irish goods to its Malthusian conclusion: not only could Irish food-stuffs be gobbled up, but the Irish themselves could literally be eaten. The hungering body of the hungerstrikers also inverted this relationship of consumption. The image of the predatory British who 'eat' the Irish is turned upside down: here the Irish are literally eating themselves, committing political autophagy, and leaving nothing to be consumed by the state. Discussing the relation of writing, starvation, and imprisonment, Ellman writes, *"autonomy* is represented as *autophagy* , because [the] starving artists eat *themselves* in the absence of any nourishment from their societies" (1993:70. Italics in original). Because food has been a historic means of British political domination, hungerstrikes have become a form of resistance.

In a similar interpretive vein, the hungerstrikes can also be interpreted in the language of law: "although the hunger strike and the war of positions it engendered were profoundly political, at another level, they were deeply legal" (Devlin 1993). In other words, the hungerstrikes illuminated the legal violence practiced by the state, and in this sense were about the fundamental nature of law. In Ireland, law not only assured political stability of the colonial regime, but implicitly legitimated that authority. Legal imperialism (simply, using law to induce political order) was the basis for colonial and economic expansion in Ireland in 1607 (Pawslisch 1985),[202] and contributed to more modern British counterinsurgency. Political violence was 'criminalized' in 1975 (Hadden, et al. 1988) following the publication of the Gardiner Report, which recommended the removal of 'special status' for those convicted of political offenses. Lord Gardiner wrote: "...terrorists who break the law...are not heroes but criminals; not the pioneers of change but its direst enemies" (1975:5). As part of a drastic policy overhaul, Lord Gardiner's recommendations were incorporated into the Northern Ireland (Emergency Provisions) Act of 1978, which drastically altered criminal procedure in Northern Ireland. Despite the specialized legal processes to which Republicans were subject, they were not considered special prisoners.

Counter-insurgency operations in the legal arena use law "as just another weapon in the government's arsenal" (Kitson 1971:69) to enforce a socio-military order. Legal violence used to enforce political order has resulted in a separation of law from justice in the eyes of

Republicans. "From the point of view of the victims of legal violence, any theoretical distinction between the authoritarian and the normal state is irrelevant: from the bottom, violence is violence" (Devlin 1993:174). Republicans, in response to criminalization, consistently refused to acknowledge the authority of British courts in Northern Ireland.[203] The refusal to recognize was a direct challenge to the legitimacy of the law. "Before the IRA can compete with the ideology of the law it must establish that it is challenging the legitimacy of the law and not just breaking the law. Consequently, the IRA's resistance to the imputation that their activity is merely criminal is the first stage in their own struggle for legitimacy" (Burton 1978:128). Legitimacy claims and counter-claims are, thus, framed in language of jurisprudence.

When "Widespread portions of the public feel "outside the law" because the law is no longer seen as an instrument of justice but as a tool of oppression...Social justice and state law are in antithesis, and people turn to symbolic representations of justice outside the law..." (Kooistra 1989:11). Republicans identified 'justice outside the law' as a component of their own ancient legal history. The hungerstrikes defied the expedient, malleable law of counterinsurgency by invoking antiquated, indigenous justice as the source of legitimacy. Hungerstriking as a form of litigation predates, and is in structural opposition to, British legal imperialism. Hungerstriking "was the espousal of cultural difference, the exposition of a jural other and the assertion of a legal right" (Devlin 1993:173). The hungerstrike is codified in the defunct medieval Irish legal code, the Senchus Mor. This document defines two types of fasting: *troscud*, fasting against a person for redress or *cealachan*, achieving justice by starvation. *Cealachan*, fasting for political purposes, is distinct from troscud, fasting for legal redress, although ultimately, the purpose is the same: "to exert moral pressure on a more powerful adversary" (Kelly 1988:182). Typically, the less powerful fasted against the powerful to redress a perceived injustice (O'Malley 1990:25).

The functional basis of the hungerstrike as litigation is the existence of a moral obligation between the ruler and the ruled. Hungerstrikes will not work where moral pressure fails to induce shame. Law, under these conditions, reflects a code of social justice based in reciprocity. By invoking ancient customary law, the hungerstrikes reintroduced the concept of justice as part of a reciprocal social and political contract into legal practice. "The hunger strike is, in many ways, the apotheosis of the nexus between the ideology of law and the violence of law" (Devlin 1993:172). Hungerstriking, by claiming a mandate in atavistic legal practice, sought to reunite law with justice.

spectacular death

"The gravest danger in post-monarchic society, "according to Virilio and Lotringer, "is the concealment of death" (1983:124). Counterinsurgency seeks to destroy public knowledge of insurgent political aims by rendering the opposition invisible. The PTA's legislation of proscription orders forms the legal basis of this invisibility: "its original purpose was modest and short-term: to remove the IRA from public sight" (Walker 1988:64).[204] British Army members are also restrained from acknowledging death: according to one Intelligence

NCO, "I can't see why we can't admit that we are shooting these terrorists" (cited in Urban 1993:76).

Counter-insurgency conceals the destruction of the opposition behind a veil of secrecy, and must deny deadly operations by "maintaining an appearance of the rule of law" (Urban 1993:76). In contradistinction to this legalistic shield, hungerstrikes spectacularized death at the hands of the state, bringing death into the foreground. As Foucault points out in <u>Discipline and Punish</u>, by the nineteenth century the penal system had undergone a disappearance of punishment as a spectacle. "As a result, justice no longer takes public responsibility for the violence that is bound up with its practice" (1979:9). During the hungerstrikers the punishment that the hungerstrikers enacted on their own bodies became the spectacle. "He suffers alone, in darkness, yet in public. People know how he is dressed, whom he has seen, what his weight is, how near his vital systems are to collapse" (Buckley 1985:126). Dreyfus and Rabinow point out that "the atrocity of torture was an enactment of power that also revealed truth" (1983:146). The atrocity of self-torture by Sands and nine other men inverted the panoptical structure of the prison: the invisibility of incarceration was rendered public, bringing the practices at the state's inner sanctum to light.

> The way the Hunger Strike worked was that the British sat and recorded every move the hunger striker made, everything he said, everything he done....a whole in depth study. The doctors were doing it, the screws were doing it, the governors of the prison were doing it, the priests were doing

it, and we were doing it. Everybody was into sussing it out (PIRA member cited in Feldman 1991:251).

To observers, the Republican focus on death appears as a morbid aberration, a "cult of death... in the political realm, where powerful symbolic performance and emotional display is...a feature of dying, wakes and funerals" (Taylor 1989:175). The death-based nationalism of "mystical Republicanism" (O'Malley 1990) apparently

> ...receives its most important revitalization in the ritualized political death. On those occasions the appropriate sacrificial victim, one who embodies the values of the struggle, can be used symbolically to call into existence the 'Catholic (or Republican) community' of nationalist discourse. Such occasions also invoke the memories and narratives which support the cult of nationalist death (Taylor 1989:184).[205]

This cult of self-sacrifice, grounded in a theology of violence (Sweeney 1993), has not escaped criticism within the Republican movement, itself. The 'blood-sacrifice' rhetoric associated with nineteenth century figures, especially Pearse (1991 fieldnotes) aggravates more pragmatic Republicans. Countess Constance Markievicz, for example, wrote from her prison cell that she "often longed for the peace of the Republican [Burial] Plot" (1987: xxviii). For Markievicz, and other early Fenians, political allegiance to revolutionary nationalism was often phrased in terms of loyalty to the dead (1987:xxix).

The exploitation of death is, in fact, a common feature of nationalism. Death is a "destruction...tantamount to a sacrilege" against the social order (Hertz 1907, 1960:77), which mortuary rites must reaffirm by disaggregating the individual from the collectivity. The state of the corpse is, thus, often linked to the fate of the soul after death (Huntington and Metcalfe 1979:14).[206] Nationalism merely extends this symbolism, linking the corpse of the martyr with the soul of the nation. Where the decaying corpse is connected to ritual sacrifice (Hubert and Mauss 1899, 1964), and embodies corporeal destruction as a prelude to spiritual immortality (Huntington and Metcalfe 1979:15), the martyr's body asserts national immortality. Anderson, in <u>Imagined Communities,</u> uses the example of the cenotaph — the tomb of the unknown soldier — to examine the linkage of "the cultural roots of nationalism with death, as the last of a whole gamut of fatalities" (1983:15). Nationalism transforms battle-field fatality into political continuity, such that "dying for one's country, which usually one does not choose, assumes a moral grandeur..." (1983:132). While nation-states may be seen as 'new', the nations to which they give political expression are born out of the past. Primordial links with the dead generations are expressed through language: "nothing connects us affectively to the dead more than a language" (1983:132).

Pearse orated at the grave of O'Donavan Rossa in 1915: "Life springs from death: and from the graves of patriot men and women spring living nations" (cited in Coogan 1980:17). Thus, the nation is defined not only by its living members, but by the political martyrdom of dead Republican heroes. The 'dead generations' are a central fact of nationhood. As Anderson has pointed out,

"If nationalness has about it an aura of fatality, it is nonetheless a fatality embedded in history" (1983:133). The National Graves Association continually updates the roster of the Republican dead (1932; 1976; 1985) and tends the Republican burial plots around Ireland. The Phoblacht na n-Eireann, the Proclamation of the Irish Republic, drafted in Easter 1916, begins: "In the name of god and *the dead generations from which she receives her old tradition of nationhood*, through us, summons her children to her flag and strikes for her freedom..."[207] Death-based nationalism (if there is such a thing) represents a highly refined and logical extension of Irish political geography. Irish nationalism, prevented from full geographical expression, can only be embodied by the dead generations of Republican martyrs. The body politic of Ireland is, in a sense, the sum total of the political bodies of the Republican martyrs. Reincorporation of the dead into the continuum of political martyrs (Van Gennep 1960) is also a feature of Republican cosmology. Bourdieu writes that "the return of the dead, that is, resurrection, is called for by every aspect of symbolism" (1977:138). The phoenix, a symbol of resurrection, is an official insignia of the Provisional Irish Republican Army. The hungerstriker contains "the spirit of the risen people, who he represents" (Sands cited in Collins 1986:113), made literal when Bobby Sands' was elected to Westminster Parliament as the member for Fermanagh and South Tyrone.[208]

Both the death of Irish culture (Scheper-Hughes 1979) and the death centered nature of Irish (Republican) culture have been subjected to considerable ethnological scrutiny (O' Suilleabhain 1979; Taylor 1989). Anthropological necrography of Ireland — 'salvage' anthropology — identifies the imminent death of the culture and seeks to preserve in documentary form its material, visual and oral culture. Scheper-Hughes, a self-identified necrographer of Ireland, argued that an obsessional focus on death was one sign of a dying culture. Irish Republicanism is clearly an example of "societies which place death at their center" (Virilio and Lotringer 1983:133).

The cultural centrality of death raises an interesting issue: whether death represents freedom, or whether death is a site of political struggle. Foucault argued that "death is power's limit, the moment that escapes it" (1990:138). In death, the body is no longer subjected to political control,[209] because no matter how panoptic, power-knowledge systems cannot penetrate the opacity of corporeal death. Baudrillard, on the other hand, referred to a heightened level of cultural control exerted on corpses as 'leuchtimisation': cultural production and consumption of death are linked to the intensity of controlling processes: "...the control problem is not one of surveillance, propaganda or paranoia. It is one of subjective influence, consent and extension to all possible spheres of life. The incorporation of the code into the corpse itself" (SPK 1983:103). When cultural leuchtimisation occurs, the body — even in death — cannot escape the controlling codes of power relations.[210]

Control of the corpse during death rituals has been vital for the construction of the Republican political community. Frank Stagg's funeral erupted into a grotesque scuffle between PIRA and members of his family over the control of his dead body. Stagg died on 12 February 1976 after hungerstriking for sixty days in Wakefield prison for repatriation to an Irish jail. PIRA accused the British of murdering Stagg, while certain members of the Stagg family blamed the Provisionals for his death. Stagg's family refused a Republican funeral and refused to allow the cadaver to be buried in the Republican plot. Although scheduled to arrive in Dublin, the body had to be flown to Shannon instead because Joe Cahill and Ruadhri O'Bradaigh were waiting at Dublin to hi-jack the body for burial in the Leigue cemetery in Ballina, County Mayo. Under Irish Gardai (police) protection, Stagg's corpse was buried beneath a concrete slab. Meanwhile, PIRA held a separate ceremony in the cemetery and O'Bradaigh swore to the 1,000 assembled mourners that the body would eventually be buried in the Republican plot. A police-guard stood watch over Stagg's grave for six months to prevent the Provisionals from swiping the body, but "when the police left in November, the IRA dug under the cement, hauled out the corpse and transferred it to the republican plot where they fired a single shot over it as a mark of respect" (Clarke 1987:85-6).

This emphasis on reclaiming the corpse for the Republican plot must be understood as an aspect of the militarization of death in Northern Ireland. Funerals are the primary site for the enactment of an ideology of military professionalism for both the British Army and the Provisionals. On 9 February, 1971 the funerals processions for James Suanders and Bernard Watt through Belfast drew huge crowds. The Army stood quietly, and many saluted as the funeral cortege passed. An IRA honor guard fired over the coffins, and Unionist politicians demanded that the Army rush in and arrest the Honor Guard. Faulkner and the Army demurred: "It is a military custom to salute funerals," explained one British Army officer to Chichester-Clark, "I would salute Hitler on his way to his last resting place" (cited in Sunday Times Insight Team 1972:246). Honoring death is the respect that soldiers accord to one another, and to their profession. During the early funerals, such as that of Charlie Hughes, British soldiers paid their respect to the IRA dead by saluting the corpses. The RUC, however, began to systematically disrupt Republican funerals in the late 1970's. As late as 1983, the funeral of PIRA Volunteer Colm McGirr, conducted under the auspices of the Army rather than the RUC, was unmolested. This, of course, shows the difference between how police and soldiers view paramilitaries. Disruptions of funerals by the RUC, such as snatching the cap and gloves from the tops of the coffins, and refusing marching privileges to funerals if the Tri-Color was draped on the coffin, struck symbolically at *military* symbols used in Republican paramilitary funerals.

This symbolic attack goes to the heart of the issue. Republican paramilitary funerals sustain Republican culture by providing the symbols and substance to the idea that these are war deaths. In other words, Republican funerals militarize death:

> The routinization of imminent violent death is a determining condition of paramilitary life....The paramilitary funeral has gradually become an extremely rationalized ceremonial response to violent death.....the politicization of his death in the funeral is tempered by the fact that the same ceremony ritualizes the inevitability of that death (Feldman 1991:106).

Funerals normalize death as a part of life and by doing so, affirm the political survival of Republican culture, which is based on death.

Martyrs sustain culture; this recognition became the basis of security policy focused on controlling Republican corpses. At the end of February 1971, the Special Powers Act outlawed the concealment of dead or injured persons (Barzilay 1973:23). Funerals, such as that of Larry Marley, became scuffles between mourners and security forces over the corpses of paramilitaries.[211] In order to preserve forensic evidence, the Security Forces always try to remove PIRA dead from the scene of gunfights, before they could be hi-jacked by PIRA. Corpses of hungerstrikers were objects of wrath. Marks were found on Frances Hughes' face which "may have resulted from a number of kicks directed at the corpse before it left the prison" (Clarke 1987:162-3). The corpse of hungerstriker Patsy O'Hara was apparently also mutilated before the departure from the prison for burial: "the face had been beaten, the nose was broken, and cigarette burns were visible on the body" (Feldman 1991:302fn.).

The policy seems to have shifted at some point from controlling actual corpse to controlling the production thereof. As early as 1979, security directives from Lisburn advised soldiers to avoid the use of lethal force, not because it was technically illegal, but to prevent dead terrorist suspects from becoming PIRA martyrs (Urban 1993:83). Militarized death, because it sustained culture, was dangerous and could only meted out sparingly.[212] The best policy was to ignore it: by 1987, security forces had stopped intervening in funerals. Paramilitary funerals were allowed to become the site of sectarian violence.[213]

silent bodies, warring bodies

Until the full-blown advent of counterinsurgency in Northern Ireland, paramilitary careers had ended either with death or imprisonment. Since the mid-1970, the high concentration of smart, motivated Republican activists in the prisons in Northern Ireland had transformed the prisons into what the British called "Universities of Terror." Rather than weakening the Republican movement, the prisons provided a space for the radicalization and re-Gaelicization of Republican activists, what Republicans called "Universities of Freedom". "Ninety-nine percent of the people on the Blanket came into the jail when they were nineteen. You had very few over that age. We were the second generation of IRA people who had joined in '74 and '75. There was no old IRA, Catholic Nationalist on the blocks..." (PIRA member cited in Feldman 1991:222). A counterinsurgency strategy was introduced which targeted the destruction of the Republican movement from within the prisons. This could be accomplished by individualizing the prisons, producing apathy amongst them, and stripping them of a cultural identity as Republicans.

> Bobby Sands made an impassioned sort of effort to make us understand his conception of the "Breaker's Yard". He would say, "No matter what we win and lose in here is not just won or lost for us. It was won and lost for the generations to come because what the Brits are into building here isn't just for breaking us. What's happening here is that the Brits

have a closely worked out psychological approach to prison. they're trying to build a 'Breaker's Yard.'" That was the term Bobby Sands used (cited in Feldman 1991:161).

Republican prison protests, beginning with the blanket protest, were a response to a counterinsurgency initiative. The prisons became the epicenter of the war. Billy McKee, who led the 1972 hungerstrike (which initially achieved special status for Republican prisoners), said "this war will be won in the prisons" (cited in Coogan 1980:14). Feldman makes essentially the same point as Clarke by pointing out that Sand's concept of the Breaker's Yard "shifted the paradigmatic topos of the Republican struggle from violence on the street to resistance in the prisons" (1991:162, 230). "We realized that the most potent weapon people had in war was organization, competent forward-moving organization....We realized that with the Blanket a new phase of the campaign had begun — that the center of the Republican movement was in the jails" (PIRA member cited in Feldman 1991:222). The blanket protest and the hungerstrike altered the inevitable: now neither death nor imprisonment ended the paramilitary career. They were only the beginning of the war effort: the war was now inside the prison and the corpse was a weapon: "Whenever [Mrs. Thatcher] used the term 'This is the IRA's last card," then we actually saw ourselves on the front line. We felt this air of importance that we were the front line in the struggle..." (cited in O'Malley 1990:85).

Even outsiders recognized the military nature of the prisoners. According to Twister McQuistan, a former UDA prisoner:

If you argued with one of them you were arguing with fifty, these guys were dedicated. They had done blanket protests, dirty protests and they had done hunger strikes. They seemed to have the attitude that they were some kind of elite regiment... (cited in Clarke 1987:202).

According to one former Maze prison officer they understood that "they were dealing with an army inside the prison...they were dealing with the army's back-up outside the prison and you're talking about the Provisional IRA outside. It wasn't any mickey-mouse stuff" (1993 BBC).

As Clark very accurately points out "it is the nature of a military organisation to accept casualties in furtherance of its objectives" (1987:172), and PIRA was more than willing to accept casualties to win their war. According to Liam McCloskey who was in the same wing with Sands,

On the night that it was announced he said that wouldn't be prepared to die for five demands only but that through the hungerstrike he would hope to gain more support for the IRA: if he died it would incense the Irish people...: if he lived it would show that Margaret Thatcher could be beaten and again draw support for the IRA. Either way the IRA won.... (cited in Clarke 1987:139)

It is also the nature of military organizations to *create* casualties in furtherance of its objectives. The deliberate deaths of the hungerstrikers can be read as an *internal pressure of a war system to produce casualties.* "We were saying, "We can go on indefinitely, we can keep sending coffins out of here every week if you give us a guarantee on the outside that you can sustain the level of the political campaign" (PIRA member cited in Feldman 1991:231). This statement displays a military cost-benefit casualty analysis, what Feldman describes as rational, calculated production of death where corpses become "political commodities" (1991:234). Benadette McAliskey (nee Devlin) said regarding the hungerstrikes:

> What separated me from him and all the rest of them was that there was no visible emotion, that total control of whatever emotion they felt was subordinate to the politics, the struggle they were involved in. They could do a balance sheet at that point in time, they could say 'Well, we lost ten volunteers. We did not get the demands but we have immeasurably strengthened the resolve of the people and built the anti-imperialist organisation nationally (cited in Clarke 1987:206-7).

The status of Republican POW's as soldiers could only be confirmed by a body count. The war-logic activated during the blanket protest demanded the production of death to confirm and validate it. The logical structure of war demands death, and in a battlespace with no enemy, it was necessary to kill themselves. McFarlane, a prison organizer said, "we are fighting a war and by choice, we have placed ourselves in the front line....we are fighting a war and we must accept that front line troops are more susceptible to casualties than anyone" (cited in O'Malley 1990:82). According to Owen Carron, "I think he [Sands] was always working on the premise that he would have to die" (Iris 1981:28). Sands death was a combat death in the context of a war. It was also, simultaneously, an autophagic regicide which expressed and confirmed a military mandate for the Republican movement.

When the hungerstrikes are understood as a military operation, what initially looks like Republican fatalism begins to resemble sometime like classical military pragmatism. Soldiers do not go to combat as a protest: they go to fight and die.[214] Sands and the other hungerstrikers understood this. According to Liam McCloskey, whose mother removed him from the hungerstrike after he went into a coma, recalled that:

> I was totally at peace with the thought of going to death....Once you were on hungerstrike I think it would have taken more courage to actually stop of your own accord than to keep going because it seemed so much like losing face and backing down when other men had died (cited in Clarke 1987:197).

The hungerstrikes were a form of combat which guaranteed 100% lethality. "When you're out on the street operating in a squad you're tight knit. Death was something that you faced, but it's not a foregone conclusion. You have a chance. In the blocks the guys who were saying "hunger strike," weren't saying, "hunger strike and you might die." They were saying, "Die! You, die for me!" (PIRA member cited in Feldman 1991:239). Strangely enough, the hungerstrikes simulate the ultra-lethality of post-modern war.

The inevitability of death made the body a source of power for the hungerstrikers. This focus on the corpse as a site of power led Patrick O'Malley, in <u>Biting at the Grave</u>, to conclude that, "from the earliest times, this tradition of passive-aggressiveness, of taking injurious action against oneself for which another was held to be responsible, was given favor" (1990:26). O'Malley, following political scientist Robert Elias, argued that Irish society is a paradigm for a "political economy of helplessness" (1990:8). The Irish, as a "victim-bonded" society, experience an "inherited sense of powerlessness" (1990:8), which results in their own self-destruction.

Contrary to psycho-political characterizations of the Irish as passive-aggressive, the hungerstrikes were a military response to the political, economic, cultural exclusions which rendered them necessary. "Rites take place," says Bourdieu, "because and only because they find their raison d'être in the conditions of existence and the dispositions of agents who cannot afford the luxury of logical speculation, mystical effusions, or metaphysical anxiety" (1977:115). In other words, rituals exist because they must exist; often the genesis of ritual is political duress. Jim Scott writes that "slavery, serfdom, the caste system, colonialism, and racism routinely generate the practices and rituals of denigration, insult, and assaults on the body that seem to occupy such a large part of the hidden transcripts of their victims. Such forms of oppression... deny subordinates the ordinary luxury of negative reciprocity: trading a slap for a slap, an insult for an insult" (1990:xi).

Direct resistance had been effectively curtailed or criminalized by counterinsurgency measures. The hungerstrikes, with their apparent obsessive focus on death, torture and starvation, resulted partially from the silence imposed by 'normalization.' British emphasis on intelligence and covert action as a containment strategy entailed a replacement of overt military spectacles with a 'normalized' police presence on the street. Although unchallenged Gramscian 'hegemony' (1971:328) was unlikely within the fractured polity of Northern Ireland,[215] a normalized state of emergency made political (now criminal) violence appear as part of a 'natural' order of things (Beetham 1991). The operations of power had, in a sense, become mute. For Bourdieu, hegemony consists of things that go without saying because they come without saying (1977:167; Comaroffs 1991:24). The state, having abandoned overt military maneuvers in favor of covert shadow operations, intended that military operations 'would go without saying'. Violence, as a banal aspect of everyday life, demanded no particular explanation. The state, having thus silenced itself, imposed a silence on its antagonists. Indeed, "power has so often been seen to lie in what it silences..." (Comaroffs 1991:23) and in Northern Ireland a *cordon sanitaire*, a strategic hamlet, keeps the war fairly quiet.

In addition to censorship of proscribed groups, silence was criminalized in the courtroom. PIRA prisoners had habitually refused to recognize the court on the grounds that any state agency deriving from partition was illegitimate. In such cases, a formal plea of not guilty was entered and the trial proceeded.

Defendants who refused to recognize the court were rarely acquitted (despite lack of evidence), since such refusal was generally taken as an admission of membership, and therefore as equivalent to a plea of guilty (Boyle, et al. 1975:111). The Right to Silence and the Criminal Evidence (NI) Act, 1988, in deviation from British Common Law practice, allowed Diplock judges to "attach whatever weight they considered proper to the fact that a suspect had remained silent under police questioning or refused to give evidence in court" (Boudin cited in Finn 1991:197). Interacting with Section 18 of the PTA, the effect of Criminal Evidence (Northern Ireland) Order 1988 was quite nefarious, and actually penalized silence (Walker 1992:138).[216]

Because secretive insurgent movements generally lack the luxury of public speech, what emerges from the barrel of the insurgent gun is open to interpretation. Eisenzweig argues that "terrorism, or if you will, the death of the author" (1988:35) is a form of utopian violence, given meaning by para-textual statements issued by the terrorists, themselves: "...the existence of the accompanying text is the very foundation of the terrorist activity, its sine qua non" (1988:33). Unanalyzed political violence often appears either as an attempt to get on the news, or as a bizarre symbolic protest against nothing in particular. The media, by leaving PIRA's rationale publicly unexplored, not only denies them speech power, but makes them appear silent by intention (Curtis 1984:151). When silence is equated with guilt in the courts, and concurrently, the political rationale of violence remains unanalyzed, minority discourse is, quite effectively, gagged. Refusal to negotiate by the heads of state meant that, for PIRA, silence was the only option. Foucault writes: "...it is probably over optimistic...to regard the silence of one's adversaries as indicative of a fear we have inspired in them. For perhaps the silence of the enemy...can also be the index of our failure to produce any such fear at all" (1984:207). The revival of the Irish language as part of a movement towards cultural separatism should be understood as a response to the imposition of political silence on them.

When silence is imposed judicially and in the public realm, the 'negative reciprocity' of direct resistance is rendered impossible. Under these conditions, the only possible form of negotiation is silence, itself. Only the body can speak without language. The body thus becomes the bearer of the 'hidden transcript' of resistance, capable of being used as a weapon and as a means of negotiation. The body, through torture (Taylor 1980) and hunger, became a site of information and control (Scarry 1985; see also Bloch 1986). For the hungerstrikers, controlling the body — stripped bare of all accoutrements — became the primary consideration. One can only renounce that over which one has control; the body was the last thing they controlled. "But there are many nuances of nothingness: and every hungerartist eats a different absence, speaks a different silence and leaves a different kind of desolation" (Ellmann 1993:113).

Where counter-insurgency sought to reduce public signs of conflict — to render invisible, to find an 'acceptable level of violence' — the hungerstrikes brought the war home, directly onto the bodies of the prisoners. An external war zone was thus internalized: "the body is...directly involved in a political field....Power relations have an immediate hold upon it; they invest it, mark it, train it, torture it, force it to carry out tasks, to perform ceremonies, to emit signs" (Foucault, cited in Dreyfus and Rabinow 1983:112). Power relations were enacted and inscribed on these human bodies: death tolls, casualty figures, mortality statistics are merely mechanical, abstract ways of calculating the consumption of bodies. "In Northern Ireland," as Feldman points out, "the practice of political violence entails the production, exchange and ideological consumption of bodies" (1991:9fn.). During the hungerstrikes, political conflict became localized in the bodies of the hungerstrikers. "The re-encoding of the body not as a criminal but as a terrain of struggle in the pursuit of freedom was considered to be the ultimate transgression of the *cordon sanitaire* of the H-Blocks" (Devlin 1993:172).

The Army Council and Sinn Fein, initially resistant to the hungerstrikes, had vetoed it on the grounds that it was un-military. Gerry Adams wrote to Sands that "we are tactically, strategically, physically and morally opposed to a hunger-strike" (cited in Reed 1984:348).[217] The Provisional IRA claimed in an official statement that they had "discouraged the prisoners on numerous occasions when they felt that a hunger-strike was the only logical means of bringing... public pressure to bear on the British" (An Phoblacht/Republican News, 5 Sept. 1981:20).[218] After the fasts had begun, the Army Council began to consider the hungerstrikes as a fundamental part of military strategy. While the no-wash protests were marked by numerous assassinations of prison warders, at the beginning of the hungerstrikes, PIRA moderated its military activity, "so that the significance of the fasting should not be obscured" (O'Bradaigh cited in Buckley 1985:85).

body-map

Cultural values, writes Bourdieu, can be "made body by the transubstantiation achieved by the hidden persuasion of an implicit pedagogy, capable of instilling a whole cosmology, an ethic, a metaphysic, a political philosophy through injunctions" (1977:94). The principles of the war were embodied in the hungerstrikers; their bodies became the battleground of the conflict. Military endurance (essentially, cultural sustainability) was 'made body', inscribed upon the bodies of the hungerstrikers, such that the fasting became a war (see Ellmann 1993). MacSwiney once said: "The contest on our side is not one of rivalry or vengeance, but of endurance. It is not those who can inflict the most, but those who can suffer the most who will conquer."[219] The hungering body was no longer simply enduring torture, but inflicting it in the capacity of a weapon. It was generally believed that the longer Sand's starved in the face of public opinion, the more pressure would be brought to bear on the British government (1991 fieldnotes). "Sand's job, then, was to survive, to last as long as possible" (Buckley 1985:126) as if he were a soldier in a very heavy firefight.

The essence of Republicanism was embodied in the hungerstrikers, whose bodies were, in turn, imprisoned. Ricoeur in <u>The Symbolism of Evil</u> suggests that the imprisonment of the body recapitulates the imprisonment of the soul within the body in Christian theology (1967:154), "for the aspect of the other world that is repeated in the body is not its divinity, but precisely its penal function" (1967:286). Ricoeur names this form of spiritual captivity the "body-prison": "But this enclosure 'in the likeness of a prison' receives from its penal character its own peculiar significance of 'alienation' in becoming an instrument of expiation, the body becomes a place of exile" (1967:283). The captivity of the body and captivity of the soul in the body-prison have ramifications for the conceptualization of subjectivity. Ellmann, after Foucault (1979), suggests that subjectivity itself may derive from incarceration: "[t]o be a person is to be a prison..." (Ellmann 1993:94). In Northern Ireland, the experience of macro-territoriality of political domain, is transferred to the micro-political domain of the body, such that siege becomes the dominant metaphor governing experiences of the physical body (Buckley 1989), and the militarization of the H-block prisoners' bodies reflects the militarization of the Irish landscape.[220]

Imprisonment contributes thus to the symbolism of the hungering body: "[h]unger strikers...resist the external limits of the institution, but only by creating an internal fortress of their flesh" (Ellmann 1993:93). The prison, the epicenter of physical State coercion, maps its geography onto the bodies of convicts. In the same way in which political authority is mapped onto a geographical terrain, the bodies of the hungerstrikers also become part of a cartographic project, part of a disputed terrain. Sands wrote: "I was thinking today about the hungerstrike. People say a lot about the body, but don't trust it." The body cannot be trusted because it is subject to the authority of the state. "I consider there is a kind of fight indeed. Firstly the body doesn't accept the lack of food, and it suffers from the temptation of food, and from other aspects which gnaw at it perpetually. The body fights back sure enough, but at the end of the day everything returns to the primary consideration, that is the mind" (1981:Day 17). Unlike their vulnerable, besieged bodies, the disembodied minds of Republican prisoners remained unassailable.

The hungerstrikers experienced their own mortal bodies as prisons, distinct from the political spirit contained in them. Referring to the physical torture undergone in Long Kesh, Sands wrote: "My spirit says, 'No, no, you cannot do what you want with me. I am not beaten...'. I refuse to be beaten" (cited in O'Malley 1990:52). Although the mortal body was vulnerable to mouth ulcers, heart arrhythmia and extreme nausea, the political body was indestructible. "The prisoners' worlds had somehow deconstructed during the years of being on the blanket and the dirty protest, of having endured physical deprivation to the point where their reaction to it had become existential: the physical self was something that existed outside the real self, and thus was something that could be discarded" (O'Malley 1990:111). "Death," reminds Foucault, "is power's limit, the moment that escapes it" (1990:138).

Under military regimes, subaltern popular memory is mapped on both bodies and landscapes (Swedenburg 1991); traditions of nationalist resistance are encoded in geographical space. The hungerstrikers' subjective experience as besieged bodies, whose freedom could only be achieved through the disembodiment experienced during death, was replicated at a macro-level in the geography of imagined nationalism. The Irish nation which the hungerstrikers represented was, itself, disembodied. The idealized nation, composed of the Republican community (living and dead), does not exist as a state. Viewed through a Republican lens, the occupation of Ulster prevents any territorial manifestation of statehood. Divided between North and South, 'Ireland' is sundered from its geographic 'body', and therefore cannot represent the nation. Ireland unfree is Ireland geographically divided. The separation of the hungerstriker from his vulnerable body and the separation of the Irish nation from its natural geography occur under adversity, specifically the occupation of territory, whether bodily or geographic. The spirit of the Irish people, entombed in the imprisoned body of the dying hungerstriker, cannot be occupied because it exists in a metaphysical realm. Death is thus equated with freedom and peace. Padraig Pearse's oratory at O'Donovan Rossa's grave neatly sums up the complex web of representation, morbid nationalism and political occupation: "They have left us our Fenian dead, and while Ireland holds these graves, Ireland unfree shall never be at peace."

The human body often seems to function as an organic metaphor of the state, political relationships often being mapped onto actual human bodies. The 'body politic' in royalist cosmology is, as Walzer (1974) and others have pointed out (Hayden 1987; Cannadine 1983, 1987), an extension of anthropomorphic theology. Ernst Kantorowitz in The King's Two Bodies, discusses the legal and cultural fiction of the King's 'body natural' and 'body politic' as a distinctive feature of Christian political theology. While the body natural is subject to decay and death, the body politic is eternal and sacred. In an Elizabethan legal dispute, the judges wrote: "...the Body natural and the Body politic are consolidated into one, and the Body politic wipes away every Imperfection of the other body..." (1957:11). The state, vested and personalized in the political body of the king, is similar to the Irish 'nation' contained in the political body of the hungerstriker. "The state in theory was the king's body politic," (Walzer 1974:13), just as the hungerstriker contained "the spirit of the risen people, who he represents..." (prison comm. from Sands, 27 Jan., 1981, cited in Collins 1986:113). This representation was institutionalized when Bobby Sands, imprisoned in Long Kesh, was elected to Westminster Parliament as the member for Fermanagh and South Tyrone.[221]

The hungerstrike, like the monarchy, has death at its center. "Regardless of the dogmatic unity of the two bodies, a separation of one from the other was nevertheless possible, to wit, that separation which, with regard to common man, is usually called Death" (Kantorowitz 1957:13). Nevertheless, the kingship endures the death of the king; Republicanism endures the death of its martyrs. Padraig Pearse, hero of the Easter Rebellion, is often quoted by Republicans as saying, "you can kill the revolutionary, but you cannot kill the revolution" (1991 fieldnotes). Sands, likewise, wrote that "I may die, but the Republic of 1916 will never die" (1981:Day 9). This ethic may be transformed into a military theory of viability.

That monarchies and revolutions are strengthened by, or even dependent on, the death of their figureheads is not too farfetched. Certainly, British strategic analysts believed that because martyrdom increased PIRA's community support, it should be avoided (Urban 1993:83). Martyrdom was seen as one cultural basis of the Irish Republican tradition, with positive consequences for political regeneration and cultural solidarity. Sand's death, for example, produced an excess of eager recruits, most of whom were rejected. Martyrdom works to affirm living communities because at one level, it demonstrates that "to die is not separation but re-incorporation ...personal separation and...political separation...will be resolved by a martyr's death..." (Taylor 1989:184). In another sense, regicides, and political martyrs, become a "symbolic paradigm for our own deaths and for the meaning of death itself" (Huntington and Metcalf 1979:182-3). The primacy of the immortal kingship in political discourse, and the function of kings as natural symbols of the political order, led Hocart to conclude that the first kings must have been dead kings (1954). Because the role exceeds the individual (Gluckman 1967), both kings and hungerstrikers are relatively interchangeable, a sentiment which is perfectly expressed in the toast, 'the king is dead, long live the king'.

The relationship between hungerstrikes and monarchy may be pressed even further. Yeat's play of 1904, The King's Threshold, represents hungerstriking as a form of publicly enacted political reciprocity:

He has chosen death:
Refusing to eat or drink, that he may bring
Disgrace upon me: for there is a custom,
An old and foolish custom, that if a man
Be wronged, or think that he is wronged,
and starve
Upon another's threshold till he die,
The Common People, for all time to come,
Will raise a heavy cry against that threshold
Even though it be the King's.

Hungerstrikes are founded on political reciprocity, a mystical relationship of obligation between the most powerful and between the least powerful. "In the darkest region of the political field," writes Foucault, "the condemned man represents the symmetrical, inverted figure of the king" (1979:29).

Because hungerstriking emphasizes the moral component of power relationships between ruler and subject, success is predicated on the existence of a unified body politic. The hungerstrikers were not nihilistically seeking the destruction of power relationships; on the contrary, the hungerstrikers were working from within an imaginary political contract. "Because its secret is to over power the oppressor with the spectacle of disempowerment, a hungerstrike is an ingenious way of *playing* hierarchical relations rather than abnegating their authority" (Ellmann 1993:21. Italics in original). The silent starving body of the hungerstrikers had "spoken to power" (Scott 1990:166), an address to the crown. The hungerstrikers refused all third-party mediation.[222] The symbolic exchange entailed by a hungerstrike depends on this primary relationship between ritual counterparts. In the absence of a functional monarchy to receive bodily discourse, Bobby Sands was "dying for a forgone conclusion; he is negotiating his life without anyone to negotiate with. He is speaking with his bones to someone who does not answer, or who answers only with insults" (Buckley 1985:119).

Father James Healy said "A hunger strike is an interaction between two parties. What is so distinctive when this interaction is between Irish and British? Perhaps it is... an unshakable conviction that any death which occurs will be the responsibility of the other party" (cited in O'Malley 1990:25). Sands is reported to have said "'Look, if I die and anything happens, the British are responsible for that and not me" (Fr. Patrick Rogan citing Sands, cited in O'Malley 1990:63). Hungerstrikes involve a transfer of moral accountability, such that "[t]heir suicide is murder by proxy" (Ellmann 1993:54). Republicans hoped that Sand's election would force this accountability shift; "Margaret Thatcher would not allow a sitting member of the Westminster Parliament to starve himself to death in the face of world opinion" (O'Malley 1990:74). Despite international protest, the British government refused public negotiation.[223]

According to Ellmann, Thatcher misread the hungerstrikers demands, she "occupied the place of blindness in the system of exchange" (1993:104). And perhaps "to the British, such hunger-strikes are merely bizarre, grotesque" (Buckley 1985:170). One soldier reported that

> Giant murals of the dead hunger strikers went up almost overnight...: 'You can kill the revolutionary but you can't kill the revolution' was one of the most famous on the Falls Road. The people there classed them as real martyrs. Myself, I thought whoever starves themselves to death for a cause must be a bit weird, but it's an Irish thing, fasting as protest (Corporal, 45 Commando, cited in Arthur 1987:145).

According to one commentator, Thatcher "did not see Sands and the others as persons, but as disgusting terrorists-trying-to-be-martyrs, insisting on dying on her doorstep" (Buckley 1985:131). The public spectacle of starvation was seen as a "calculated piece... of moral blackmail" (The Protestant Church of Ireland Primate, Most Rev. John Armstrong, cited in Buckley 1985:139) intended to assign blame to the British government.

Despite the distaste which the hungerstrikes may have provoked, British intractability resulted from their *very* accurate reading of the nature of the hungerstrikers demands. The demands themselves were not so important, at issue were the principles defining political violence: "we simply cannot give way on the essential principle... murder is murder. Murder is a crime. It is not to be excused because the motive is political" (Northern Ireland Secretary Humprey Atkins, cited in Irish Times, 7 May 1981:8). According to Thatcher, "[Her Majesty's government] is not prepared, through the granting of political status, to legitimize criminal acts undertaken in pursuit of political ends" (The Times, 15 May 1981:1). And again: "there is no such thing as political murder" (Thatcher cited in the Irish Times 27 November 1980). But as Adams pointed out, "No one, despite British protestations to the contrary, presumes that any of these prisoners are criminal" (letter to the editor, The Guardian, 21 April, 1981).

republican mandates

COIN seeks to abolish the capability of the war system to reproduce itself by interfering with cultural regeneration: because martyrdom gives cultural power, preventing martyrdom became a central part of security strategy. Recognition of this underlying logic of Republicanism by Security Forces shifted security strategy from 'shoot to kill' to covert 'clean kills'. Were the Republicans then deliberately producing martyrs in an attempt to sustain the culture? If this proposition is true, the hungerstrikes were not a political protest, but a mass suicide. This was the view of certain clerics.

After the British government suggested that PIRA were ordering men to die, McFarlane wrote to Adams: "with the authorities still adamant, more deaths would start to look like suicide, which would be damaging for the movement" (cited in O'Malley 1990:72). Statements from religious authorities that the hungerstrikers were "an act of violence" (Archbishop Runcie of the Anglican Church, cited in Buckley 1985:127) echoed Thatcher's refrain, "a crime is a crime is a crime" (21 April 1981 statement on BBC, cited in Buckley 1985:137), but failed to establish against whom the alleged crime was being perpetrated. Characterized as 'suicide', which in the Catholic tradition is a sin, the hungerstrikes were a crime against God, a neat extension of criminalization into the symbolic realm. The Catholic ecclesiastical hierarchy in the form of Cardinal Basil Hume, archbishop of Westminster, agreed. The Irish Catholic Church, on the other hand, did not consider the deaths as suicide, rather as "a responsible protest though of a sacrificial nature" (Father Denis Faul, cited in O'Malley 1990:178). The Protestant Church of Ireland Primate, Most Rev. John Armstrong said that the hungerstrikes were "one of the most calculated pieces of moral blackmail in recent times" (Buckley 1985:139). Bishop Edward Daly of [London]Derry asserted that Sand's death was categorically "not suicide". The sole neutral party was the Belfast city coroner, who refused to examine motives, ruled the deaths as due to starvation, and promptly resigned.

The issue of hungerstrikers deaths as self-sacrifice or as suicide became central to the negotiation of ideological legitimacy. "Hunger-striking, when taken to the death, has a sublime quality about it; in conjunction with terrorism it offers a consummation of murder and self sacrifice which in a sense can legitimize the violence which precedes and follows it" (Beresford 1987:25). By turning the terrorist into a victim, the hungerstrikes translated a 'bad' violence into a 'good' violence. Rene Girard in <u>Violence and the Sacred</u> writes that, "this obscurity [of ritual] coincides with the transcendental effectiveness of a violence that is holy, legal, and legitimate successfully opposed to a violence that is unjust, illegal and illegitimate" (1972:23). As a ritual, the hungerstrikes claimed that a crime was not a crime was not a crime, and never had been.

The language of martyrdom and sacrifice is a common feature of death-based nationalism. European nation-states tend to identify national causes with Christian sacrifice, which is partially the result of the "fatal capacity that Christian language has for redeployment in warfare" (Sykes 1991:90). The death of Christ, which in turn serves as a metaphor for the deaths of martyrs, is the primary example of the metaphorical extension of Christian symbolic language. The "theological subversion of the Christian language of sacrifice" is most powerful in the context of war" (Sykes 1991:95). In order to redeem the 'bad' violence of terrorism, and animate the narrative of nationalist martyrdom during war, the hungerstrikes had to be characterized as self-sacrifice rather than suicide: this is the language of war. While the self-abnegation of martyrdom is altruistic, suicide destroys without principle, affirming and negating the primacy of the individual in the same moment (Durkheim 1970). 'Taking one's own life' hinges on the concept of the life-force itself as private property, belonging to the individual and disposable at will. A Catholic priest stated that "from reading a great deal about Irish hungerstrikers I am convinced that any suggestion that they are taking, rather than giving, their lives would sound absurd to them..." (cited in O'Malley 1990:25).

The hungerstrike was seen as self-sacrifice of a prime order because it valued the continuity of Republican community above the value of the individual lives that were sacrificed. The sacrifice is "rather...a substitute for all the members of the community, offered up by the members themselves" (Girard 1972:8). Bobby Sands wrote: "I am dying not just to attempt to end the barbarity of H-block or to gain the rightful recognition of a political prisoner but primarily because what *is lost in here is lost for the Republic*" (cited in O'Malley 1990:57. Italics are mine). The logic of sacrifice maintains that a life must be relinquished if life is to be attained (Bloch and Parry 1982:8; Strathern 1982). By giving their lives to the Republic, the hungerstrikers would become part of a continuum of martyrdom; their deaths would be legitimated by future-time history. Gilbert argues that "sacrifices renew ancestral power, and the latter is dependent upon those who 'serve' in the sense of being killed" (1987:307). By allowing themselves to be killed by proxy, the hungerstrikers assured that the Republic would live. "The cult is that of Ireland," informs Conor Cruise O'Brien, "as a sentient being called for human sacrifice — and also of those who heard that call, and died. And killed" (cited in the Observer, 10 May 1981:14). Precisely, though O'Brien probably has little understanding what his statement implies.

Behind the idea of death as a soldier in the service of the Republic lies the concept of *mandate*. Only a legitimate state has the authority to send soldiers to death. Sand's autophagic regicide expressed and confirmed a military mandate for the Republican movement: that is, the legitimate right to use force for the furtherance of political objectives. Sand's autophagic regicide also expressed and confirmed a cultural mandate for the Republican movement: that is, the legitimate right to choose the time and the place of one's own death. The hungerstrikes thus express two primary messages: the right to kill, and the right to die.

Sacrifice, especially self-sacrifice, is a 'good' death because it represents the capacity to choose: "Both the impulse to determine the time and place of death, and the dissociation of social death from the termination of bodily function, clearly represent an attempt to control the unpredictable nature of biological death and hence dramatize the victory of order over biology" (Bloch and Parry 1982:15). 'Good' sacrificial death, by mastering the arbitrariness of biology, results in social regeneration. Sacrificial death becomes a central mechanism in the establishment of social order (see Mack and Hamerton-Kelly 1987) and certainly, the hungerstrikes did regenerate the Republican movement. Girard sees "the concept of sacrifice as... an instrument of prevention in the struggle against violence" (1972:17). Sacrifice, or judicial scapegoating, transforms chaotic violence into a rational appeasement of bloody desire. Thus, "carnivorous sacrifice is essential to the structure of subjectivity" (Derrida 1992:18-19). While Girard argues that judicial systems can be seen as substitutes for sacrificial rites (1972:18-19), the hungerstrikes also suggest that sacrificial rites can function as substitute for justice.

Sacrifice as a *chosen* death links up with arguments made earlier in this chapter. Irish ethnography, popular culture, and political discourse invoke a number of common tropes about the "death of the Celtic fringe" and the "death of the Irish language". These "deaths" are treated anthropomorphically in the ethnographic literature and in popular discourse: cultural mortality thus appears to be an inevitable conclusion to 'living' culture. The causes of culture death escape examination. Cultures, of course, do not die magically of their own accord, but must be actively killed (through deprivation of resources, dispossession from arable land, outlawing language and religious expression, appropriation of foodstuffs, and so on). Republicans and many other Irish people believe that Irish culture has undergone a systematic extermination by British imperialism. While death might be inevitable, murder is intentional. The hungerstrikes force us to conduct a forensic pathology of culture by positing the question: if an Irish hungerstriker dies, who is responsible? if Irish culture dies, who is responsible? For Republicans, the answer to both questions is: the British.

Republicans read criminalization as yet another attempt at destroy not just Irish nationalism, but Irish culture. Counterinsurgency was understood as an attempt to murder Irish culture, and to pretend that it wasn't happening, that everything was normal in Ulster. If death is inevitable, isn't it better to choose how one dies? War gives men the opportunity for a meaningful death in an otherwise banal, empty world. The preference for a meaningful death may be difficult to understand if one has been taught to value life — any sort of life — above all. In West Belfast, "any sort of life" is hardly any life at all: daily life is incomprehensibly hard. Joining the Provisionals is the highest possible achievement not despite, but *because* membership will probably result in death. Death for paramilitaries represents the capacity to choose in conditions when choice is the last luxury. Death represents the survival of the culture at the expense of the self.[224] Death represents political autonomy when the other option (life) is the worst kind of slavery. From within the war-logic of the Provisionals, starving to death in the service of Republicanism — a heroic act — was preferable to being degraded by a foreign uniform and living to tell about the experience. Sacrifice made during times of war activates and affirms a number of transcendental principles which are commonly invoked as the "noble sentiments" of war: glory, compassion, and honor. Suicide, through a careful selection of the time and place of death, expresses a preference for death before dishonor. Cultural *seppuku*, an honorable autophagic suicide, was preferable to the dishonor of culture-murder in a Breaker's Yard.

The right to die raises certain aspects of royalist cosmology: the death of the king reaffirms the life of the state. At the heart of every theology of war is a "fatal coupling" which Ricoeur calls the "King-Enemy" relation: "for Sands, the battle in the H-blocks wasn't so much about conditions in the jail, as it was two ideologies fighting each other. It was two cultures fighting each other. It was the crystallization of the whole struggle" (Jake Jackson cited in O'Malley 1990:53). This King-Enemy relationship is genesis of social order: "...every historical drama, every historical conflict, must be attached by a bond of *re-enactment* of the cultural-ritual type to the drama of creation" (Ricoeur 1967:191). Cultural rituals, by excluding time and history, re-enact the creation myth. Ritual praxis often centers on the form of the King, who mediates between the gods and man. Royalty thus ties history to the cult. The center of this mythic narrative is "a theology of war founded on the identification of the Enemy with the powers that the god has vanquished ...the mythological type of the drama of creation is marked by the *King-Enemy* relation, which becomes the political relation *par excellence*" (Ricoeur 1967:197-198). The foundation of all human political relationships lies in the exclusion of the "Enemy-Other". Evil is structurally associated with the enemy, so that "in the final analysis, evil is not an accident that upsets a previous order; it belongs constitutionally to the foundation of order" (Ricoeur 1967:198). "By negation, order reaffirms itself" (1967:43). Because sacrifice is part of a system of exchange, the logic of sacrifice in the interest of order demands that chaos be banished: "we must therefore first kill the gift..." (Leach 1976:83).[225]

This raises the second issue: the right to kill. At the root of Republicanism is an ideology of the political efficacy of violence, a conviction that the application of force will bring England to the bargaining table. This is the central fact of Republicanism: if you don't believe in the right of the Provisionals to use force, you are not a Republican. "For the Provos, force was the first resort and electoral politics only arose when force was seen to flounder. For this reason it could command the support of only a small and fluctuating minority, yet because it was the uniting factor in the movement it could not be rationally debated, much less abandoned..." (Clarke 1987:225). The diplomatic failure of the hungerstrikes to win concessions is hardly surprising, since they were not performed from within the tradition of diplomatic politics, but as a military operation.[226] The final breakdown of the hungerstrike is evidence of a lack of any institutionalized political control over a military action.[227]

There are strong anti-Clausewitzian aspects in hungerstrikes as paramilitary practice. PIRA claims an historical mandate as the legitimate government of the Republic. No political organization has this authority —

it is the prerogative of the Army Council itself. Every Republican owes allegiance to the Army Council, of which Sinn Fein is merely the proxy. The training manual of the Provisionals, The Green Book, states that "The leadership of the IRA is the lawful government of the Irish Republic" (cited in Clarke 1987:2). Compare this with clause twenty of the Constitution of the Irish Republican Brotherhood:

> The Supreme Council of the Irish Republican Brotherhood is hereby declared in fact, as well as by right, the sole Government of the Irish Republic. Its enactments shall be the Laws of the Irish Republic until Ireland secures absolute National Independence, and a permanent Republican Government be established.

These two Republican documents claim legitimacy in almost identical language. "The IRA in any event sees itself as the legal successor of the 'Second Dail', and as such the government of the Irish Republic" (Cronin 1980:208).[228] As the legitimate government of Ireland, the Provisionals have a legitimate mandate to declare war against Great Britain. This is exactly what they believe.

> If one accepts the Provisionals' premise that they are fighting a war, that their instruments of violence are not always controllable, that to be militarily significant they must carry out certain operations, with insufficient safety precautions, it must be conceded that they have not been totally irresponsible in their conduct or choice of

target. If one rejects their premise, or their right to wage war, the argument changes: there is no justification for anything they do (Cronin 1980:207).

From where does this distrust of politics, this structural and ideological primacy of the military above the political, arise? Some Irish history may, at this point, be necessary. The Proclamation of the Irish Republic on Easter, 1916 in Dublin constituted the first Dail Eireann. The Dail Eireann was the parliament of the Irish Republic. The IRB, rechristened by the Dail to the IRA, took oaths to "support and defend the Irish Republic and the Government of the Irish Republic, which is Dail Eireann, against all enemies, foreign and domestic." In December 1921, the Dail Eireann signed a treaty with England which gave them the status as a nation-state within the dominion of the British Empire, and which ended the Anglo-Irish War. Sinn Fein split into two factions in the Dail: those favoring the Treaty, and those opposed. "But when the Dail 'betrayed the Republic' by accepting the Treaty, the IRA, in the spring of 1922, withdrew its allegiance and fell back on its own convention for authority" (Cronin 1980:132). The Anti-Treaty faction called themselves Republicans, since they believed that the Anglo-Irish War had been fought in order to found an independent Republic, outside of any political influence of England. The Irish Free State came into existence in 1922, and the Republican anti-Treaty forces began a guerrilla war against it.[229]

These historical facts have informed Republican political thinking for over fifty years, and still influence the daily functioning of the organization. The refusal to negotiate during the hungerstrikes (not to mention the breakdown of the more recent talks between 51/55 Falls Road and Whitehall), was not simply due to mistrust of 'perfidious Albion.' Sinn Fein's electoral abstentionism, the Blanketmen's refusal to wear uniforms, and the equation of political compromise with treachery in the Republican movement, result directly from the Republican belief that their legitimacy derives from a military mandate, which is merely confirmed by violent praxis. "The reasoning in the IRA goes back to Rory O'Connor, who occupied the Four Courts in 1922 and repudiated all political authority because Dail Eireann had betrayed the Republic by endorsing the Treaty" (Cronin 1980:209).

The subordination of the political to military control within the Republican movement brings Ludendorff to mind: "Both warfare and politics are meant to serve the preservation of the people, but warfare is the highest expression of the national 'will to live', and politics must, therefore, be subservient to the conduct of war" (1936:24). Although it may seem that the tail is wagging the dog, the Provisional IRA's physical force tradition not only established the Republic of Ireland (Eire) which today has the status of a state, but has assured the continuity and preservation of the movement — its viability, if you will — for at least a century (or eight centuries, depending on when one begins counting). The ideology of Republicanism is also remarkably consistent: Seamas Twomey, former commander of the Belfast Brigade of PIRA in the 1970's and Gerry Adam's predecessor as Chief of Staff, summed up the basic ideology of Republicanism very simply: "Our first prime and main objective is the unification of our country. This means getting the British out of the occupied part of the country. After that the whole system in North and South would have to be changed" (cited in Cronin 1980:214). For Republicans, the presence of the British Army in Ulster recapitulates a prior myth of *occupation*. British reasons for the troop presence (e.g., peacekeeping) ring false in Republican ears, and fundamentally, the reasons just don't matter. Republicans believe they have the right to kill, and die, in the service of the Republic. The presence of troops, which activates the mandate, obligates them to do so. But as Churchill said while Colonial Secretary: "We do not recognize the Irish Republic. We have never recognized it, and never will recognize it" (cited in Cronin 1980:148). Despite perpetual hope of a political solution, the problem is really about the legitimacy of armed force, and can only be solved by military means. This view is nothing more than pragmatic.

Chapter Seven:

In Conclusion: Thing-killing Lethality

But when we're duly white-washed, being dead,
The race will bear Field-Marshal God's inspection.
—Wilfred Owen, Inspection (1917)

War in the age of intelligent machines depends primarily on machines, rather than human beings, for the production, analysis and distribution of death. Decentralization of command and control schemes, the development of 'smart' weapons, video combat simulation, and other technological manifestations, increasingly remove human beings from the lethality of war. And war is becoming increasingly lethal. Lethality indices for weapons (based on characteristics such as rate of fire, potential targets per strike, accuracy, reliability, range of action, etc.) which calculate the theoretical maximum killing capacity per hour show a logarithmic increase in lethality in the evolution of machine system s from the smooth-bore artillery to the ballistic missile (Bellamy 1990:46).

Dispersion of combat troops on the battlefield has increased proportionally. "In ancient times, one man probably occupied 10 square meters of battlefield...In the American Civil War one man occupied rather over 200 square metres; in World War One, over 2,000 and in World War Two, over 20,000" (Bellamy 1990:47).[230] In low-intensity conflict, however, battlespace is very compact, and limited by geography. Northern Ireland is only 5,238 square miles. Belfast is only 25 square miles. Dispersion of troops in the restricted, compact theater of Northern Ireland is simply not possible. The *cordon sanitaire* restricts expansion of the war.

Nevertheless, the geospace of Ulster is full of things designed to kill other things. Increased metalization has been required for armored personnel carriers (APCs), also known as 'Pigs', to reduce their vulnerability to PIRA's 7.62mm armor-piercing rounds. PIRA has almost certainly now acquired heat-seeking surface-to-air guided missiles, and in response to this threat, most RAF helicopters have been fitted with heat shrouds over their engine exhausts, and other electronic countermeasures too classified to mention. In South Armagh, the 'Bandit Country' along the border of the Irish Republic, sixteen 20 meter high observation towers monitor roads, border crossings and cow pastures. These towers are staffed by men who live in underground bunkers. Men and supplies are flown in by helicopter because even armored vehicles are vulnerable to land mines and improvised explosive devices (IEDs). Road traffic throughout Northern Ireland is controlled by permanent vehicle checkpoints (PVCPs), with automatic barriers and intercom systems through which soldiers communicate with drivers. Soldiers manning PCVPs are armed with .50 cal Browning heavy machine guns and French manufactured Luchaire Close Light Assault Weapons (CLAW). The CLAW rocket grenades are tipped with HEAT warheads and can be fired from SA-80 rifles, the standard British Army weapon. HEAT

> refers to the relatively new High Energy Anti-Tank warhead, a round utilising a shaped charge that is ignited from the rear at the very instant the conically hollowed front of the charge comes in contact with the armour. This combination of directed and kinetic energy focuses a jet of hot gas capable of penetrating several hundred centimetres of the toughest armour and squirting molten metal throughout the turret of a tank (Walker 1986:33).

HEAT warheads are a 'thing-killing' weapon. In the world of weapons R&D, man-killing is passé; missiles, tanks, bombs and artillery are designed for 'thing-killing'. The weapons are nominally designed to seek out and annihilate non-human targets; yet they are, of course, exceptionally lethal to human beings, as anyone who watched Gulf TV is well aware. DeLanda (1991) suggests that the development of 'non-lethal' technologies—'thing-killing' as opposed to man-killing weapons (see also Keegan 1976) — brings weapons systems lethality out of control (1991:18-25). Any moral inhibitions against barbarities in design are displaced by the abstraction of warheads designed to pierce armored tanks (see also Gabriel 1987; Best 1980, on attempts to establish limits to weapons lethality). Hyper-lethal armor piercing war heads, for example, are designed to penetrate and attack tanks. Actually, HEAT rounds kill human beings, although this happens merely *as a by-product of their primary tank-killing task.* They destroy the weakest link in the chain of the weapons system, which is the human soldier.[231]

In any war machine, the human operator is one aspect of a more complex unit, which includes actual hard technology (for example, a tank) and the skills or techniques needed to manipulate the technology (what buttons to push to drive the tank). Human operators, like tanks, are interchangeable because they are merely carriers of tank-driving skills. The system can continue to function as long as tanks and soldiers with the necessary skills to operate them are available. The system, if it is functioning properly with the proper technology, techniques and operators, will have what is known as a rate of 'survivability'. Simply, survivability means the ability of a weapons system to survive attack; to take a hit and continue to function. 'Survivability' as a military concept does not apply to human life *per se*, but rather the survivability of the system as a whole. Because human soldiers, regardless of how well-disciplined or automatized, are the weakest, most vulnerable component, keeping them in the loop can threaten the survivability of the total weapons system. Human survival is clearly not the priority here: tanks are infinitely more valuable and not so easy to replace.

The vulnerability of humans as biological organisms threatens the systemic survivability of weapons systems. Human beings are obviously much easier to kill than tanks. The quantum lethality of high-intensity combat using 'thing-killing' weapons presents a number of options: either humans can be replaced by machines — taken out of the loop—, or a cybernetic man/machine combination can be devised (Arnett 1992). Post hoc analysis of the 1991 war in Iraq has stirred up a controversy regarding the relationship of men and machines in future war (see Arnett 1992). While weapons scientists and civilian strategists are advocating the development of autonomous 'brilliant' weapons, military thinkers prefer to keep men 'in the loop'. The ethos of war probably requires the participation of men: combat by machines being perceived as mere cowardice.

But what happens to the ethos of war when combat lethality reaches 100%? According to British Army Brigadier Richard Simpkin (1982), in <u>Human Factors in Mechanized War,</u>

> Sure, microprocessor-based control engineering enables semi-skilled individuals of limited intellect to select and initiate complex programs. At the same

time the very nature of these advanced systems combines with the operators lack of understanding of them to take the man out of the primary control loop and so prevent him from 'pulling something out the bag in an emergency' (8).

The deployment of weapons designed to destroy other weapons, and human beings only incidentally, increases the possibility of 100% human casualty rate during combat. Human soldiers are psychologically vulnerable and cannot be trusted to continue to fight in battlefield conditions where death is taken for granted.

Problems concerning lethality and human vulnerability will be compounded for any army which employs *Auftragstaktik*. While decentralized tactical schemes absorb and maximize local chaos, they also "stretch the chain of command by allowing more local initiative. This increases the reliance of the war machine...of its human element...[T]rust...is expensive for State war machines" (DeLanda 1991:79). Trust is also extremely difficult to engineer as a quality in human soldiers, who, after all, are not machines just yet. Combat reliability of soldiers may be maximized by the production of sociopathic personalities, simultaneously obedient and super-violent (Gabriel 1987), by the use of stress reduction drugs, by discipline, etc.[232] Cybernetic technology can also be used to substitute for soldiers in the feedback loop (DeLanda 1991:79).

"The human must be modified if it is not to be the weakest link in an integrated weapons system" (Gray 1989:59). Soldiers are therefore being (re)constructed and (re)programmed to fit integrally into weapons systems...the human body is the site of these modifications— whether it is of the 'wetware' (the mind and hormones), the 'software' (habits, skills, disciplines) or the 'hardware' (the physical body). To overcome the limitations of yesterday's soldiers, as well as the limitations of automation as such, the military is moving towards a more subtle man/machine integration: a cybernetic organism ('cyborg') model of the soldier, that combines machine-like endurance with a redefined human intellect subordinated to the overall weapons system (1989:43).

The biotechnological aspects of armored war-fighting have been thoroughly explored by the military establishment. Richard Simpkin writes: "a crewman's station should be seen as an extension of his body— his nervous system and his muscle power" (1982:102). Simpkin advocated that particular concern should be paid to the somatic nervous system (103), which is the subconscious programmatic responses to stimuli during combat operations, which are more easily manipulated and trained than frontal cortex responses. Simpkin wrote Human Factors in Mechanized War (1982) to correct for the emphasis on pure machine technology in Antitank (1982). This suggests an understanding that humans should not be replaced with machines, but unified and integrated. This process is already quite advanced: the personalities and thoughts of soldiers are often hardly more than the by-product of interacting with machines.

The real problem for Simpkin and other military theoreticians seeking to improve troop performance in high-intensity mechanized infantry combat, is no longer how to manipulate, desensitize and program soldiers, *but how to conserve individuality and avoid dehumanization* (29). Intense inter-personal bonds between soldiers, in other words, extremely humane expressions of empathy, have historically contributed to the sustainability of combat. Human qualities are so now eroded by machine oriented cybercombat, and soldiers are so prone to become extensions of the deadly machines which they operate, that prevention, conservation and maintenance of the humanity of infantry soldiers is the priority. Humanity must be systematically reprogrammed into cyborg personalities, rather than programmed out.

Removing humans from the loop, constructing cyborg soldiers, dehumanization through the application of discipline, the introduction of AI, and even the use of drugs, represent possible solutions to the conundrum of how to maximize the capacity of human soldiers to operate in a battlespace characterized by thing-killing lethality. On the other hand, the persistent humanization of death in counterinsurgency makes these solutions problematic. In low-intensity conflict, despite the presence of lethal machines, the human factor predominates. To explore this, we will compare military subjectivity in cyberwar and in Northern Ireland.

derealization and hyper-vigilant paranoid inversion

The central tenet of every military strategy is: *know your enemy*. Cyberwar enemies are no longer identified by actual visual clues, such as shields, helmets, or banners. Battle-management systems select and identify enemies mechanically. Autonomous predatory machines deploying 'thing-killing' weapons prevent conceiving of enemies as human. The abstraction and distancing of the enemy in military visioning systems, what Shapiro calls 'derealisation', began in Vietnam, with target selection made by pilotless planes. Imaging devices created data, which were electronically transmitted to computers, so that enemies remained invisible to commanders. Relying on electronic information, operators struck at symbols rather than actual bodies. Thus, "the visual rift between adversaries, who appear to each other on the video scopes of the hi-tech weapons of modern warfare, amounts to an obscuring...of the objects of lethal violence" (Shapiro 1993:120). Enemies are now very largely abstract:

> during large-scale hostilities the enemy/ objects of violence are familiar neither to the antagonistic populations nor the combatants, in modern warfare, the visioning and weapons technologies render the antagonists even less familiar be derealising or dematerialising them , i.e. by apprehending and targeting them primarily through remote visioning devices (Shapiro 1993:114).

The visual rift of derealization shifts the moral norms of combat: weapons requiring direct contact and physical expenditure of energy now appear 'savage'. Physical proximity in battle no longer signifies bravery, but savagery.[233]

This abstraction prevents the production of enemies through political deliberation: "electronic friend and foe identification systems (IFF)...represent the disappearance of political space. Spy satellites and electronic ship- and aircraft carried information systems, which not only determine threats but also automatically send hostile responses, constitute what Paul Virilio has called 'the intelligence of war that eludes politics" (Shapiro 1993:116). Logistically oriented decision making procedures, reliant on electronic information systems, for example the US Navy's Aegis battle management system (Shapiro 1993:117), now determine enemies automatically. Indeed, future technology is tasked to identify, detect, and distinguish friends and foes: in the era of covert combat X-ray machines and people 'sniffers' have more applications than tanks and bombers (Van Creveld 1992:63).

The normalization of military lethality, and of panoptical surveillance, has made 'deep black' surveillance rather mundane. "Because the modern notion of national security is linked with a militarisation of the globe, imaging from space satellites and high altitude reconnaissance...is a continuous, every-day phenomenon" (Shapiro 1993:118). Surveillance is also the normal, continual, unexceptional social condition in the urban battlespace of Northern Ireland. Identifiable signs of military presence are indistinguishable from the general militarized conditions of West Belfast. Because defence is now primarily electronic and invisible, military landscapes need not be characterized by bunker architecture.[234] Any landscape can be a military landscape. "Demilitarization of the core" creates the impression or normality (Jarman 1993:117). The transparent, electronic battlefield is populated by sensing devices, night-seeing automata, and sentient computers, which create an electromagnetic environment in which nothing hostile can survive. Combat areas are under twenty four hour surveillance. The electronic battlefield is an epicenter of surveillance, blanketed by "...a wide range of equipment and sensors in the military arsenal... monitoring...the battlefield by day and by night in all weathers..." (Gudgin 1989:138). Technostrategic panopticism occurs continuously at every level.

Remote visioning devices, thermal sensors and electronic eavesdropping contribute to the transparency of the electronic battlefield in the urban and rural spaces of Ulster. In low-intensity conflict, however, derealization of the enemy can never be as explicit as in a high-intensity cyberwar. The primacy of the human factor in counter-insurgency, despite the bevy of electronic warfare devices being deployed, means that violence is always to some degree personal, killing is done face-to-face — point blank — by soldiers and 'volunteers' who very often know one another by name, and have memorized each other's photographs and dates of birth from snapshots tacked to ops room walls. Republican paramilitaries choose specific targets for assassination (for example, prison warders and UDR members), and are very familiar with the individual British regiments assigned to patrol urban neighborhoods. The face-to-face interactions on city streets personalize the enemy for British soldiers and for Republican paramilitaries, and guarantee that the 'war of gazes' is intense.

Surveillance by intelligence organizations and soldiers on the ground — "a kind of state-sanctioned voyeurism" (Der Derian 1993:25)[235] — was advocated as the most effective means of countering terrorism (Styles 1975:165; Evelegh 1978). Surveillance was at "the heart of the terrorist-defeating system" (Evelegh 1978:131), electronically linking private and governmental information into the Security Forces' intelligence data system (Evelegh 1978:131). Security Forces on the streets of Belfast and Derry observe and patrol as a 'real time' task. Observation and patrolling are the key to "domination on the ground" (Dewar 1985:180) because "the rule of law cannot be maintained without regular visits from those upholding the law" (Dewar 1985:182). Observation posts were established in Republican areas in which soldiers were required to sit and watch. In one case, soldiers sat in a cramped OP for six weeks before shooting a gunman (Hamill 1985:216).

The persistent humanization of death in low-intensity conflict through point blank killing and surveillance on-the-ground has implications for military subjectivity. Soldiers who have been trained to target X's on a map, and accustomed to counting kills in distant puffs of smoke are thrust into a power relationship of observing and being observed within a very limited geographical space. During extended tours of duty, this power relationship of the observing and the observed is inverted; the state eye penetrating the terrorists' coal bin, laundry hamper and kitchenette, becomes objectified by the very thing it seeks to observe.[236]

Soldiers often experienced themselves as objects of surveillance: "you can't relax because that's another of their weapons. They wait for the time when you are relaxed, and they're watching and waiting all the time, and as soon as you relax they hit you. Then you're fucked" (Corporal, 45 Commando, cited in Arthur 1987:166). The small, mobile British Army patrols "...were taught that they were watched all the time" (Hamill 1985:141). And indeed, they were. "You never see the person who throws it — but as soon as it hits you, probably a thousand voices start cheering. You never see a curtain twitch but they're all watching and they know you can't do a lot about it" (Marine, 45 Commando, cited in Arthur 1987:167). On approaching a road block in Northern Ireland, drivers are always signaled by soldiers to turn off their headlights, as the illumination provided by headlights and by streetlights affords PIRA snipers excellent targets. "To be in a street where gunmen operate in uniform and to be illuminated by street lights or car headlights while the surrounding areas are dark is a truly horrible experience" (Evelegh 1978:84).

The function of the Fianna Eireann, the youth wing of PIRA (who actually have a charter from the Boy Scouts, secured by Countess Markievicz during the Anglo-Irish war), is to observe, scout and report on the patterns of Army/RUC patrols. In Republican areas, housewives, mom and pop groceries, taxi drivers and bricklayers keep their eyes open and pass on information about the doings of the Army. Surveillance of the British Army by PIRA has also been performed electronically. According to one Lisburn staff officer, "Our telephone was actually tapped by the Provisional IRA at one stage. Really! you'd have thought you could trust you own phone in a headquarters..." (cited in Hamill 1985:177). In 1971, OIRA hacked into the teleprinter network linking Army HQ with all battalion tactical HQ's and were thus able to access all daily intelligence summaries and battalion commanders requests for specific raids (Doherty 1983:117-123).

Der Derian has written that cybernetic surveillance regimes display a pathological hyper-vigilance which resembles paranoia (1993:33). The paranoid pathology of the national-security state manifests in a very different form of paranoia for men 'on the ground' in a low-intensity conflict:

..I felt their eyes following me and measuring my every step. I knew that I was becoming paranoid, but I was unable to prevent it. The very streets, the derelict houses, the shop-signs and the graffiti breathed a feotid breath of hostility. The city was literally alive with hatred. When I walked through the streets, I felt like a victim in a Disney movie, where the houses developed fanged mouths and demon eyes and watched me gloatingly, waiting to pounce (Asher 1991:262).

The permanence of war, and the normalization of violence, force spectators and participants to avoid acknowledging or internalizing the horror of the spectacle of daily life. This results in a weird, ambivalent indifference to death:

Card lay on his left side, bleeding heavily from a hole the size of a florin in the side of his neck. He was conscious and groaning loudly... The unseen gunman fired intermittently just over our heads and we could not stand, or even kneel upright....we waited, pressed flat to the earth. Other platoons were deploying quickly to cordon the area and flush the gunman, the doctor was on his way and we were stuck in that damnable patch of grass, unable to move without provoking a bullet from the hidden sniper....In the bedroom of one of the nearby houses a man was getting dressed in a calm unhurried manner. He buttoned his shirt and put on his tie, watching our small group with *the same casual disinterest* that he might have shown to a group of sparrows clustered round some crumbs in his garden" (Major Gary Johnson of the Royal Green Jackets, now Major-General Johnson, Defence Staff, Ministry of Defence, cited in Dewar 1985:71, first published in Greenjacket Chronicle vii, 1972. Italics are mine).

military compassion and anti-politics

This weird ambivalence to death is a necessity for professional soldiers. As a job, war requires the suppression of humane compassion. According to a Platoon Commander:

I don't think that the men felt the same sense of shame or intrusion which I felt when going into someone else's home. I could never come to terms with bursting in at four in the morning... screaming children!...women!...It was absolutely hateful and I loathed it....In those days we didn't knock but went straight through the door...I can remember a sense of revulsion — of almost physical sickness. Not because of the dirt and squalor. It was an emotional reaction to invading someone else's house...seeing adults and kids, bleary-eyed and scruffy and dirty...and then there was no one we wanted in that house! Ninety-nine times out of a hundred there never was! (cited in Hamill 1985:75).

A Sergeant from 40 Commando reported:

> My saddest moment in Northern Ireland wasn't being shot at, or bombed, or attacked by rioting crowds, but the first time I was spat at. That was the biggest shock, just being spat on, by an extremely pretty girl. If you're shot at, it's detached. They're doing it for military advantage....there's no personal contact. But if someone spits at you it's hate,...and that's a very strong emotion to inflict on someone... (cited in Arthur 1987:126).

War requires that soldiers bear the burden of hate. The longevity and endurance of the institution depends on the existence of a capable, professional class of soldiers who can tolerate hatred and suspend compassion. These men view war as a job:

> And I never felt that I was being taken advantage of by the government or the system, being told to stand there and take the hail of bottles and bricks....I just felt I was doing a job, and it was a bit more interesting than some of the other things I might have been doing (Subaltern, 1 King's Own Scottish Borders, cited in Arthur 1987:23).

Although war may be undertaken for political aims in the national interest, and the military professionals who organize and fight wars may subscribe to a particular ideology, the fundamental requirement of war that *soldiers follow orders* demands that politics be excluded from the praxis of war. Clausewitz said "patrols do not make their rounds from political considerations" (1987:404). Politics must be excluded from praxis.

Military professionalism may thus be correlated with moral ambivalence. In counter-insurgency, wrote Kitson, "as in other forms of conflict, there is some right and some wrong on both sides" (1971:8). Moral ambivalence forms the basis of soldier's compassion.

> I felt sorry for the people there...and acutely sorry for them because of the conditions in which they were living. The Lower Falls was like Coronation Street: no baths, outside loos, houses that had never been repaired, deterioration everywhere — a terrible, filthy place. In those days we used to do the search at two o'clock in the morning, at random: we didn't have to have proof, we just searched a house for no better reason than it hadn't been done before. And if the door wasn't answered within thirty seconds, we kicked it in...It was nothing to find four or five people sleeping in a double bed; women who you thought were fantastically attractive in the day time, you'd see them in a nightie with no shoes on and notice how black their feet were. It undermined their self-respect for you to see them like that (Sergeant, 1 Royal Green Jackets, cited in Arthur 1987: 90-91).

There are professionals on both sides; Republican culture, as a military culture, produces good soldiers. According to Future Terrorist Trends: "PIRA will probably continue to recruit the men it needs. They will still be able to attract enough people with leadership talent, good education and manual skills to enhance their all-round professionalism."

The compassion born of military professionalism allows soldiers of opposing armies to recognize professional similarities with one another: "...the opposition you are up against are one of the most professional groups in the Western world. They are very committed....I grew up with Catholics who have joined the IRA. I have a healthy respect for them and they for me..." (Major, UDR, cited in Arthur 1987:229). Soldiers identify with each other as soldiers, and understand that the discrete components of antagonism — they, themselves— are interchangeable parts of military systems.

> If I'd been born in the Falls Road I know I'd be in the IRA. They want a free Ireland. They see the British Army as oppressors....I could see that those IRA guys are really dedicated to their cause, prepared to die for it, and I respect them for that (Corporal, 1 Welsh Guards cited in Arthur 1987:247).

An Army Intelligence officer said regarding PIRA Volunteer Thomas McElwaine, "[h]e was an extraordinary bloke who would have been in the SAS if he was in the Army. It is just as well he is dead" (cited in Urban 1993:219). McElwaine was born into the war, and did not have the luxury of choice; neither do British soldiers.

Compassion fundamentally contradicts the art of war, which is founded on atrocity, death, and obedience. Thus, "those for whom knowing the enemy is very professionally relevant — are likely to be those who are least well equipped to engage in the activity [of empathy]" (Booth 1979: 103). The conduct of war demands dehumanization.

> The Provisional IRA...thought of British soldiers as 'uniforms'...If they had regarded us as human beings, they would never have been able to kill us. Our worship of violence insulated us from guilt or fear. It prevented us from remembering that the IRA were human beings like us. The ability of people to dehumanize anyone outside their own 'tribe' is the sole cause of war. It is also, paradoxically, the reason for human survival. The world exists in harmony, but everywhere nature is in conflict, tooth and claw. Conflict is what gives nature its structure (Asher 1991:120).

War ends, and soldiers become incapable of protracted combat, when compassion and identification with the enemy, supersede the ambivalence instilled by professionalism.

> I rapped on the door. 'Shut up!" I said. Donnelly turned and looked at me. No other word was spoken between us, but that look said it all. In those dark eyes I saw reflected

a picture of myself, sitting here in the uniform of an oppressor....I saw something else too: this Donnelly was like me. He came from a long line of Republicans, stretching back generations....If I had been here to protect my wife and family, I might have felt differently. But I was here from choice (Asher 1991:268).

And from the other side, the view looks very much the same:

The hate, I found out what hatred was....Looking at them ones [warders] as parts of a machine, like robots. If we can't overcome that, we're nothing (PIRA member cited in Feldman 1991:197).

The culture of war may very well convince soldiers that they choose death, *as if they had a choice*. The illusion of choice is a device which allows us to believe that we control war. The war system perpetually reproduces itself, as an autonomous cultural form beyond the boundary of the state, by normalizing death. Death, actual somatic death, cannot be disavowed, simulated, denied, or abstracted. "A bullet in the stomach is a bullet in the stomach, whether it comes from a member of a political group with a long name operating in an internal or 'unconventional interstate' conflict, or whether it comes from a member of the 3rd Shock Army" (Bellamy 1990:10). The bullet in the ambush is no simulacra to its human victim: the praxis of violence is extremely concrete.

Appendix 1

principle security legislation
(excluding annual renewals of emergency legislation)

1969 Ulster Defence Regiment Act
1970 Criminal Justice (Temporary Provisions) Act (NI)
      Public Order (Amendment) Act (NI)
      Police Act (NI)
1972 Prosecution of Offences (NI) Order
      Detention of Terrorists (NI) Order
      Northern Ireland Act
1973 Northern Ireland (Emergency Provisions) Act
1974 Prevention of Terrorism (Temporary Provisions) Act
1975 Criminal Jurisdiction Act
      Northern Ireland (Emergency Provisions)
      (Amendment) Act
1976 Prevention of Terrorism (Temporary Provisions) Act
1977 Police (NI) Order
1978 Northern Ireland (Emergency Provisions) Act
1984 Prevention of Terrorism (Temporary Provisions) Act
1987 Northern Ireland (Emergency Provisions) Act
      Public Order (NI) Order
      Police (NI) Order
1988 Criminal Evidence (NI) Order
1989 Prevention of Terrorism (Temporary Provisions) Act

Appendix 2

victorian small wars (1837-1901)

Anti-colonial Revolt in Canada, 1837

The Capture of Aden and Operations against the Persians, 1838

The First Afghan War, 1838-42

The War in the Levant, 1840

The War in China, 1840-1

The Conquest of Sind, India, 1843

The Gwalior War, India, 1843

The First Sikh War, India, 1845-6

Campaigns against the Boers in South Africa, 1838-52

The Second Sikh War, India, 1848-9

North-West Frontier of India, 1847-54

The Second Burmese War, 1852

The Eureka Stockade, Australia, 1854

The War with Persia, 1856-7

The Storming of the Taku Forts, China, 1859-60

The Maori Wars, 1861-4

Operations in Sikkim, India 1861

Expeditions on the North-West Frontier of India, 1858-67

Japan, 1864

The Expedition to Abyssinia, 1868

The Red River Expedition, Canada, 1870

The Ashanti War, West Africa, 1874

The Expedition to Perak, Malaya, 1875-6

Campaign against the Galekas and Gaikas, Cape Colony, 1877

North-West Frontier, India, 1878-9

The Second Afghan War, 1878

The Third Afghan War, 1879

The Zulu War, 1879

Operations on the North-West Frontier, India, 1880-4

The First Boer War, 1881

The Bombardment of Alexandria, 1882

The Expedition to the Sudan, 1884-5

The Third Burmese War, 1885

Battles around Suakin, Sudan, 1885

The End of the Nile Campaign, 1885

Minor Operations on the North-West Frontier of India, 1888-92

Minor Operations in India, 1888-94

The Siege and Relief of Chitral, India, 1895

The Mashonaland Rising, East Africa, 1896

The Re-Conquest of Egypt, 1896-8

The Operations of the Tirah Expeditionary Force, India, 1897-8

Operations on the North-West Frontier, 18976-8

The Boxer Rising, China, 1900-1

Appendix 3

brush fire wars

Greece, 1944-7
Palestine, 1945-8   (223 dead, 478 wounded)
Vietnam, 1945
Indonesia (Java), 1945-6
Aden, 1947
Ethiopia (Eritrea), 1948-51
Malaya, 1948-60   (489 dead, 961 wounded)
Kenya 1952-6   (12 dead, 69 wounded)
Cyprus 1954-9   (79 dead, 414 wounded)
Suez 1956   (12 dead, 63 wounded)
Aden (border), 1955-60
Togoland 1957
Muscat and Oman, 1957-9
Malaysia, 1962-6   (59 dead, 123 wounded)
Brunei, 1962
Aden 1963-8   (92 dead, 510 wounded)
Oman (Dhofar), 1965-?
Northern Ireland, 1969-present

Note: casualties are British. Figures for dead and wounded
from Stanhope (1979:14)

Appendix 4

security statistics

Table 1

Deaths, August 1969-December 1988

Year	RUC	RUCR	Army	UDR	Civs.	Total
1969	1	0	0	0	12	13
1970	2	0	0	0	23	25
1971	11	0	43	5	115	174
1972	14	3	103	26	321	467
1973	10	3	58	8	171	250
1974	12	3	28	7	166	216
1975	7	4	14	5	217	247
1976	13	10	14	15	245	297
1977	8	6	15	14	69	112
1978	4	6	14	7	50	81
1979	9	5	38	10	51	113
1980	3	6	8	9	50	76
1981	13	8	10	13	57	101
1982	8	4	21	7	57	97
1983	9	9	5	10	44	77
1984	7	2	9	10	36	64
1985	14	9	2	4	25	54
1986	10	2	4	8	37	61
1987	9	7	3	8	66	93
1988	4	2	21	12	54	93
Total	168	89	410	178	1,866	2,711

Source: Northern Ireland: A Political Directory, 1968-88, WD. Flackes and S. Elliott, eds. Belfast: The Blackstaff Press. Note: Figures for civilians include terrorist suspects.

Table 2

Incidents

Year	IED (bombs)	Explosions	Shootings
1969	1	9	73
1970	17	153	213
1971	493	1022	1756
1972	471	1382	10630
1973	542	978	5018
1974	428	685	3206
1975	236	399	1803
1976	426	766	1908
1977	169	366	1081
1978	178	455	755
1979	142	422	728
1980	120	280	642
1981	131	398	1142
1982	118	220	547
1983	101	266	424
1984	55	193	334
1985	65	136	226
1986	82	173	385
1987	148	236	600
1988	62	56	143
Total	398	8603	3170

Source: Army Internal Document. Lisburn July 1986, Amended Jan 1987, LHL (Pol. Col.) P 3025

Appendix 5

glossary and list of abbreviations

AI	— Artificial Intelligence
APC	— Armoured Personnel Carrier
ASU	— Active Service Unit
BAOR	— British Army of the Rhine
CID	— Criminal Investigation Division
CLAW	— Close Light Assault Weapons
CLF	— Commander Land Forces
CO	— Commanding Officer
DPP	— Director of Public Prosecutions
EPA	— Emergency Provisions Act
FOFA	— Follow-On Forces Attack
GOC	— General Officer Commanding
HEAT	— High Energy Anti-Tank
HMP	— Her Majesty's Prison
HQ	— Headquarters
HUMINT	— Human Intelligence
ICRC	— International Committee of the Red Cross
IED	— Improvised Explosive Device
IEW	— Intelligence Electronic Warfare
IFF	— Identification Friend or Foe
INLA	— Irish National Liberation Army
IRB	— Irish Republican Brotherhood
IRSP	— Irish Republican Socialist Party
IS	— Internal Security
JIC	— Joint Intelligence Committee
LIC	— Low Intensity Conflict
LMP	— London Metropolitan Police
LRP	— Long Range Penetration
MI	— Military Intelligence
MI5	— Military Intelligence 5 (the Security Service)
MOD	— Ministry of Defence
MP	— Member of Parliament
MRF	— Mobile Reconnaissance Force
NCO	— Non-Commissioned Officer
NICRA	— Northern Ireland Civil Rights Association

NIO	— Northern Ireland Office
NITAT	— Northern Ireland Training and Advisory Team
NLM	— National Liberation Movement
NORAID	— Irish Northern Aid
OC	— Officer Commanding
OIRA	— Official Irish Republican Army
OP	— Observation Post
PARAs	— Parachute Regiment
PIRA	— Provisional Irish Republican Army
PLO	— Palestine Liberation Organization
POW	— Prisoner of War
PSF	— Provisional Sinn Fein
PTA	— Prevention of Terrorism Act
PVCP	— Permanent Vehicle Checkpoint
QM	— Quartermaster
RAC	— Relatives Action Committee
RIC	— Royal Irish Constabulary
RUC	— Royal Ulster Constabulary
SAS	— Special Air Service
SDLP	— Social Democratic and Labour Party
SOE	— Special Operations Executive
UDA	— Ulster Defence Association
UDR	— Ulster Defence Regiment
USC	— Ulster Special Constabulary
UVF	— Ulster Volunteer Force
UXB	— Un-exploded Bomb

bibliography

Ackroyd, Carol, Karen Margolis, Jonathan Rosenhead and
    Tim Shallice
1980 The Technology of Political Control. 2nd edition.
    London: Pluto Press.

Adams, Gerry
1982 Falls Memories. Dingle: Brandon Books.
1986 The Politics of Irish Freedom. Dublin: Brandon
    Books.
1988 A Pathway to Peace. Cork: Mercier Press.

Adorno, Theodore
1973 Negative Dialectics. New York: Continuum Books.

Ahmed, Akbar S.
1976 Models and Method in Anthropology, Millennium
    and Charisma among the Pathans: A Critical Essay
    in Social Anthropology. London: Routledge and
    Kegan Paul.

Aldred, Margaret
1993 Britain's MoD links with the Academic Community.
    Army Quarterly and Defence Journal 123(3):261-270.

Allen, Charles
1990 The Savage Wars of Peace: Soldiers' Voices, 1945-89.
    London: Michael Joseph.

Alves, Maria Helena Moriera
1985 State and Opposition in Military Brazil. Austin:
    University of Texas Press.

American Joint Chiefs of Staff
1987 Dictionary of Military and Associated Terms. Washington, DC: US Government Printing Office.

Amis, Kingsley
1966 The Anti-Death League. London: Penguin Books.

Amnesty International
1975 Report of an Inquiry into Allegations of ill-treatment in Northern Ireland. London: Amnesty International.

Anderson, Benedict
1983 Imagined Communities: Reflections on the Origin and Spread of Nationalism. London and New York: Verso.

Anderson, David M.
1991 Policing, prosecution and the law in colonial Kenya, c. 1905-39. In Policing the Empire: Government, Authority and Control, 1830-1940. D. Anderson and D. Killingray, eds. Pp. 183-202. Manchester and New York: University of Manchester Press.

Anderson, David M. and David Killingray
1991 Consent, Coercion and Colonial Control: Policing the Empire 1830-1940. In Policing the Empire: Government, Authority and Control, 1830-1940. D. Anderson and D. Killingray, eds. Pp. 1-18. Manchester and New York: University of Manchester Press.
1992 An Orderly retreat? Policing the end of empire. In Policing and Decolonisation: Politics, Nationalism and the Police, 1917-1965. D. Anderson and D.

Killingray, eds. Pp. 1-22. Manchester and New York: Manchester University Press.

Appadurai, Arjun
1990 Disjuncture and Difference in the Global Cultural Economy. Public Culture 2(2):1-24.

Applegate, Major RAD and JR Moore
1990 Warfare — an Option of difficulties: An Examination of Forms of war and the Impact of Military Culture. RUSI Journal 135 (3):13-20.

Apter, David
1991 A View from the Bogside. forthcoming.

Ardrey, Robert
1961 African Genesis. New York: Atheneum.
1966 The Territorial Imperative. New York: Atheneum.

Arensburg, Conrad M. and Solon T. Kimball
1940 Family and Community in Ireland. Cambridge: Harvard University Press.
1968 Family and Community in Ireland. 2nd edition. Cambridge: Harvard University Press.

Army Magazine
1988 How the Army Works: Tour of Duty. Army Magazine, Part 1. London: Orbis Publications.

Arnett, Eric
1992 Welcome to Hyperwar. The Bulletin of Atomic Scientists 48(7):14-22.

Arquilla, John and David Ronfeldt
1993 Cyberwar is Coming! Comparative Strategy 12:141-165.

Arthur, Max
1988 Northern Ireland: Soldiers Talking. London: Sidgwick & Jackson.

Asad, Talal
1975 Two European Images on Non-European Rule. *In* Anthropology and the Colonial Encounter. T Asad, ed. London and Atlantic Highlands, NJ: Ithaca Press and Humanities Press.
1972 Market Model, Class Structure and Consent: A Reconsideration of Swat Political Organization. Man 7(1):74-95.

Asher, Michael
1991 Shoot to Kill: A Soldier's Journey through Violence. London: Penguin Books.

Asmal, Kader, ed.
1985 Shoot to Kill? International Layer' Inquiry into the Lethal Use of Firearms by the Security Forces in Northern Ireland. Cork and Dublin: Mercier Press.

Aulich, James
1991 Framing the Falklands: Nationhood, Culture and Identity. Buckingham: Open University Press.

Austin, John
1961 The Province of Jurisprudence Determined. 2nd edition. 3 vols. London: John Murry.

Axtell, James
1985 The Invasion Within: The Conquest of Cultures in Colonial North America. New York: Oxford University Press.

Babington, Anthony
1990 Military Intervention in Britain: From the Gordon Riots to the Gibraltar Incident. London: Routledge.

Balandier, Georges
1986 An anthropology of violence and war. International Social Science Journal 110: 499-511.

Baldwin, Robert
1982 Police Powers and Politics. London: Quartet.

Barfield, Thomas J.
1994 The Devil's Horsemen: Steppe Nomadic Warfare in Historical Perspective. *In* Studying War: Anthropological Perspectives. S.P. Reyna and R.E. Downs, eds. Pp. 157-185. Langhorne, Penn.: Gordon and Breach Science Publisher.

Barker, Dennis
1981 Soldiering On: An Unofficial Portrait of the British Army. London: Andre Deutsch.

Barkun, Michael
1968 Law without Sanctions. New Haven: Yale University Press.

Barnet, Richard J.
1988 The Costs and Perils of Intervention. *In* Low Intensity Warfare: Counterinsurgency, Proinsurgency and

Antiterrorism in the Eighties. Klare and Knornbluh, eds. Pp. 207-223. New York: Pantheon Books.

Barnett, Donald and Karari Njama
1966 Mau Mau from Within: Autobiography and Analysis of Kenya's Peasant Revolt. Letchworth and London: MacGibbon and Kee.

Barnett, S.A.
1968 On the Hazards of Analogies. *In* Man and Aggression. M.F. Ashley Montagu, ed. Pp. 18-27. Oxford: Oxford University Press.

Barritt, Denis P., and Charles F. Carter
1972 The Northern Ireland Problem. 2nd edition. London: Oxford University Press.

Barth, Fredrick
1959 Political Leadership among Swat Pathans. New York: Humanities Press.

Barzilay, David
1973 The British Army in Ulster, vol. 1. Belfast: Century Services Limited.

Bateson, Gregory
1942 Morale and National Character. *In* Society for the Psychological Study of Social Issues, second yearbook: Civilian Morale. G. Watson, ed. Pp. 71-91. Boston: Houghton Mifflin Company.

Baudrillard, Jean

1990 Fatal Strategies. Translated by Philip Beitchman and W.G.J. Niesluchowski. Jim Fleming, ed. New York: Semiotext(e)/Pluto.

Beckett, Ian F.W.

1982 Guerrilla Warfare: Insurgency and Counter-insurgency since 1945. *In* Warfare in the Twentieth Century: Theory and Practice. C.D. McInnes and G.D. Sheffield, eds. Pp. 194-213. London: Unwin and Hyman.

Beckett, Ian F.W., ed.

1988 The Roots of Counter-Insurgency: Armies and Guerrilla Warfare, 1900-1945. London: Blandford Press.

Beckett, Ian F.W., and Keith Simpson, eds.

1985 A Nation in Arms: A Social History of the British Army in the First World War. Manchester: Manchester University Press.

Beckett, Ian F.W., and John Pimlott, eds.

1985 Armed Forces and Modern Counter-Insurgency. London and Sydney: Croom Helm.

Beetham, David A.

1991 The Legitimation of Power. Atlantic Heights: The Humanities Press International.

Behagg, Colonel A. MBE

1993 Increasing Tempo on the Modern Battlefield. *In* The Science of War: Back to First Principles. B. H. Reid, ed. Pp. 110-131. London and New York: Routledge.

Belfrage, Sally
1987 The Crack: A Belfast Year. London: Andre Deutsch.

Bellamy, Christopher
1983 Heirs of Genghis Khan: The Influence of the Tartar Mongols on the Imperial Russian and Soviet Armies. RUSI 128(1):52-60.
1990 The Evolution of Modern Land Warfare: Theory and Practice. London and New York: Routledge.

Bell, G.H. (Mrs.)
1923 The Veiled Wives of Indian Soldiers: A Problem. Army Quarterly VII: 70-86.

Benedict, Ruth
(1946) 1989 The Chrysanthemum and the Sword: Patterns of Japanese Culture. Boston: Houghton Mifflin.

Benjamin, Walter
1978 Critique of Violence (Zur Kritik der Gewalt). *In* Reflections: Essays, Aphorisms, Autobiographical Writings. Pp. 277-301. New York and London: Harcourt Brace Jovanovich.

Bentham, Jeremy
1970 Of Laws in General. London: Athlone Press.

Beresford, Peter
1987 Ten Men Dead: The Story of the 1981 Irish Hungerstrike. New York: Atlantic Monthly Press.

Best, Geoffrey
1980 Humanity in Warfare. New York: Columbia University Press.

Bew, Paul, Peter Gibbon and Henry Patterson
1979 The State in Northern Ireland: 1921-1972. Manchester: Manchester University Press.

Bew, Paul and Henry Patterson
1985 The British State and the Ulster Crisis: From Wilson to Thatcher. London: Verso.

Bidwell, Shelford and Dominick Graham, eds.
1982 Firepower — British Army Weapons and Theories of War, 1904-45. London: George, Allen and Unwin.

Blaxland, Gregory
1971 The Regiments Depart. London: William Kimber.

Bloch, Jonathan and Peter Fitzgerald
1983 British Intelligence and Covert Action: Africa, Middle East and Europe since 1945. London: Junction Books.

Bloch, Maurice
1986 From Blessing to Violence: History and Ideology in the Circumcision Ritual of the Merina of Madagascar. Cambridge: Cambridge University Press.

Bloch, Maurice, and Jonathan Parry, eds.
1982 Death and the Regeneration of Life. Cambridge: Cambridge University Press.

Boal, F.W. and Russell Murray
1977 A city in conflict. Geographical Magazine 44:364-71.

Boal, F.W., and J. Neville H. Douglas
1982 Integration and Division: Geographical Perspectives on the Northern Ireland Problem. London: Academic Press.

Bodley, John
1992 Anthropology and the Politics of Genocide. *In* The Paths to Domination, Resistance, and Terror. C. Nordstrom and J. Martin, eds. Pp. 37-55. Berkeley and Los Angeles: University of California Press.

Bohannon, Paul
1965 The Differing Realms of Law. *In* The Ethnography of Law. L. Nader, ed. American Anthropologist 67(6) part 2:33-43.

Bohannon, Paul, ed.
1980 Law and Warfare: Studies in the Anthropology of Conflict. Austin: University of Texas Press.

Bond, Brian
1977 Liddell Hart: A Study of His Military Thought. London: Cassell.

Bond, Brian, and Ian Roy, eds.
1975 War and Society: A Yearbook of Military History. London: Croom Helm.

Booth, Ken
1979 Strategy and Ethnocentrism. New York: Holmes and Meier Publishers.

Boserup, Anders
1972 Contradictions and Struggles in Northern Ireland. Socialist Register. Pp. 157-92.

Bourdieu, Pierre
1990 Outline of a Theory of Practice. 2nd edition. Cambridge: Cambridge University Press.

Boyle, Kevin, Tom Hadden, and Paddy Hillyard
1975 Law and State: The Case of Northern Ireland. London: Martin Robertson and Amherst: University of Massachusetts Press.
1980 Northern Ireland: The Communal Roots of Violence. New Society 6 November: 270-1.

Bracken, Paul
1985 The Command and Control of Nuclear Forces. New Haven: Yale University Press.

Bradby, Barbara
1975 The Destruction of Natural Economy. Economy and Society 4:127-61.

Brewer, John
1989 The Sinews of Power, War, Money and the English State, 1688-1783. New York: Alfred A. Knopf.

Brodie, Bernard
1959 Strategy in the Missile Age. Princeton: Princeton University Press.

Broeker, Galen
1970 Rural Disorder and Police Reform in Ireland. London: Routledge & Kegan Paul.

Brown, James and William Snyder, eds.
1985 The Regionalization of Warfare: The Falklands/ Malvinas Islands, Lebanon, and the Iran-Iraq Conflict. New Brunswick and Oxford: Transaction Books.

Buckley, Vincent
1986 Memory Ireland. Australia: Penguin Books.

Bufwack, Mary S.
1982 Village without Violence. Cambridge: Schenkman Publishing Company, Inc.

Bunyan, Tony
1976 The History and Practice of the Political Police in Britain. London: Julian Friedmann Publishers.

Burk, James
1993 Morris Janowitz and the Origins of Sociological Research on Armed Forces and Society. Armed Forces and Society 19(2):167-187.

Burman, Sandra B., and Barbara Harrell-Bond, eds.
1979 The Imposition of Law. New York: Academic Press.

Burrows, William E.
1986 Deep Black: Space Espionage and National Security. New York: Random House.

Burton, Frank
1978 The Politics of Legitimacy: Struggles in a Belfast Community. Routledge and Kegan Paul.
1979 Ideological social relations in Northern Ireland. British Journal of Sociology 30:61-80.

Burton, Frank and Pat Carlen
1979 Official Discourse. London: Routledge & Kegan Paul.

Bynum, Caroline
1987 Holy Feast and Holy Fast: The Religious Significance of Food to Medieval Women. Berkeley: University of California Press.

Callaghan, James
1973 A House Divided: The Dilemma of Northern Ireland. London: Collins.

Callwell, Col. Charles E.
1906 Small Wars: Their Principles and Practice. 3rd edition. London: Printed for His Majesty's Stationery Office by Harrison and Sons.

Calverley, John
1985 Country Risk Analysis. London: Butterworths.

Campbell, Kenneth J.
1993 Intelligence, Human (HUMINT). International Military and Defense Encyclopedia. Col. T.N. Dupuy, ed. Pp. 1335-1341. Washington and London: Brassey's (US), Inc.

Cannadine, David

1987 Introduction: The Divine Rights of Kings. *In* Rituals of Royalty: Power and Ceremonial in Traditional Societies. D. Cannadine and Price, eds. Pp. 1-20. Cambridge: Cambridge University Press.

1983 The Context, Performance and Meaning of Ritual: the British Monarchy and the 'Invention of Tradition. *In* The Invention of Tradition. E. Hobsbawm and T. Ranger, eds. Cambridge: Cambridge University Press.

Canny, Nicholas P.

1976 The Elizabethan Conquest of Ireland: A Pattern Established, 1565-76. Sussex: The Harvester Press.

Cavendish, Anthony

1990 I nside Intelligence. London: Collins.

Chagnon, Napoleon

1974 Studying the Yanomamo. New York: Holt, Rinehard and Winston.

Charters, David

1979 Special Operations in Counter-insurgency: the Farran Case, Palestine 1947. RUSI 124(2):56-61.

1989 From Palestine to Northern Ireland: British Adaptation to low-intensity operations. *In* Armies in Low-Intensity Conflict: A Comparative Analysis. D.

Charters and M. Tugwell, eds. Pp. 169-251. London: Brassey's Defence Publishers.

Chatterjee, Partha
1986 Nationalist Thought and the Colonial World: A Derivative Discourse. London: Zed Books.

Chomsky, Noam and Edward S. Herman
1988 Manufacturing Consent: The Political Economy of the Mass Media. New York: Pantheon Books.

Chowdhury, Subrata Roy
1989 The Rule of Law in a State of Emergency: The Paris Minimum Standards of Human Rights Norms in a State of Emergency. London: Pinter Publishers.

Cincinnatus [pseudonym]
1981 Self Destruction: The Disintegration and Decay of the US Army during the Vietnam War. New York: W.W. Norton.

Clarke, A.F.N.
1983 Contact. London: Martin Secker & Warburg Ltd.

Clarke, Liam
1987 Broadening the Battlefield: the H-Blocks and the Rise of Sinn Fein. Dublin: Gill and Macmillan.

Clarke, Brigadier M.H.F., Lt. Col. T. Glynn, and Lt. Col. A.P.V. Rogers
1989 Combatant and Prisoner of War Status. *In* Armed Conflict and the New Law: Aspects of the 1977 Geneva Protocols and the 1981 Weapons Convention. M. Meyer, ed. Pp. 107-137. London: British Institute of International and Contemporary Law.

Clastres, Pierre
1977 Society Against the State. Translated by Robert Hurley and Abe Stein. Oxford: Blackwell.

Clausewitz, Carl von
1823, 1976, 1987 On War. New York: Penguin Books.

Clayton, Anthony
1976 Counter-Insurgency in Kenya: A Study of Military Operations Against Mau Mau. Nairobi: Transafrica Publishers.

Clifford, James
1983 Power and Dialogue in Ethnography: Marcel Griaule's Initiation. *In* Observers Observed: Essays on Ethnographic Fieldwork, History of Anthropology, Volume 1. G. Stocking, Jr., ed. Pp. 121-156. Madison: University of Wisconsin Press.

Clifford, James and George E. Marcus
1986 Writing Culture: The Poetics and Politics of Ethnography. Berkeley: University of California Press.

Clode, Charles M.
1869 Military Forces of the Crown: Their Administration and Government. 2 Vols. London: Murray.

Clutterbuck, Richard
1974 A Third Force? Army Quarterly 104 (1):22-8.
1975 Living with Terrorism. London: Faber and Faber.
1978 Britain in Agony: The Growth of Political Violence. London: Faber and Faber.

Cohen, Ronald
1986 War and War Proneness in Pre- and Postindustrial States. *In* Peace and War: Cross-Cultural Perspectives. M.L. Foster and R. Rubinstein, eds. New Brunswick: Transaction Books.

Cole, Major D.H. and Major E.C. Priestly
1936 An Outline of British Military History, 1660-1936. London: Sifton Praed.

Collier, John G.
1991 The Legal Basis of the Institution of War. *In* The Institution of War. R. Hinde, ed. Pp. 121-133. London: Macmillan.

Collins, Tom
1986 The Irish Hungerstrike. Dublin: White Island Book Company.

Comaroff, John and Jean
1991 Of Revelation and Revolution: Christianity, Colonialism and Consciousness in South Africa, vol. 1. Chicago and London: The University of Chicago Press.

Conroy, John
1988 War as a Way of Life: A Belfast Diary. London: Heinemann.

Coogan, Tim Pat
1980 On the Blanket: The H-Block Story. Dublin: Ward River Press.

Cookridge, E.H.
1966 Inside SOE: The Story of Special Operations in Western Europe, 1940-45. London: Arthur Baker, Ltd.

Copet-Rougier, Elisabeth
1986 'Le Mal Court': Visible and Invisible Violence in an Acephalous Society — The Mkako of Cameroon. *In* The Anthropology of Violence. D. Riches, ed. Pp. 50-70. London: Basil Blackwell.

Corbin, John
1986 Insurrections in Spain: Casa Viejas 1933 and Madrid 1981. *In* The Anthropology of Violence. D. Riches, ed. Pp. 28-50. London: Basil Blackwell.

Cornell, Drucilla
1992 The Philosophy of the Limit. New York: Routledge.

Crapanzano, Vincent
1985 Waiting: the Whites of South Africa. New York: Random House.

Cronin, Sean
1980 Irish Nationalism: A History and its Roots and Ideology. Dublin: The Academy Press.

Crowder, Michael, ed.
1978 West African Resistance: The Military Response to Occupation. London: Hutchinson.

Crozier, Brian
1960 The Rebels: A Study of Post-War Insurrections. London: Chatto & Windus.

Cunningham, Cyril
1972 International Interrogation Techniques. RUSI 117:31-34.

Cunningham, Michael J.
1991 British Government Policy in Northern Ireland, 1969-1989: Its Nature and Execution. Manchester: Manchester University Press.

Curtis, Liz
1984 Ireland: The Propaganda War: The Media and the 'Battle for Hearts and Minds'. London: Pluto.

Dandeker, Christopher
1990 Surveillance, Power and Modernity: Bureaucracy and Discipline from 1700 to the Present Day. Cambridge: Polity Press.

Darby, John
1976 Conflict in Northern Ireland: The Development of a Polarised Community. Dublin: Gill and Macmillan.
1986 ntimidation and the Control of Conflict in Northern Ireland. Dublin: Gill and Macmillan.

DeBaroid, Ciaran
1990 Ballymurphy and the Irish War. London: Pluto Press.

DeLanda, Manuel
1991 War in the Age of Intelligent Machines. New York: Swerve Editions/Zone Books.

Delueze, Gilles and Felix Guattari

1986 Nomadology: The War Machine. Translated by Brian Massumi. New York: Semiotext(e).

1992 Treatise on Nomadology — The War Machine. *In* A Thousand Plateaus: Capitalism and Schizophrenia. Translation and Foreword by Brian Massumi. Pp. 351-423. London: The Athlone Press.

Denning, Major BC

1927 Modern Problems of Guerrilla Warfare. Army Quarterly and Defence Journal 13:347-54. London: W. Clowes & Sons, Ltd.

DePaor, Liam

1970 Divided Ulster. Harmondsworth: Penguin.

Der Derian, James

1992 Antidiplomacy: Spies, Terror, Speed, and War. Cambridge: Blackwell Publishers.

Derrida, Jacques

1992 Force of Law: The 'Mystical Foundation of Authority'. *In* Deconstruction and the Possibility of Justice. D. Carlson, D. Cornell, and M. Rosenfeld, eds. Pp. 3-68. New York and London: Routledge.

Detter DeLupus, Ingrid

1987 The Law of War. Cambridge: Cambridge University Press.

Devlin, Richard

1993 The Rule of Law and the Politics of Fear: Reflections on Northern Ireland. Law and Critique 4(2):155-185.

Dewar, Michael
1984 Brush Fire Wars: Minor Campaigns of the British Army Since 1945. New York: St. Martin's Press; London: Robert Hale Limited.
1985 The British Army in Northern Ireland. London: Arms and Armour Press.

Dicey, A.V.
(1885) 1959   Introduction to the Study of the Law of the Constitution. 10th edition. London: Macmillan.

Dillon, Martin
1989 The Shankill Butchers, a Case Study of Mass Murder. London: Hutchinson.

Dixon, Norman
1976 On the Psychology of Military Incompetence. London: Jonathan Cape.

Dockery, J.T. and A.E.R. Woodcock, eds.
1993 The Military Landscape: Mathematical Models of Combat. Cambridge: Woodhead Publishing, Ltd.

Dodd, Norman. L
1976 Corporals War: Internal Security Operations in Northern Ireland. Military Review, 56(7):58-68.
1979 COIN and I.S. Operations. Defence 10(5):318-323.

Doherty, Frank
1983 SIGINT used by Anti-State Forces: A Case Study of PIRA ops. *In* War and Order. C. Bledoska, ed. London: Junction Books.

Doob, Leonard W., and William J. Foltz

1973 The Belfast Workshop: An Application of Group Techniques to a Destructive Conflict. Journal of Conflict Resolution 17(3):489-512.

Dorrill, Stephan

1993 The Silent Conspiracy: Inside the Intelligence Services in the 1990's. London: Heinemann.

1991 Smear! Wilson and the Secret State. London: 4th Estate.

Douglas, Mary

1966 Purity and danger; an analysis of concepts of pollution and taboo. London: Routledge and Kegan Paul.

Draper, G.I.A.D.

1979 Wars of National Liberation and War Criminality. *In* Restraints on War: Studies in the Limitation of Armed Conflict. M. Howard, ed. Pp. 135-163. Oxford: Oxford University Press.

Drewry, Brigadier C.F.

1993 The Lessons of the 1920s and Modern Experience. *In* The Science of War: Back to First Principles. B. H. Reid, ed. Pp. 12-24. London and New York: Routledge.

Dreyfus, Hubert L., and Paul Rabinow

1983 Michel Foucault: Beyond Structuralism and Hermeneutics. 2nd edition. Chicago: University of Chicago Press.

Duffy, Christopher
1981 Russian's Military Way to the West. London: Routledge & Kegan Paul.

Duggan, G.C.
1968 The Royal Irish Constabulary: Forgotten Force in a Troubled Land. In *1916*: The Easter Rising. O. Dudley Edwards and F. Pyle, eds. Pp. 91-99. London: MacGibbon and Kee.

Dupre, G., and P.P. Rey
1978 Reflections on the Relevance of a Theory of the History of Exchange. *In* Relations of Production: Marxist Approaches to Economic Anthropology. Translated by Helen Lackner. D. Seddon, ed. pp. 171-208. London: Cass.

Durkheim, Emile
1933 The Division of Labor in Society. New York: The Free Press.
1970 Suicide: A Study in Sociology. London: Routledge and Kegan Paul.

Durrant, Major G.R. MBE
1978 The Evolution of the Covert Combat. British Army Review 60:54-59.

Edwards, Paul N.
1989 The closed world: Systems discourse, military policy and post- World War II US historical consciousness. *In* Cyborg Worlds: The Military Information Society. L. Levidow and K. Robins, eds. Pp. 135-159. London: Free Association Books.

Ehrlich, Eugen
1936 Fundamental Principles of the Sociology of Law. Cambridge: Harvard University Press.

Eisenzweig, Uri
1988 Terrorism in Life and in Real Literature. Diacritics 18(3):32-43.

Eliot, Philip
1977 Reporting Northern Ireland. In Ethnicity and the Media: An Analysis of Media Reporting in the U.K., Canada and Ireland. Pp. 263-275. Paris: UNESCO.

Ellis, John
1975 The Social History of the Machine Gun. London: Croom Helm.

Ellmann, Maud
1993 The Hunger Artists: Starving, Writing, and Imprisonment. Cambridge: Harvard University Press.

Enloe, Cynthia
1980 Ethnic Soldiers: State Security in Divided Societies. Athens: University of Georgia Press.

Evans-Pritchard, E.E.
1940 The Nuer: A Description of the Modes of Livelihood and Political Institutions of a Nilotic People. Oxford: Clarendon Press.

Evelegh, Robin
1978 Peace-Keeping in a Democratic Society: The Lessons of Northern Ireland. Montreal: McGill-Queens University Press.

Faligot, Roger
1983 Britain's Military Strategy in Ireland: The Kitson Experiment. London: Zed Press.

Falk, Richard, and S. Kim, eds.
1980 The War System: an Interdisciplinary Approach. Boulder: Westview Press.

Falk-Moore, Sally
1978 Law as Process. London: Routledge & Kegan Paul, Ltd.

Fall, Bernard
1966 Vietnam Witness. New York: Praeger Books.

Farago, Ladislas
1942 German Psychological Warfare. New York: Arno Press.

Faris, James
1973 'Pax Britannia' and the Sudan: S.F. Nadel. *In* Anthropology and the Colonial Encounter. T. Asad, ed. London and Atlantic Highlands, NJ: Ithaca Press and Humanities Press.

Farrell, Michael

1983 Arming the Protestants: The Formation of the Ulster Special Constabulary and the Royal Ulster Constabulary, 1920-1927. London: Pluto Press.

1988 Twenty Years On. Dingle: Brandon Books.

Featherstone, Donald

1973 Colonial Small Wars, 1837-1901. Newton Abbot: David & Charles.

Feld, Maury D.

1977 The Structure of Violence: Armed Forces as Social Systems. Beverly Hills and London: Sage Publications.

Feldman, Allen

1991 Formations of Violence: The Narrative of the Body and Political Terror in Northern Ireland. Chicago: University of Chicago Press.

Fergusson, R. Brian and Neil L. Whitehead, eds.

1992 War in the Tribal Zone: Expanding States and Indigenous Warfare. Santa Fe, New Mexico: School of American Research Press.

Ferguson, R. Brian, ed.

1984 Warfare, Culture and Environment. Orlando: Academic Press.

Ferguson, R. Brian

1989 Anthropology and War: Theory, Politics, Ethics. *In* The Anthropology of War and Peace: Perspectives on the Nuclear Age. P. R. Turner and D. Pitt, eds.

Pp. 141-159. South Hadley, Massachusetts: Bergin & Garver Publishers, Inc.

1990 Explaining War. *In* The Anthropology of War. J. Haas, ed. Pp. 27-55. Cambridge: Cambridge University Press.

Fergusson, Sir James
1964 The Curragh Incident. London: Faber and Faber.

Fields, Rona
1977 Society under Siege: A Psychology of Northern Ireland. Philadelphia: Temple University Press.

Finn, John E.
1991 Constitutions in Crisis: Political Violence and the Rule of Law. Oxford: Oxford University Press.

Fisk, Robert
1978 The Effect of Social and Political Crime on the Police and British Army in Northern Ireland. In International Terrorism in the Contemporary World. M.H. Livingston, L.B. Kress and M.G. Wanek, eds. Pp. 840-94. Westport, CT: Greenwood Press.

Fitzpatrick, Peter
1992 The Mythology of Modern Law. London and New York: Routledge.

Foley, Conor
1992 Legion of the Rear Guard: The IRA and the Modern Irish State. London: Pluto Press.

Foot, Paul
1990 Who Framed Colin Wallace? London: Pan Books.

Foot, M.R.D
1973 The IRA and the Origins of SOE. *In* War and Society: Historical Essays in Honour and Memory of J.R. Western, 1928-71. M.R.D. Foot, ed. Pp. 57-71. London: Pal Ellele.
1986 SOE. New York: University Publications of America.

Fortes, Meyer, and E.E. Evans-Pritchard, eds.
1940 Introduction. *In* African Political Systems. Pp. 1-25. Oxford: Oxford University Press.

Foster, Mary LeCron, and Robert A. Rubinstein
1986 Introduction. *In* Peace and War: Cross Cultural Perspectives. New Brunswick: Transaction Books.

Foucault, Michel
1979 Discipline and Punish: The Birth of the Prison. New York: Vintage Books.
1984 The Juridical Apparatus. *In* Legitimacy and the State: Readings in Social and Political Theory. W. Connolly, ed. Pp. 201-222. London: Basil Blackwell.
1990 The History of Sexuality, Volume I: An Introduction. New York: Vintage Books.

Fox, Captain K.O.
1974 Public Order: The Law and the Military. Army Quarterly and Defence Journal 104(3):295-308.

Fraser, Morris
1973 Children in Conflict. London: Secker and Warburg.

French, David
1990 The British Wary in Warfare, 1688-2000. London: Unwin Hyman.

Fried, Morton, Marvin Harris, and R. Murphy, eds.
1968 War: the Anthropology of Armed Conflict and Aggression. Garden City: The Natural History Press.

Fuller, JCF
1925 Foundations of the Science of War. London: Hutchinson.
1942 Machine Warfare. An Enquiry into the Influences of Mechanics on the Art of War. London: Hutchinson.

Furniss, Edgar
1964 De Gaulle and the French Army: A Crisis in Civil-Military Relations. New York: Twentieth Century Fund.

Gabriel, Richard
1987 No More Heroes: Madness and Psychiatry in War. New York: Hill and Wang.

Gallaher, Frank
1957 The Indivisible Island: the History of Partition of Ireland. London: Gollancz.

Galvin, General John R.
1991 Uncomfortable Wars: Toward a New Paradigm. *In* *Uncomfortable* Wars: Toward a New Paradigm of Low Intensity Conflict. M. Manwaring, ed. Pp. 9-19. Boulder: Westview Press.

Gardner, Robert, and Karl G. Heider
1968   Gardens of War: Life and Death in the New Guinea
       Stone Age. New York: Random House.

Gazit, Schlomo and Michael Handel
1980   Insurgency, Terrorism and Intelligence. *In*
       Intelligence Requirements for the 1980: Counter-
       intelligence. R. Godson, ed. Pp. 125-148.
       Washington: National Strategy Information Center,
       Inc.

Geary, Roger
1985   Policing Industrial Disputes 1893-1985. Cambridge:
       Cambridge University Press.

Geertz, Clifford
1980   Negara: The Theater State in Nineteenth Century
       Bali. Princeton: Princeton University Press.
1983   Local Knowledge: Further Essays in Interpretive
       Anthropology. New York: Basic Books.
1988   Works and Lives: The Anthropologist as Author.
       Stanford: Stanford University Press.

Gellner, Ernest
1991   An Anthropological View of War and Violence. *In*
       The Institution of War. R.A. Hinde, ed. Pp. 62-81.
       London: Macmillan.

Geraghty, Tony
1980   Who Dares Wins: The Story of Special Air Service,
       1950-1980. London: Arms and Armour Press.

Geschiere, Peter
1985   Imposing Capitalist Dominance through the State.
       In Old Modes of Production and Capitalist

Encroachment. W. Van Binsbergen and P. Geschiere, eds. London: KPI.

Gibson, James
1986 Perfect War: Technowar in Vietnam. Boston: Atlantic Monthly Press.

Giddens, Anthony
1987 The Nation-State and Violence: Volume Two of A Contemporary Critique of Historical Materialism. Berkeley and Los Angeles: University of California Press.

Gilbert, Michelle
1987 Ritual and Power in a Ghanaian State. *In* Rituals of Royalty: Power and Ceremonial in Traditional Societies. D. Cannadine and S.R.F. Price, eds. Pp. 298-331. Cambridge: Cambridge University Press.

Girard, Rene
1977 Violence and the Sacred. Baltimore: The Johns Hopkins University Press.

Gluckman, Max
1940 The Kingdom of the Zulu. *In* African Political Systems. E.E. Evans-Pritchard and M. Fortes, eds. Pp. 25-56. Oxford: Oxford University Press.
1960 Custom and Conflict in Africa. Oxford: Basil Blackwell.
1967 Judicial Process Among the Barotese of Northern Rhodesia. 2nd edition. Manchester: Manchester University Press.

Gonzales-Casanova, Pablo
1965 Internal Colonialism and National Development. Studies in Comparative International Development 1(4):27-37.

Gorer, Geoffrey
1943 Themes in Japanese Culture. Transactions of the New York Academy of Sciences, Ser. 2, 5(5): 106-124.
1968 Man Has No 'Killer' Instinct. *In* Man and Aggression. M.F. Ashley Montagu, ed. Pp. 27-37. Oxford: Oxford University Press.

Gramsci, Antonio
1971 Selections from the Prison Notebooks. Edited and translated by Q. Hoare and G. Nowell Smith. New York: International Publishers.

Gray, Chris Hables
1989 The Cyborg Soldier: The US Military and the Post-Modern Warrior. In Cyborg Worlds. L. Levidow and K. Robins, eds. Pp. 43-73. London: Free Association Books.

Greenhouse, Carol J.
1986 Fighting For Peace. *In Peace* and War: Cross-Cultural Perspectives. M.L. Foster and R. Rubinstein, eds. New Brunswick: Transaction Books.

Greenwood, Christopher
1991 In Defence of the Laws of War. *In* The Institution of War. R. Hinde, ed. Pp. 133-148. London: Macmillan.

Greer, Steven
1988 The Supergrass System. *In* Justice Under Fire: The Abuse of Civil Liberties in Northern Ireland. A. Jennings, LL.B., ed. Pp. 73-104. London: Pluto Press.

Griffith, Kenneth, and Timothy E. O'Grady
1982 Curious Journey: An Oral History of Ireland's Unfinished Revolution. London: Hutchinson.

Gudgin, Peter
1989 Military Intelligence: The British Story. London: Arms and Armour Press.

Guelke, Adrian
1988 Northern Ireland: The International Perspective. Dublin: Gill and Macmillan.

Gwynn, Major-General Sir Charles W.
1934 Imperial Policing. London: Macmillan and Co., Limited.

Hackett, Sir John
1983 The Profession of Arms: The History and Role of the Military Profession over 4,000 Years. London: Sidgwick and Jackson.

Hadden, Tom, Kevin Boyle, and Colm Campbell
1988 Emergency Law in Northern Ireland: The Context. *In* Justice Under Fire: The Abuse of Civil Liberties in Northern Ireland. A. Jennings, LL.B., ed. Pp. 1-27. London: Pluto Press.

Hall, Peter
1988   The Prevention of Terrorism Acts. *In* Justice Under Fire: The Abuse of Civil Liberties in Northern Ireland. A. Jennings, LL.B., ed. Pp. 144-191. London: Pluto Press.

Halstock, Max
1981   Rats: the Story of a Dog Soldier. London: Gollancz.

Hamerton-Kelly, Robert G. ed.
1987   Violent Origins: Ritual Killing and Cultural Formation. Stanford: Stanford University Press.

Hamill, Desmond
1985   Pig in the Middle: The Army in Northern Ireland, 1969-1984. London: Methuen.

Hanle, Donald J.
1989   Terrorism: The Newest Face of Warfare. New York: Pergamon-Brassey's International Defense Publishers, Inc.

Hancock, W.K.
1937   Survey of British Commonwealth Affairs, Vol. I: Problems of Nationality, 1918-1936. Oxford: Oxford University Press.

Harris, Marvin
1975   Cows, Pigs, Wars, and Witches: the Riddles of Culture. New York: Vintage Books.

Harris, Rosemary
1972 Prejudice and Tolerance in Ulster: A Study of Neighbours and 'Strangers' in a Border Community. Manchester: Manchester University Press.

Hattendorf, John B., and Malcolm H. Murfett, eds.
1990 The Limitations of Military Power: Essays Presented to Professor Norman Gibbs on his Eightieth Birthday. Houndmills and London: Macmillan Academic and Professional, Ltd.

Hawkins, Richard
1991 The 'Irish Model' and the Empire: A Case for Reassessment. *In* Policing the Empire: Government, Authority and Control, 1830-1940. D. Anderson and D. Killingray, eds. Pp. 18-33. Manchester and New York: University of Manchester Press.

Hayden, Ilsa
1987 Symbol and Privilege: The Ritual Context of British Royalty. Tucson: University of Arizona Press.

Heaney, Seamas
1984 Station Island, canto IX. *In* Station Island. London: Faber and Faber.

Hechter, Michael
1977 Internal Colonialism: The Celtic Fringe in British National Development, 1536-1966. Berkeley and Los Angeles: University of California Press.

Hegel, Georg Wilhelm Friedrich
1977 Phenomenology of Spirit. Translated by A.V. Miller. Oxford: Clarendon Press.

Heilbrunn, Otto and Brigadier C. A. Dixon
1954 Communist Guerrilla Warfare. London: George Allen and Unwin Ltd.

Herman, Edward and Gerry O'Sullivan
1989 The 'Terrorism' Industry: The Experts and Institutions That Shape our View of Terror. New York: Pantheon Books.

Hertz, Robert
1960 Death and the Right Hand. Translated by Rodney and Claudia Needham. London: Cohen & West.

Heskin, Ken
1980 Northern Ireland: A Psychological Analysis. Dublin: Gill and Macmillan.

Hill, Richard S.
1991 The Policing of Colonial New Zealand: from informal to formal control, 1840- 1907. *In* Policing the Empire: Government, Authority and Control, 1830-1940. D. Anderson and D. Killingray, eds. Pp. 52-71. Manchester and New York: University of Manchester Press.

Hill, Stuart and Donald Rothchild
1992 The Impact of Regime on the Diffusion of Political Conflict. *In* The Internationalization of Communal Strife. M.I. Midlarsky, ed. Pp. 189-209. London: Routledge.

Hillyard, Paddy
1983 Law and Order. *In* Northern Ireland: the Background to the Conflict. J. Darby, ed. Pp. 32-61. Belfast: Appletree Press.
1988 Political and Social Dimensions of Emergency Law in Northern Ireland. *In* Justice Under Fire: The Abuse of Civil Liberties in Northern Ireland. A. Jennings, LL.B., ed. Pp. 191-213. London: Pluto Press.

Hillyard, Paddy, and Janie Percy-Smith
1988 The Coercive State. London and New York: Pinter Publishers.

Hilsman, Roger
1967 To Move a Nation: The Politics of Foreign Policy in the Administration of J.F.K.

Hobbes, Thomas
1968 Leviathan. C.B. Macpherson, ed. Harmondsworth: Penguin.

Hobhouse, Leonard Trelawny, Gerald C. Wheeler, and Morris Ginsberg
1915 The Material Culture and Social Institutions of the Simpler Peoples: An Essay in Correlation. London: Chapman Hill.

Hocart, Arthur M.
1931 Warfare in Eddystone of the Solomon Islands. Journal of the Royal Anthropological Institute 61:301-324.
1954 Social Origins. London: Watts.

Hoebel, E.A.
1954 The Law of Primitive Man. Cambridge: Harvard University Press.

Hoffman, Bruce, and Jennifer M. Taw
1991 Defense Policy and Low-Intensity Conflict: The Development of Britain's 'Small Wars' Doctrine During the 1950's. Santa Monica: The RAND Corporation.

Hogan, Gerard, and Clive Walker
1989 Political Violence and the Law in Ireland. Manchester and New York: Manchester University Press.

Holmes, Richard
1987 Firing Line. Harmondsworth: Penguin.

Holroyd, Fred, with Nick Burbridge
1989 War Without Honor. Hull: Medium.

Holsti, Rudolf
1914 The Relation of War to the Origin of the State. Annales Academic Scientarium Fennicae XIII. Helsingfors, Finland.

Howard, Michael Sir
1957 Introduction: The Armed Forces as a Political Problem. *In* Soldiers and Governments: Nine Studies in Civil-Military Relations. M. Howard, ed. Pp. 9-25. London: Eyre and Spottiswoode.
1975 The British Way in Warfare: A Reappraisal. London: Jonathan Cape.

1983 The British Way in Warfare: A Reappraisal. *In* The Causes of Wars. M. Howard, ed. Pp. 169-188. London: Unwin Counterpoint.

1986 Men Against Fire: The Doctrine of the Offensive in 1914. *In* Makers of Modern Strategy. P. Paret, ed. Princeton: Princeton University Press.

Howard, Michael, ed.

1979 Restraints on War: Studies in the Limitation of Armed Conflict. New York: Oxford University Press.

1984 The Causes of Wars and Other Essays. Cambridge: Harvard University Press.

Hubert, H., and Marcel Mauss

(1899) 1964 Sacrifice: Its Nature and Function. Chicago: University of Chicago Press.

Hunt, Albert

1981 The Language of Television. London: Eyre Methuen.

Huntington, Richard and Peter Metcalf

1979 Celebrations of Death: The Anthropology of Mortuary Ritual. Cambridge: Cambridge University Press.

Huntington, Samuel P.

(1957) 1985 The Soldier and the State: The Theory and Politics of Civil-Military Relations. Cambridge and London: The Belknap Press of Harvard University Press.

1968 Political Order in Changing Societies. New Haven: Yale University Press.

1993 The Clash of Civilizations? Foreign Affairs (72)3:22-50.

Hutchinson Simon
1969 The Police Role in Counter-Insurgency Operations. RUSI Journal 114(4):56-61.

Irvine, D.D.
1938 The Origin of Capital Staffs. Journal of Modern History 10(2):161-179.

Jack, Ian
1988 Gibraltar. Granta 25:13-87.

Jackson, Robert
1991 The Malayan Emergency: the Commonwealth's Wars, 1948-66. London: Routledge.

Jackson, William
1986 Withdrawal from Empire: A Military View. New York: St. Martin's Press.

Janowitz, Morris
1960 The Professional Soldier. Glencoe, Ill: the Free Press.

Jarman, Neil
1993 Intersecting Belfast. In Landscape: Politics and Perspectives. B. Bender, ed. Pp. 107-138. Providence and Oxford: Berg.

Jefferies, Charles Sir
1952 The Colonial Police. London: Max Parrish.

Jeffery, Keith

1984 The British Army and the Crisis of Empire, 1918-1922. Manchester: Manchester University Press.

1987 Intelligence and Counterinsurgency operations-- some reflections on the British experience. Intelligence and National Security 2(1):118-150.

1988 Colonial War, 1900-1939. *In Warfare* in the Twentieth Century: Theory and Practice. C.D. McInnes and G.D. Sheffield, eds. Pp. 24-51. London: Unwin and Hyman.

Jenkins, Brian

1975 International Terrorism: A New Mode of Conflict. *In* International Terrorism and World Security. International School on Disarmament and Research on Conflicts, 5th Course. Pp. 13-50. New York: John Wiley and Sons.

Jennings, Anthony

1988 Shoot to Kill: The Final Courts of Justice. *In* Justice Under Fire: The Abuse of Civil Liberties in Northern Ireland. A. Jennings, LL.B., ed. Pp. 104-131. London: Pluto Press.

Johnson, Douglas

1991 From Military to Tribal Police: Policing the Upper Nile Province of the Sudan. *In* Policing the Empire: Government, Authority and Control, 1830-1940. D. Anderson and D. Killingray, eds. Pp. 151-168. Manchester and New York: University of Manchester Press.

Jones, Timothy Llewellyn
1991 The Development of British Counter-insurgency Polices and Doctrine, 1945-52. Ph.D. Dissertation. King's College, University of London. Unpublished.

Judd, Denis
1973 Someone Has Blundered: Calamities of the British Army in the Victorian Age. London: Arthur Barker Limited.

Caldor, Mary
1991 Do Modern Economies Require War or Preparations for Warfare? *In The* Institution of War. R. A. Hinde, ed. Pp. 178-192. London: Macmillan.

Kantorowicz, Herman
1958 The Definition of Law. Cambridge: Cambridge University Press.

Kantorowicz, Ernst H.
1957 The King's Two Bodies: A Study in Mediaeval Political Theology. Princeton: Princeton University Press.

Kaplan, Robert D.
1994 The Coming Anarchy. Atlantic Monthly (Feb): 44-76.

Keegan, John
1976 The Face of Battle. London: Cape.

Keen, Sam
1986 Faces of the Enemy. San Francisco: Harper and Roe.

Kelly, Fergus
1988 A Guide to Early Irish Law. Dublin: Mt. Salus Press.

Kenyatta, Jomo
1938 Facing Mt. Kenya: the tribal life of the Gikuyu. With an introduction by B. Malinowski. London: Secker and Warburg.

Kerruish, Valerie
1991 Jurisprudence as Ideology. London and New York: Routledge.

Kertzer, David I.
1988 Ritual, Politics and Power. New Haven: Yale University Press.

Khong, Yuen Foong
1992 Analogies at War: Korea, Munich, Dien Bien Phu and the Vietnam Decisions of 1965. Princeton: Princeton University Press.

Kidron, Michael and Dan Smith
1983 The War Atlas: Armed Conflict-Armed Peace. London and Sydney: Pan Books.
1991 The New State of War and Peace: An International Atlas. New York: Simon and Schuster, Inc.

Killingray, David

1991   Guarding the Extending Frontier: Policing the Gold
       Coast, 1865-1913. *In* Policing the Empire:
       Government, Authority and Control, 1830-1940. D.
       Anderson and D. Killingray, eds. Pp. 106-126.
       Manchester and New York: University of
       Manchester Press.

Kitson, Frank

1971   Low Intensity Operations: Subversion, Insurgency,
       Peace-keeping. Harrisburg: Stackpole Books.
1977   Bunch of Five. London: Faber and Faber.
1987   Warfare as a Whole. London: Faber and Faber.

Kitson, Frank

1989   Directing Operations. London: Faber and Faber.

Klare, Michael T. and Peter Kornbluh

1988   The New Interventionism: Low-Intensity Warfare
       in the 1980's and Beyond. *In* Low Intensity
       Warfare: Counterinsurgency, Proinsurgency and
       Antiterrorism in the Eighties. M. Klare and P.
       Kornbluh, eds. Pp. 3-21. New York: Pantheon
       Books.

Kooistra, Paul

1989   Criminals as Heroes: Structure, Power, and
       Identity. Bowling Green, Ohio: Bowling Green
       State University Press.

Krepinevich, Andrew F., Jr.

1986   The Army and Vietnam. Baltimore: Johns Hopkins
       University Press.

Krigsman, Henry Azel, Jr.
1993 Internal Security and Armed Forces. International Military and Defense Encyclopedia. Col. T.N. Dupuy, ed. Pp. 1335-1341. Washington and London: Brassey's (US), Inc.

Kuper, Adam
1988 The Invention of Primitive Society: Transformations of an Illusion. London and New York: Routledge.

Kuper, Hilda Beemer
1937 The development of the military organization in Swaziland. Africa 10:55-74, 176-205.

Lake, Carney
1990 Reflected Glory. Devon: Otter Books.

Lan, David
1985 Guns and Rain: Guerillas and Spirit Mediums in Zimbabwe. London: James Currey; Berkeley and Los Angeles: University of California Press.

Lawrence, T.E.
1920 The Evolution of a Revolt. Army Quarterly 1(1): 55-69.
(1926) 1949 Seven Pillars of Wisdom: A Triumph. London: Jonathan Cape.
1927 Revolt in the Desert. London: Jonathan Cape.

Leach, Edmund R.
1954 Political Systems of Highland Burma. Boston: Beacon Press.

1968 Don't Say 'Boo' to a Goose. *In* Man and Aggression. M.F. Ashley Montagu, ed. Pp. 65-74. Oxford: Oxford University Press.

1976 Culture and Communication: The Logic by which Symbols are Connected. Cambridge: Cambridge University Press.

Lee, Richard
1984 The Dobe !Kung. New York: Holt Rinehart and Winston.

Lee, W.L. Melville
1901 A History of the Police in England. London: Methuen.

Leeman, Richard W.
1991 The Rhetoric of Terrorism and Counterterrorism. New York: Greenwood Press.

Levi-Strauss, Claude
1966 The Savage Mind. Chicago: University of Chicago Press.

Leyton, Elliott
1974 Opposition and Integration in Ulster. Man 9:185-98.

Liddell Hart, Basil H.
1927 Jenghiz Khan and Sabutai. *In Great* Captains Unveiled. Pp. 1-35. Edinburgh and London: William Blackwood and Sons.

1935 The British Way in Warfare: Adaptability and Mobility. Harmondsworth: Penguin.

Lienhardt, Godfrey
1961 Divinity and Experience: the religion of the Dinka. Oxford: Clarendon Press.

Lindblom, Charles and David Cohen
1979 Usable Knowledge: Social Science and Social Problem Solving. New Haven: Yale University Press.

Lorenz, Konrad
1966 On Aggression. Harcourt Brace & World.

Lowe, T.A. Major
1922 Some Reflections of a junior commander upon the campaign in Ireland, 1920 and 1921. Army Quarterly 5:50-57.

Ludendorff, General Eric Von
1935(6) Der Totale Kreig. Translated as The Nation at War. Translated by A.S. Rappoport. London: Hutchinson & Co.

Luttwak, Edward N.
1987 Strategy: the Logic of War and Peace. Cambridge: The Belknap Press of Harvard University Press.

MacDonald, Michael
1986 Children of Wrath: Political Violence in Northern Ireland. Cambridge: Polity Press.

Mack, Burton, and Robert G. Hamerton-Kelly, eds.
1987 Violent Origins: Ritual Killing and Cultural Transformation. Stanford: Stanford University Press.

MacKenzie, Compton
1954   Water on the Brain. 2nd edition. London: Chatto and Windus Publishers.

MacKenzie, John, ed.
1992   Popular Imperialism and the Military 1850-1950. Manchester: Manchester University Press.

MacLeod, William Christie
1967   Celt and Indian: Britain's Old World Frontier in Relation to the New *In Beyond* the Frontier: Social Process and Cultural Change. P. Bohannon and F. Plog, eds. Pp. 25-43. Garden City, New York: The Natural History Press.

MacMahon, Major M.
1993   The Tenets of Manoeuvre Warfare and their Application to Low Intensity Conflict. British Army Review 105:25-31.

MacStiofain, Sean
1979   Memoirs of a Revolutionary. Letchworth: Free Ireland Book Club.

Maine, Sir Henry Sumner
1906   Ancient Law: Its Connection with the Early History of Society and its Relation to Modern Ideas. New York: Henry Holt and Company.

Malinowski, Bronislaw
1922   War and Weapons among the natives of the Trobriand Islands. Man 20:10-12.
1926   Crime and Custom in Savage Society. London: Kegan Paul, Trench and Trubner.

1932 Crime and Custom in Savage Society. London: Routledge & Kegan Paul, Ltd.

1938 Introduction. *In* Facing Mt. Kenya: the tribal life of the Gikuyu. By Jomo Kenyatta. Pp. vii-xiii. London: Secker and Warburg.

1945 The Dynamics of Culture Change: An Inquiry into Race Relations in Africa. New Haven: Yale University Press.

Mallory, Keith and Arvid Ottar
1973 The Architecture of Aggression: A History of Military Architecture in N.W. Europe, 1900-1945. London: The Architectural Press, Ltd.

Manwaring-White, Sarah
1983 The Policing Revolution: Police Technology, Democracy and Liberty in Britain. Brighton, Sussex: The Harvester Press.

Markievicz, Constance Countess
1987 The Prison Letters of Countess Markievicz. London: The Virago Press, Ltd.

Marks, John D. and Victor E. Marchetti
1974 The CIA and the Cult of Intelligence. New York: A.A. Knopf.

Marvin, Garry
1986 Honour, Integrity and the Problem of Violence in the Spanish Bullfight. *In* The Anthropology of Violence. D. Riches, ed. Pp. 118-136. London: Basil Blackwell Ltd.

Marx, Karl and Frederick Engels
1971 On Ireland. London: Lawrence and Wishart.
1972 Ireland and the Irish Question. New York: International Publishers.

Marx, Karl
1956 Selected Writings in Sociology and Social Philosophy. T.B. Bottomore and M. Rubel, eds. London: Watts.

Mauss, Marcel
1954 The Gift. London: Cohen & West.
1967 The Gift, Forms and Functions of Exchange in Archaic Societies. New York and London: W.W Norton & Company.
1906 Les Variations saisonnieres des societies esquimaux. Annee Sociologiques, vol. IX.

McCann, Eamonn
(1974) 1980 War and an Irish Town. London: Pluto.

McCauley, Clark
1990 Conference Overview. *In* The Anthropology of War. J. Haas, ed. Pp. 1-25. Cambridge: Cambridge University Press.

McCuen, Lt. Col. John J.
1966 The Art of Counter-Revolutionary War: The Strategy of Counter-Insurgency. Harrisburg, Penn.: Stackpole Books.

McGuffin, John
1973 Internment. Tralee, Ireland: Anvil Press.

McKeown, Michael
1989 Two Seven Six Three: An Analysis of fatalities attributable to civil disturbances in Northern Ireland in the twenty years between July 13, 1969 and July 12, 1989. Belfast: Murlough Press.

McKinley, Michael
1987 Ireland, Britain, and the "Torture Case". Conflict 7(3):249-83.

McKnight, G.
1974 The Mind of the Terrorist. London.

Mead, Margaret
1979 Anthropological Contributions to National Policies During and Immediately After World War II. In The Uses of Anthropology. W. Goldschmidt, ed. Pp. 145-157.

Mead, Margaret, and Rhoda Metraux
1965 The Anthropology of Human Conflict. In The Nature of Human Conflict. Elton B. McNeil, ed. Pp. 116-139. Englewood Cliffs: Prentice-Hall.

Meron, Theodor
1987 Human Rights in Internal Strife: Their International Protection. Cambridge: Grotius Publications, Limited.

Metraux, Rhoda, and Nelly S. Hoyt
1953 German National Character: A Study of German Self Images, Studies in Contemporary Cultures. New York: American Museum of Natural History.

Middleton, John
1966 Resolution of Conflict In the Lugbara of Uganda. *In* Political Anthropology. M. Swartz, V. Turner, A. Tuden, eds. Pp. 141-155. Chicago: Aldine.

Midlarsky, M. I., ed.
1992 The Internationalization of Communal Strife. London: Routledge.

Mitchell, Timothy
1991 Colonising Egypt. Berkeley: University of California Press.

Mockaitis, Thomas R.
1990 British Counterinsurgency, 1919-1960. London: MacMillan.

Montagu, Ashley M.F.
1968 The New Litany of 'Innate Depravity', or Original Sin Revisited. *In* Man and Aggression. M.F. Ashley Montagu, ed. Pp. 3-18. Oxford: Oxford University Press.

Moore, Sally Falk, and Barbara G. Meyerhoff, eds.
1977 Secular Ritual. Amsterdam: Van Gorcum.

Morris, Eric
1986 Churchill's Private Armies. London: Hutchinson.

Morrison, William R.
1991 Imposing the British way: the Canadian Mounted Police and the Klondike Gold Rush. *In* Policing the Empire: Government, Authority and Control, 1830-1940. D. Anderson and D. Killingray, eds. Pp.

92-106. Manchester and New York: University of Manchester Press.

Morton, Brigadier Peter
1989 Emergency Tour: 3 Para in South Armagh. Wellingborough: William Kimber and Co.

Moxon-Browne, Edward
1983 Nation, Class and Creed in Northern Ireland. Aldershot: Gower.

Mueller, John
1991 War: Natural but not Necessary. *In* The Institution of War. R. A. Hinde, ed. Pp. 13-30. London: Macmillan.
1989 Retreat From Doomsday: The Obsolescence of Major War. New York: Basic Books.

Muenger, Elizabeth A.
1991 The British Military Dilemma in Ireland: Occupation Politics, 1886-1914. Lawrence: University of Kansas Press.

Murphy, N.T.P. and C.W. Blandy
1993 Sustainability and Viability. *In* International Military and Defense Encyclopedia. Col. T.N. Dupuy, ed. Pp. 2636-2640. Washington and London: Brassey's (US), Inc.

Nadel, Siegfried F.
1947 The Nuba: An Anthropological study of the Hill Tribes in Kordofan. London: Oxford University Press.

Nader, Laura and Duane Metzger
1963 Conflict Resolution in Two Mexican Communities. American Anthropologist 65(3):584-593.

National Graves Association (Cumann Na n-Uagheann Na`isiu`nta)
(1932) 1976 The Last Post: The Details and Stories of Republican Dead, 1913-1975, second ed. Dublin: The National Graves Association.
1985 Belfast Graves. Dublin: AP/RN Print.

Newell, Clayton R.
1991 The Framework of Operational Warfare. London: Routledge.

Nietzsche, Friedrich
1968 The Will to Power. W. Kaufman and R.J. Hollingdale, eds. New York: Random House.

Nugent, David
1982 Closed System and Contradiction: The Kachin In and Out of History. Man 17: 508-527.

O'Brien, Conor Cruise
1974 States of Ireland. London: Panther.

O'Brien, William V.
1978 The Jus in Bello in Revolutionary War and Counterinsurgency. Virginia Journal of International Law 18(2):193-245.
1979 Guidelines for Limited War. Military Review LIX(2):64-72.
1989 Counterterror, Law and Morality in LIC. In Low-Intensity Conflict: The Pattern of Warfare in the

Modern World. L. B. Thompson, ed. Pp. 187-209. Lexington: Lexington Books.

O'Duffy, Brendan
1993 Containment or regulation? The British approach to ethnic conflict in Northern Ireland. *In* The Politics of Ethnic Conflict Regulation: Case Studies of Protracted Ethnic Conflicts. J. McGarry and B. O'Leary, eds. Pp. 125-151. London: Routledge.

Offerdal, Audun and Jan O. Jacobsen
1993 *Auftragstaktik* in the Norwegian Armed Forces. Defence Analysis 9(2):211-225.

O'Malley, Padraig
1983 The Uncivil Wars: Ireland Today. Belfast: Blackstaff Press.
1990 Biting at the Grave: The Irish Hungerstrikes and the Politics of Despair. Boston: Beacon Press.

O'Neill, Bard E.
1990 Insurgency and Terrorism: Inside Modern Revolutionary Warfare. Washington and London: Brassey's.

Oraa, Jaime
1992 Human Rights in States of Emergency in International Law. Oxford: Clarendon Press.

Ortner, Sherry
1981 Theory in Anthropology since the Sixties. Comparative Studies in Society and History 26:126-166.

O' Suilleabhain, Sean
1979 Irish Wake Amusements. Dublin and Cork: The Mercier Press.

Ottenberg, Simon
1978 Anthropological Interpretations of War. *In* War: A Historical, Political and Social Study. L.L. Farrar, Jr, ed. Pp. 29-37. Studies in International and Comparative Politics #9. Santa Barbara: American Bibliographical Center-Clio Press, Ltd.

Otterbein, Keith
(1970) 1986 The Evolution of War: A Cross-Cultural Study. 2nd edition. New Haven: HRAF Press.

Overholt, William and William Ascher
1983 Strategic Planning and Forecasting: Political Risk and Economic Opportunity. New York: Wiley.

Paget, Julian
1967a Counter-Insurgency Operations: Techniques of Guerilla Warfare. New York: Walker and Company.
1967b Counter-Insurgency Campaigning. London: Faber and Faber.

Palmer, Stanley H.
1988 Police and Protest in England and Ireland, 1780-1850. Cambridge: Cambridge University Press.

Parker, Tony
1985 Soldier, Soldier. London: William Heineman, Ltd.
1993 May the Lord in his Mercy be kind to Belfast. London: Jonathan Cape.

Pawlisch, Hans S.

1985 Sir John Davies and the Conquest of Ireland: A Study in Legal Imperialism. Cambridge: Cambridge University Press.

Peace, Adrian

1983 From Arcadia to Anomie: Critical Notes on the Constitution of Irish Society as an Anthropological Object. Critique of Anthropology 9(1):89-111.

Pemberton, W.B.

1962 Battles of the Boer War. London: Batsford.

Perkins, Major General K.

1981 Counterinsurgency and Internal Security. British Army Review 69:27-33.

Pick, Daniel

1993 War Machine: The Rationalisation of Slaughter in the Modern Age. New Haven: Yale University Press.

Pimlott, John

1985 The British Army: The Dhofar Campaign 1970-75. In Armed Forces and Modern Counter-Insurgency. I. F.W. Beckett and J. Pimlott, eds. Pp. 16-46. London and Sydney: Croom Helm.

1988 The British Experience. In The Roots of Counter-Insurgency: Armies and Guerrilla Warfare, 1900-45. I.F.W. Beckett, ed. Pp. 17-40. London: Blandford Press.

Ponting, Clive

1990 Secrecy in Britain. London: Basil Blackwell.

Pospisil, Leopold
1974 The Anthropology of Law: A Comparative Theory. New Haven: HRAF Press.

Priestland, Gerald
1974 The Future of Violence. London: Hamish and Hamilton.

Prisk, Courtney E.
1991 The Umbrella of Legitimacy. *In Uncomfortable* Wars: Toward a New Paradigm of Low Intensity Conflict. M. G. Manwaring, ed. Pp. 69-93. Boulder: Westview Press.

Purdie, Bob
1990 Politics in the Streets: The Origins of the civil Rights Movement in Northern Ireland. Belfast: The Blackstaff Press.

Pye, Lucien
1963 Military Development in the New Countries. *In* Social Science Research and National Security, A Report prepared by the Research Group in Psychology and the Social Sciences, under Office of Naval Research contract No. 1354 (18), Task Number NR 170-379. deSola Pool, et al., eds. Washington, D.C.: Smithsonian Institution.
1964 Guerrilla Communism in Malaya: its source and political meaning. 2nd ed. Princeton: Princeton University Press.
1964 The Roots of Insurgency and the Commencement of Rebellions. *In* Internal War: Problems and Approaches. H. Eckstein, ed. Pp. 157-180. London: Collier and Macmillan.

Pye, Lucian W., and Sidney Verba, eds.
1965 Political Culture and Political Development. Princeton: Princeton University Press.

Rabinow, Paul
1977 Reflections on Fieldwork in Morocco. Berkeley: University of California Press.

Radcliffe-Brown, A.R.
1940 Preface. *In* African Political Systems. M. Fortes and E.E. Evans-Pritchard, eds. Pp. xi-xxiii. London: Oxford University Press.

Rawlins, Colonel P. P. MBE
1993 Economy of Effort: A Passive Principle. *In The* Science of War: Back to First Principles. B. H. Reid, ed. Pp. 49-63. London and New York: Routledge.

Reed, David
1984 Ireland: The Key to British Revolution. London: Larkin Press.

Reynolds, Charles
1989 The Politics of War: A Study of the Rationality of Violence in Inter-State Relations. New York: St. Martin's Press.

Reith, Charles
1943 British Police and the Democratic Ideal. London: Oxford University Press.

Riches, David, ed.
1986 The Anthropology of Violence. London: Basil Blackwell.

Richter, Daniel
1983 War and Culture: the Iroquois Experience. William and Mary Quarterly 3rd series (40):528-559.

Ricoeur, Paul
1967 The Symbolism of Evil. Boston: Beacon Press.
1979 The Model of the Text: Meaningful Action Considered as Text. *In* Interpretive Social Science: A Reader. P. Rabinow and W. M. Sullivan, eds. Berkeley: University of California Press.

Roberts, Adam and Richard Guelff, eds.
1989 Documents on the Laws of War. Oxford: Clarendon Press.

Rolston, Bill
1991 Containment and its Failure: The British State and the Control of conflict in Northern Ireland. *In* Western State Terrorism. A. George, ed. Pp. 155-180. New York: Routledge.

Rolston, Bill, Liam O'Dowd, Bob Miller, and Jim Smyth
1983 A Social Science Bibliography of Northern Ireland, 1945-1983. Belfast: Queen's University Press.

Rose, Richard
1971 Governing without Consensus: An Irish Perspective. London: Faber and Faber.

Roseberry, William, and Jay O'Brien, eds.
1991 Golden Ages, Dark Ages: Imagining the Past in Anthropology and History. Berkeley: University of California Press.

Rousseau, Jean-Jacques
1960 The Social Contract. *In* Social Contract: Essays by Locke, Hume and Rousseau. Sir Ernest Barker, ed. Pp. 167-307. London: Oxford University Press.

Rowe, Peter
1987 Defence: The Legal Implication: Military Law and the Laws of War. London: Brassey's.

Rowtorn, Bob, and Naomi Wayne
1988 Northern Ireland: The Political Economy of Conflict. Boulder: The Westview Press.

Rule, James B.
1973 Private Lives and Public Surveillance. London: Allen Lane.

Said, Edward
1989 Representing the Colonized: Anthropology's' Interlocutors. Critical Inquiry 15:205-225.

Sahlins, Marshall
1985 Islands of History. Chicago: University of Chicago Press.

Said, Edward
1989 Representing the Colonized: Anthropology's' Interlocutors. Critical Inquiry 15:205-225.

Salemink, Oscar
1991 Mois and Maquis: The Invention and Appropriation of Vietnam's Montagnards from Sabatier to the CIA. *In* Colonial Situations: Essays on the Contextualization of Ethnographic

Knowledge. G. Stocking, ed. Pp. 269-283. HOA 7.
Madison: University of Wisconsin Press.

Sands, Bobby
1981 The Diary of Bobby Sands. Dublin: Republican
Publications.

Sarkesian Sam C.
1985 Low-Intensity Conflict: Concepts, Principles and
Policy Guidelines. Air University Review 36(2):
4-24.

deSaussure, Ferdinand
1983 Course in General Linguistics. C. Bally and A.
Sechenaye, eds. London: Duckworth Press.

Scarry, Elaine
1985 The Body in Pain: The Making and the Unmaking
of the World. New York, Oxford: Oxford
University Press.

Schaeffer, Robert
1990 Warpaths: The Politics of Partition. New York: Hill
and Wang.

Scheper-Hughes, Nancy
1979 Saints, Scholars, and Schizophrenics: Mental Illness
in Rural Ireland. Berkeley: UC Press.

Schlesinger, Philip, Graham Murdock and Philip Eliot
1983 Televising Terrorism. London: Comedia.

Schmid, Alex P. and Janny DeGraf
1982 Violence as Communication: Insurgent Terrorism and the Western Mass Media. London: Sage.

Scott, James
1990 Domination and the Arts of Resistance. New Haven: Yale University Press.

Sereseres, Caesar D.
1985 Lessons from Central America's Revolutionary Wars, 1972-1984. *In* The Lessons of Recent Wars in the Third World, vol. 1. R. Harkavy and S. Newman, eds. Pp. 161-189. Lexington: Lexington Books.

Seymour, William
1985 British Special Forces. London: Sidgwick and Jackson.

Shafer, D. Michael
1988 Deadly Paradigms: The Failure of U.S. Counterinsurgency Policy. Leicester: Leicester University Press.

Shapiro, Michael J.
1993 That obscure object of violence: Logistics and Desire in the Gulf War. *In* The Political Subject of Violence. D. Campbell and M. Dillon, eds. Pp. 114-137. Manchester and New York: Manchester University Press.

Shaw, R. Paul, and Yuwa Wong
1989 The Genetic Seeds of Warfare: Evolution, Nationalism and Patriotism. Boston: Unwin Hyman

Shy, John
1993 The Cultural Approach to the History of War. Journal of Military History 57(5):13-27.

Simmel, Georg
1955 Conflict. Glencoe, Ill.: The Free Press.

Simpkin, Richard
1982 Human Factors in Mechanized Warfare. Oxford: Brassey's.
1982 Antitank: An Air-Mechanized Response to Armoured Threats in the 90's. London: Pergamon-Brassey's.
1985 Race to the Swift: Thoughts on Twenty-First Century Warfare. London: Brassey's Defence Publishers.
1987 Deep Battle: The Brainchild of Marshall Tukhachevskii. London: Brassey's.

Simpson, Henry J.
1937 British Rule and Rebellion. London: William Blackwood.

Sinn Fein Education Department
1990 The History of Republicanism: 1798-1990. Basic Education Programme, Section 1, Part 1.

Slocum, Sally

1975   Woman the Gatherer: Male Bias in Anthropology.
       In Towards an Anthropology of Women. R. Reiter,
       ed. Pp. 36-51. New York: Monthly Review Press.

Sluka, Jeffrey A.

1989   Hearts and Minds, Water and Fish: Support for the
       IRA and INLA in a Northern Irish Ghetto.
       Greenwich, Conn: JAI Press.

Smith, Mike

1991   The Role of the Military Instrument in Irish
       Republican Strategic Thinking: An evolutionary
       Analysis. King's College, University of London.
       D.Phil. Dissertation. Unpublished.

Snyder, Jack

1984   The Ideology of the Offensive: Military Decision
       Making and the Disaster of 1914. Ithaca: Cornell
       University Press.

Spenser, Edmund

1949   A Brief Note of Ireland. *In* The Works of Edmund
       Spenser, Vol. 9: The Prose Works. Baltimore: The
       Johns Hopkins University Press.

Spjut, R.J.

1986   The 'Official' Use of Deadly Force by the Security
       Forces against Suspected Terrorists. Public Law
       38:38-66

SPK

1983   The Post-Industrial Society. *In* The Industrial
       Culture Handbook. San Francisco: Re Search.

Spivak, Gayatri Chakravorty
1988 Can the Subaltern Speak? *In* Marxism and the Interpretation of Cultures. Nelson and Grossberg, eds. Pp. 271-317. Basingstoke: Macmillan Education.

Sterling, Claire
1981 The Terror Network: The Secret War of International Terrorism. London: Weidenfeld and Nicholson.

Stewart, Brig.-Gen. John F.
1988 Military Intelligence Operations in Low-Intensity Conflict: An Organizational Model. Military Review LXVIII: 17-28.

Stocking, Jr., George W.
1987 Victorian Anthropology. New York: The Free Press.

Stockwell, A.J.
1992 Policing during the Malayan Emergency, 1948-60: Communism, Communalism And Decolonisation. *In* Policing and Decolonisation: Politics, Nationalism and the Police, 1917-65. D. Anderson and D. Killingray, eds. Pp. 105-127. Manchester and New York: Manchester University Press.

Stoler, Laura Ann
1991 Carnal Knowledge and Imperial Power: Gender, Race and Morality in Colonial Asia. *In* Gender at the Crossroads of Knowledge: Feminist Anthropology in the Postmodern Era. M. diLeonardo, ed. Pp. 1-51. Berkeley and Los Angeles: University of California Press.

Strachan, Hew
1983 European Armies and the Conduct of War. London: George Allen and Unwin.

Stanhope, Henry
1979 The Soldiers: An Anatomy of the British Army. London: Hamish Hamilton.

Strathern, Andrew, ed.
1979 Ongka: A Self-Account by a New Guinea Big-Man. London: Duckworth.

Strathern, Andrew
1982 Witchcraft, greed, cannibalism and death: some related themes from the New Guinea Highlands. *In* Death and the Regeneration of Life. M. Bloch and J. Parry, eds. Pp. 111-134. Cambridge: Cambridge University Press.

Styles, George Lieut.-Colonel G.C.
1975 Bombs have No Pity: My War Against Terrorism. London: William Luscombe.

Swedenburg, Ted
1991 Popular Memory and Palestinian National Past. In Golden Ages, Dark Ages: Imagining the Past in Anthropology and History. W. Roseberry and J. O'Brien, eds. Pp. 152-180. Berkeley and Los Angeles: University of California Press.

Sunday Times Insight Team
1972 Ulster. London: Penguin.

Sun Tzu
1983 The Art of War. New York: Delacorte Press.

Suter, Keith
1984 An International Law of Guerrilla Warfare: The Global Practice of PeaceMaking. London: Francis Pinter.

Swedenburg, Ted
1991 Popular Memory and the Palestinian National Past. *In* Golden Ages, Dark Ages: Imagining the Past in Anthropology and History. W. Roseberry and J. O'Brien, eds. Pp. 152-180. Berkeley: University of California Press.

Sweeney, George
1993 Self-Immolation in Ireland: Hungerstrikes and Political Confrontation. Anthropology Today 9 (5): 10-14.

Swift, Jonathan
1969 A Modest Proposal. Columbus, Ohio: Charles E. Merrill.

Sykes, Stephen
1991 Sacrifice and the Ideology of War. *In* The Institution of War. R.A. Hinde, ed. Pp. 87-99. London: Macmillan.

Szwed, John
1975 Race and the Embodiment of Culture. Ethnicity(2): 19-33.

Taber, Robert
1965 The War of the Flea: Guerrilla Warfare: Theory and Practice. Paladin Press.

Taussig, Michael
1987 Shamanism, Colonialism and the Wild Man. Chicago: University of Chicago Press.
1992 The Nervous System. New York and London: Routledge.

Taylor, Lawrence
1989 Bas InEireinn: Cultural Construction of Death in Ireland. Anthropology Quarterly 64(2):175-187.

Taylor, Maxwell
1991 The Fanatics: A Behavioural Approach to Political Violence. London: Brassey's (U.K.).

Taylor, Peter
1980 Beating the Terrorists?: Interrogation in Omagh, Gough and Castlereagh. New York: Penguin Books.

Taylor, Trevor and Keith Hayward
1989 The UK Defence Industrial Base: Development and Future Policy Options. London: Brassey's.

Thompson, Loren B.
1989 Low-Intensity Conflict: An Overview. In Low-Intensity Conflict: The Pattern of Warfare in the Modern World. L. B. Thompson, ed. Pp. 1-27. Lexington: Lexington Books.

Thompson, Sir Robert

1966 Defeating Communist Insurgency: Experiences from Malaya and Vietnam. London: Chatto & Windus.

1969 No Exit From Vietnam. London: Chatto and Windus.

1970 Revolutionary War in World Strategy, 1945-1970. London: Secker and Warburg.

Thompson, William Irwin

1967 The Imagination of an Insurrection, Dublin, Easter 1916: A Study of an Ideological Movement. New York: Harper Colophon Books.

Thronton, Thomas P.

1964 Terror as a Weapon of Political Agitation. *In* Internal War: Problems and Approaches. H. Eckstein, ed. Pp. 71-100. New York: The Free Press.

Tilly, Charles

1990 Coercion, Capital, and European States, AD 990-1990. London: Basil Blackwell.

Tilman, Robert O.

1966 The Non-Lessons of the Malayan Emergency. Military Review XLVI(12):62-72.

Townshend, Charles

1975 The British Campaign in Ireland, 1919-1921. Oxford: Oxford University Press.

1982 Martial Law: Legal and Administrative Problems of Civil Emergency in Britain and the Empire, 1800-1940. The Historical Journal 25(1):167-95.

1983   Political Violence in Ireland. Oxford: Oxford University Press.

1986   Britain's Civil Wars: Counterinsurgency in the Twentieth Century. London: Faber and Faber.

1992   Policing Insurgency in Ireland 1914-23. *In* Policing and Decolonisation: Politics, Nationalism and the Police, 1917-1965. D. Anderson and D. Killingray, eds. Pp. 22-42. Manchester and New York: Manchester University Press.

Tugwell, Maurice

1989   Adapt or Perish: The Forms of Evolution in Warfare. *In* Armies in Low-Intensity Conflict: A Comparative Analysis. D. A. Charters and M. Tugwell, eds. Pp. 1-19. London: Brassey's Defence Publishers.

Turner, Victor W.

1969   The Ritual Process: Structure and Anti-Structure. London: Routledge, and Kegan Paul.

1967   The Forest of Symbols: Aspects of Ndembu Ritual. Ithaca: Cornell University Press.

Urban, Mark

1993   Big Boys Rule's: The SAS and the Secret Struggle Against the IRA. London: Faber and Faber.

Van Creveld, Martin

1992   High Technology and the Transformation of War, Part II. RUSI Journal 137(6):61-64.

Van Gennep, Arnold

1960   The Rites of Passage. Chicago: University of Chicago Press.

Vaughn, Megan
1990 Curing Their Ills: Colonial Power and African Illness. Stanford: Stanford University Press.

Vetschera, Heinz
1993 Low-intensity Conflict: Theory and Concept. *In* The International Military and Defense Encyclopedia. Col. T.N. Dupuy, ed. Pp. 1378-1582. Washington and London: Brassey's (US), Inc.

Vincent, Joan
1990 Anthropology and Politics: Visions, Traditions and Trends. Tucson: University of Arizona Press.

Virilio, Paul
1986 Speed and Politics: An Essay on Dromology. Translated by Mark Polizzotti. New York: Semiotext(e).
1986b Popular Defense and Ecological Struggles. New York: Semiotext(e) Foreign Agent Series.

Virilio, Paul, and Sylvere Lotringer
1983 Pure War. New York: Semiotext(e).

Vogler, Richard
1991 Reading the Riot Act: The Magistry, the Police and the Army in Civil Disorder. Philadelphia: Open University Press.

Volkan, Vamik D.
1988 The Need to Have Enemies and Allies: From Clinical Practice to International Relationships. Northvale, New Jersey, and London: Jason Aronson, Inc.

Wakin, Eric
1992 Anthropology Goes to War: Professional Ethics and Counterinsurgency in Thailand. Madison: University of Wisconsin Press.

Walker, Clive
1992 The Prevention of Terrorism in British Law. 2nd edition. Manchester and New York: Manchester University Press.

Walker, Paul F.
1986 Emerging Technologies and Conventional Defence. *In* Emerging Technologies and Military Doctrine: A Political Assessment. F. Barnaby and M. ter Borg, eds. Pp. 26-43. London: Macmillan Press.

Wallace, Anthony F.C.
1968 Psychological Preparations for War. *In* War: the Anthropology of Armed Conflict and Aggression. M. Fried, M. Harris and R. Murphy, eds. Pp. 173-183. Garden City NY: The Natural History Press.

Walsh, Dermot P.J.
1985 The Use of the Acts in Northern Ireland. *In* The New Prevention of Terrorism Act: The Case for Repeal. 3rd edition. C. Scorer, S. Spencer, and P. Hewitt, eds. London: NCCL.
1988 Arrest and Interrogation. *In* Justice Under Fire: The Abuse of Civil Liberties in Northern Ireland. A. Jennings, LL.B., ed. Pp. 27-47. London: Pluto Press.

Walzer, Michael
1974 Regicide and Revolution: Speeches at the Trial of
      Louis XVI. London: Cambridge University Press.
1978 Just and Unjust Wars: A Moral Argument with
      Historical Illustrations. New York: Basic Books.

Warner, W. Lloyd
1931 Murngin Warfare. Oceania 1:457-483.

Washburn, Sherwood and Lancaster, C.
1968 The Evolution of Hunting. *In* Man the Hunter. R.B.
      Lee and I. DeVore, eds. Pp. 293-304. Chicago:
      Aldine.

Waugh, Evelyn
1952 Men at Arms. Middlesex: Penguin Books.

Weber, Max
1984 Legitimacy, Politics and the State In Legitimacy
      and the State: Readings in Social and Political
      Theory. William Connolly, ed. Pp. 32-63. Oxford:
      Basil Blackwell.

Weitzer, Roland John
1990 Transforming Settler States: Communal Conflict
      and Internal Security in Northern Ireland and
      Zimbabwe. Berkeley and Los Angeles: University
      of California Press.

Welch, Stephen
1993 The Concept of Political Culture. Basingstoke:
      Macmillan.

White, Hayden
1987 The Content of the Form: Narrative Discourse and Historical Representation. Baltimore and London: The Johns Hopkins University Press.

Whitfield, Stephen J.
1991 The Culture of the Cold War. Baltimore: Johns Hopkins University Press.

Whyte, John
1991 Interpreting Northern Ireland. Oxford: Claredon Press.

Wilkinson, Paul
1977 Terrorism and the Liberal State. New York: Halsted Press.
1986 Terrorism and the Liberal State. 2nd edition. New York: New York University Press.

Wilkinson, Peter and Joan Bright Astley
1993 Gubbins and SOE. London: Leo Cooper.

Williams, Maxine
1989 Murder on the Rock: How the British Government Got Away with Murder. London: Larkin Publications.

Wilson, Edward O.
1975 Sociobiology: The New Synthesis. Cambridge: Harvard University Press.
1976 Sociobiology: A New Approach to Understanding the Basis of Human Nature. New Scientist 70(1000):342-9.

Wilson, Harold
1971 The Labour Government: 1964-1970. London: Weidenfeld and Nicholson, and Michael Joseph.

Wilson, Heather A.
1988 International Law and the Use of Force by National Liberation Movements. Oxford: Clarendon Press; New York: Oxford University Press.

Wilson, Thomas M.
1984 From Clare to the Common Market: Perspectives in Irish Ethnography. Anthropology Quarterly 57(1): 1-15.

Wimberly, Captain D.
1933 Bertrand Steward Army Quarterly Prize Essay. Army Quarterly 26:209-232.

Wolf, Eric
1969 Peasant Wars of the Twentieth Century. New York: Harper & Row.
1982 Europe and the People Without History. Berkeley: University of California Press.
1987 Cycles of Violence: the Anthropology of War and Peace. *In* Waymarks: the Notre Dame inaugural lectures in anthropology. K. Moore, ed. Pp. 127-150. Notre Dame: University of Notre Dame.

Woodham-Smith, Cecil
1980 The Great Hunger: Ireland 1845-1849. New York: E.P. Dutton.

Worsley, Peter

1986 The Superpowers and the Tribes. *In* Peace and War: Cross-Cultural Perspectives. M.L. Foster and R. Rubinstein, eds. New Brunswick: Transaction Books.

Wright, Frank

1987 Northern Ireland: A Comparative Analysis. Dublin: Gill and Macmillan.

Wright, Peter, with Paul Greengrass

1987 Spycatcher. London: Viking.

Wyllie, James H.

1984 The Influence of British Arms: An Analysis of British Military Intervention Since 1956. London: George Allen and Unwin.

Yarborough, William P.

1993 Low-Intensity Conflict: The Military Dimension. *In* International Military and Defense Encyclopedia. Col. T.N. Dupuy, ed. Pp. 1575-1578. Washington and London: Brassey's (US), Inc.

Yeats, William Butler

1982 The King's Threshold. *In* Collected Plays. London: Macmillan.

other sources

[The Baker Report]
Review of the Operation of the Northern Ireland (Emergency Provisions) Act 1978 (Cmnd. 9222, 1984. London: HMSO).

[The Bennett Report]
Report of the Committee of Inquiry into Police Interrogation Procedures in Northern Ireland (Cmnd. 7497, 1979. London: HMSO).

[The Cameron Report]
Report of the Commission on Disturbances in Northern Ireland (Cmnd. 532. Belfast: HMSO).

[The Compton Report]
Report of the Enquiry into the Allegations against the Security Forces of Physical Brutality in Northern Ireland, Arising out of Events on 9th August, 1971 (Cmnd. 4823, 1971. London: HHSO).

[The Diplock Report]
1972 Report of the Commission to consider legal procedures to deal with terrorist activities in Northern Ireland (Cmnd. 5185. London: HMSO).

[Gardiner Report]
1975 Report of a committee to consider, in the Context of Civil Liberties and Human Rights, Measures to deal with Terrorism in Northern Ireland (Cmnd. 5847. London: HMSO).

[Hunt Report]
1969 Report of the Advisory Committee on Police in Northern Ireland (Cmnd. 535, Lord Hunt, Chair. Belfast: HMSO).

[Parker Report]
1972 Report of the Commission of the Privy Councillors Appointed to Consider Authorized Procedures for the Interrogation of Persons Suspected of Terrorism (Cmnd. 4901. London: HMSO).

[Widgery Report]
1972 Report of the Tribunal appointed to inquire into the events of Sunday 30 January 1972, which led to the loss of life in connection with the procession in Londonderry on that day (H.L. 101/H.C. 220) (April 1972, London).

newspapers and magazines

An Phoblacht/Republican News
15 June 1977
9 December 1978
18 October 1980.
5 September 1981.

The Guardian
21 April 1981

Iris Magazine
Vol. 1, No. 2, November 1981

Irish Times
27 April, 1978
27 November 1980
21 March, 1981
13 April 1981
14 April 1981
7 May 1981

Newsweek
10 April 1981

New York Times
29 November 1993

The Observer
May 10 1991

Soldier Magazine
13 June 1988 44(2)

The Times
1 March 1981 (Sunday)
15 May 1981

Visor: Weekly Report for Soldiers in Northern Ireland
21 November 1974. Serial No. 40

decisions of United Kingdom and Irish Courts

R v. MacNaughton [1975] N.I. 203 67, 82-3
The State (O'Connor) v. O'Caomhanaigh [1963] I.R. 112 20
R. (Ronayne and Mulcahy) v. Strickland [1921] 2 I.R. 333
Higgins v. Willis [1921] 2 I.R. 386

Re: Jones (deceased) All England Law Reports [1981] 1 All
   ER
Anderson, In the Will of [1958] 75 WN (NSW) 334

acts, by year

Tumultuous Risings (Ireland) Act 1831
Capital Punishment (Ireland) Act 1842
Crime and Outrage Act of 1847
Peace Preservation Act in 1856
Criminal Law and Procedure (Ireland) Act 1887
Defence of the Realm Acts 1914-1918
Restoration of Order Act 1920
Government of Ireland Act 1920
Criminal Procedure Act (Northern Ireland) 1922
Civil Authorities (Special Powers) Acts (Northern
   Ireland) 1922-23
Public Safety Act 1927
New Whiteboy Act 1931
Constitution (Amendment No. 17) Act 1931
Offenses against the State Act 1939
Emergency Powers Acts 1940-43
Army Act 1955
Criminal Law Act (Northern Ireland) 1962
Criminal Law Act 1967
Northern Ireland (Emergency Provisions) Acts 1972, 1978
Prevention of Terrorism Acts
Right to Silence and Criminal Evidence Act (Northern
   Ireland) 1988

government papers and archival sources

House of Commons Debates, vol. 834, col. 240, 28 March
72
House of Commons Debates, vol. 882, 25 November 1974
House of Commons Debates, vol. 73, col. 100
House of Commons Debates, vol. 70, col. 575, 20
December 1974

Jamaica rebellion file (6 Jan. 1838), (WO 32 6235)

Ministry of Defence
1972 Notes and Information on Training Matters. Army
Code No. 9565(45).
1973 Notes and Information on Training Matters. Army
Code No. 9565(46).

unattributed dossier
1972 Incursions of the I.R.A. from Bases in the Irish
Republic.

Army Doctrine Publication, Volume One, Operations,
Final Draft, June 1993

PRO WO 35/90 May 22, 1920 and July 4, 1920

Jeudwine Papers 72/82/2, correspondence from H.Q.
November, 23, 1920. Held at the Imperial War
Museum. London, England.

Templer papers 7410-29-1-9. Held at the National Army
Museum. London, England.

Duties in Aid of the Civil Power, 1923

Notes on Imperial Policing, 1934

Imperial Policing and Duties in Aid of the Civil Power, 1949

Staff College Counter-Revolutionary Warfare Handbook, 1986

Northern Ireland: Future Terrorist Trends, 2 November 1978

Proceedings of the Low-Intensity Warfare Conference, Jan. 14-15 1986. Sponsored by the Secretary of Defense at Fort McNair, Washington, D.C.

Keeping the Peace (Duties in Support of the Civil Power), 1957

Northern Ireland Information Service
Parliamentary Statement: Messages Between the IRA and the Government. 29 November 1993.

pamphlets and other sources

Information on Ireland
[undated] British Soldiers Speak Out on Ireland. Pamphlet No. 1. London.

Faul, Fr. Denis, and Fr. Raymond Murray
1979H-Block X-Mas (pamphlet). Belfast.

Sinn Fein
1980Smash H-Block (pamphlet). Dublin.
1990Republican Resistance Calendar.

1988 Death of a Terrorist. PBS documentary.
1993 Hungerstrike: The Inside Story. BBC 2 documentary,
    aired on (10/13/93).

PIRA GHQ The Staff Report, 1977.

Major-General Michael Willcocks, speech at RUSI, Feb.
    1994.

The End

endnotes

[1]Urban (1993) disputes that Lance-Corporal Jones was a member of the SAS: "his name does not appear on the plaque on the clock tower at Hereford" where the names of SAS dead are inscribed. He was listed in Pegasus, the journal of the airborne Forces as a member of 3 Para. Urban believes that Jones was probably a member of the Close Observation Platoon of the 3rd Battalion, Parachute Regiment.

[2]This assumes a Weberian definition of the state as "a human community that successfully claims the monopoly on the legitimate use of physical force within a given territory" (1984:33). The state is then the sole source of the right to use violence.

[3]Actually, it was Maine in Ancient Law who said, "nobody cares about criminal law, except theorists and habitual criminals" (1906:102-3)

[4]The AAA Statement on Problems of Anthropological Research and Ethics states: "Academic institutions and individual members of the academic community, including students, should scrupulously avoid both involvement in clandestine intelligence activities and the use of the name of anthropology, or the title of anthropologist as a cover for intelligence activities" (1967).

[5]Conclusions about 'primitive' warfare, often derived from inference rather than direct observation (Ottenberg 1978), suffer from historical decontextualization. Barth's work among the Swat Pathan (1959) and Leach's work amongst the Kachin in Burma (1954) have been criticized on this point (Ahmed 1976; Asad 1975; Nugent 1982).

[6]To this end, this paper addresses not only warfare, but concepts of aggression and violence which are intimately linked in the mind of the anthropologist with warfare.

452

[7]This take on Marxism is merely a lowest-common denominator gloss on developmental models. Certainly, most Marxist interpretations are far more sophisticated.

[8]Anthropology as science' has, of course, been problematized recently by the incursion of literary criticism (Clifford and Marcus 1986).

[9]I've heard detailed discussions in military circles about the theories of Kaplan (1994) and Huntington (1993), who elaborate worst-case scenarios of socio-cultural implosion in the third world.

[10]Janowitz' The Professional Soldier (1960) was formative to the discipline of military sociology, argues that the boundaries separating military and civilian society are progressively breaking down. Berk (1993) offers an assessment of Janowitz' theoretical and institutional contributions to the field.

[11]Arensberg and Kimball's base-line 1930's model, makes any alteration in the structure of kin-based agrarian society appear as decline from an 'Arcadian' past. "Simply put, one cannot trace decline from a 'traditional' or 'stable' society (Clare of the 1930's) when it never was either" (Wilson 1984:3). Scheper-Hughes saw mental illness and "the general disinterest of the local populace in sexuality, marriage and procreation are further signs of cultural stagnation" (1979:4). In fact, the depopulation of western Ireland did not signify culture death, but reflected a stable and long-established pattern of marrying brides eastward, into the more prosperous lands of eastern and central Ireland. Thus, "the problems of population decline in Ireland are results not of poverty but of prosperity" (Arensburg and Kimball 1968:220).

[12]McPherson's Ossian cycles (1790) represented the Celts as a dying culture. McPherson introduced the 'noble suicide' trope of 18th century sentimental literature into Irish historiography. The myth of Ireland's death is perpetrated by the Irish themselves, for example by Yeats who claimed that his poetry was the beacon that started the Easter Rebellion.

[13]The myth of the death of Ireland assumes only part of the Darwinian metaphor; it does not incorporate the principle of actual genetic or cultural survival.

[14]Their purpose is to "describe how common law reasoning and empiricist social research are articulated in ideological practice when judges head official investigations into social situations (crises) which are characterized by threats to the proclaimed authority of judicial administration and the maintenance of public order" (Burton and Carlen 1979:53).

[15]Deleuze and Guattari also derive theoretical material from Paul Virilio, who identified a transition from the governance of war by a metaphor of siege, which is expressed in spatial and territorial coordinates, to the governance of war by a metaphor of emergency, which is expressed as time (1986:140). This concept, of course, can be roughly compared with Simpkin's concepts of the transition from attrition to maneuver warfighting forms (1985). Both Simpkin and Virilio were, of course, reading Sun Tzu. Virilio also influenced DeLanda's (1991) discussion of weapons evolution as a machinic phylum and Der Derian's notion of anti-diplomacy.

[16]Societies without a state, as well as societies without power, are nevertheless expressed in political form (Clastres 1977:9): "It is imperative to accept the idea that negation does not signify nothingness; that when the mirror does not reflect our own likeness, it does not prove there is nothing to perceive" (Clastres 1977:12).

[17]The basic assumption in this school of thought is that militaries should be under civilian control, at all times and at almost any cost. The military would thus be excluded from politics, no doubt as a barrier against military despotism. British dread of the army unbalancing political equilibrium no doubt derives from the experience of Cromwell. The US Constitution, by making the President the commander-in-chief of the armed services, seeks to guarantee the subordination of the military to civilian control and resides firmly in the British tradition.

[18]According to Kitson, for example, "political factors which tend to have a more direct bearing on those involved in countering insurgency than they do on soldiers involved in more conventional forms of war, have an even greater impact on peace-keeping. For these two reasons peace-keeping provides very good experience for counter-insurgency and vice versa" (1977:253).

[19]Evidence crops up to support the hypothesis that MI5 was out of control just as much as PIRA (Smith 1991) in the mid-1970s, an issue reexamined in Chapter Five.

[20]Low-intensity conflict may also referred to as 'low-intensity warfare', 'low-intensity conflict', 'low-intensity operations', 'limited engagement', 'guerrilla insurgency', etc. Low-intensity conflict is defined by the United States Joint Chiefs of Staff as "...a limited politico-military struggle to achieve political, social, economic, or psychological objectives. It is often protracted and ranges from diplomatic...pressures through terrorism and insurgency. Low-intensity conflict is often characterized by constraints on the weaponry, tactics and level of violence" (cited in Thompson 1991:3).

[21]Statistics on numerical incidence of LIC since WW II vary wildly: Serseres (1985), a US Army lieutenant-colonel, estimates that of the 125 to 150 conflicts in the past four decades, 90 percent can be characterized as 'low-intensity'. Kidron and Smith (1983) estimate that over 300 LIC 'wars' have occurred since 1945.

[22]National emergency plans are formulated in response to popular conceptions of political threats to the state. Prior to World War I, contingency plans were based on meeting a foreign enemy, while the immediate post war concerns were with 'domestic enemies'. Similarly, the contingency plans of the 1930's invoked a foreign threat. Since the end of World War II, and the dawn of the cold war, attention was once again focused on internal dissent (Bunyan 1976).

[23]'War space' is the actual space in which the military operates, which envelops the world and penetrates the surface. As of 1991, nine states occupied war spaces larger than the entire biosphere (see Kidron and Smith 1991:54-55).

[24]The notion of a Provisional Irish Republican air force was quite an amusing topic for British soldiers in the early 1970's. Seamus Twomey had been quoted in a German newspaper, Der Spiegel on November 3, 1973, as saying: "We shall fight from the air and from the ocean and we shall win. I just don't know whether I shall be alive to see it". This provoked a witty spoof in Visor (21 November 1974) on the PIRAF (the Provisional Irish Republican Air Force)! Here are some excerpts:

Well, now things looked pretty bad for us in mid 1974 and no mistake. We were more or less holding our own on the ground, where the IRA was doing a grand job in premature explosions and knee-cappings, but in the air it was a different matter. The British were after having complete supremacy, which is one way of saying they had planes and we didn't. So we put our heads together and did some thinking and when that got too hard work we did some talking and up comes Squadron Leader Barry Kaid with this great idea. 'If we haven't got the flyin' machines,' he said, 'we'll steal some and bomb the British Forces of Occupation all the way back to Kilburn!

[25]Guerrillas, too, have at least the possibility of acquiring airpower without resorting to hijacking.

[26]While the notion of a 'terrorist network' is perhaps overstated, the hardware connections between PIRA and Arab nationalists are fairly well-documented. Members of PIRA were trained near Baddawi refugee camp in the Lebanon, and acquired some RPG7 rocket launchers in 1973. Intercepted weapons consignments from the Middle East bound for Ireland 1977 would seem to suggest that there are more Kalashnikovs where those came from. The transfer of technique between rebel groups is more interesting: Guevara's Guerrilla Warfare circulated among PIRA members imprisoned in Long Kesh. Additionally, "Key figures of the present military leadership of the IRA had been heavily influenced by their prison experiences in the late 1950s and early 1960s, that is, when they met with imprisoned Cypriot insurgents and adopted General Grivas's terrorist strategy for defeating Britain in a guerrilla war" (Feldman 1991:162). General Grivas's strategy was, in fact, formulated partially as a result of having served in the British Army. Being quite familiar with its organization, structure and strategy, Grivas proved quite adept at countering it. The link between subaltern (para)military groups is, of course, nothing new: numerous Fenians fought on both sides of the American Civil War. Feldman suggests that the revolutionary cell system which replaced the old brigade system was a product of this prison association between Cypriot and Irish insurgents.

[27]Repeating to the Cabinet his warning to de Valera, Lloyd George declared

> that if disorder broke out again, the struggle would bear an entirely different character. British military commitments in different parts of the world were gradually being reduced, which had enabled the Government to concentrate their forces at home. As it was immaterial whether they were quartered in Great Britain or Ireland they would be sent to the latter country, where a great military concentration would take place with a view to the suppression of the rebellion and the restoration of order (cited in Jeffery 1984:90).

[28]Manwaring-White, for example, writes that "...the experience of governing dissenters in Ireland led to new methods of maintaining law and order in Britain; then it was the setting up of a force, now it is the strategy and technology of counting dissent that is learnt from the RUC in Northern Ireland by the mainland police" (1983:5). Military hardware (such as rifles, gas guns, water cannons and body armor), tested in Ireland, is now routinely used in mainland Britain (Manwaring-White 1983:117-154). Surplus army equipment is sometimes acquired by mainland forces and the West Midlands police recruit from amongst Army members (Hillyard and Percy-Smith 1988:247).

[29]O'Duffy (1993) argues that violence in Northern Ireland is not the cause of the conflict; rather violence is both the cause and effect of a counter-productive security policy. O'Duffy asserts that the British have, perhaps inadvertently, engineered the correct level of repression necessary to marginalize the proponents of the violence (136), just enough to make the democratic political process unfeasible.

[30]Hechter suggests in a different vocabulary an equivalent idea, that decolonization correlates with 'endo-colonization' inside state territory in the form of "internal colonialism" and the exploitation of "the periphery as an internal colony" (1977:30). He distinguishes internal colonialism, "or the political incorporation of culturally distinct groups by the core" from internal colonization, the settlement of previously unoccupied territories within state borders (1977:32 fn.).

[31]Depending very much on the political viewpoint adopted, British military power 'pointing outwards' may sometime point towards PIRA. If PIRA is considered as a nation without a territorial state, military power may simultaneously be 'pointing inwards'.

[32]Industrial capitalism was seen as essentially pacific by many nineteenth century figures such as St. Simon and Rousseau. Marx's theories, which explicitly link capitalist penetration with violence, are an obvious counterpoint.

[33]"Indeed the attempt is always made to ensure that force will appear to be based on the consent of the majority, expressed by the so-called organs of public opinion (politicians, media, etc.)" (Gramsci 1971:80) Force has never appeared as neutral in Northern Ireland.

[34]Hence the common Irish nickname for the police: 'peelers'. Robert Peel, himself was to acquire the nickname 'Orange' Peel, the 'orange' referring to his Protestant heritage.

[35]In 1867, the Fenian risings were repressed by the RIC, acting in small groups under own their own command. After this feat, the Irish Constabulary were rewarded with royal accolade and became henceforth the Royal Irish Constabulary.

[36]The 1839 Commissioners of inquiry into a police for England and Wales reported that "the Irish constabulary force is in its origination and action essentially inapplicable to England and Wales. It partakes more of the character of a military and repressive force, and is consequently required to act in greater numbers than the description of force which we consider the most applicable, as a preventive force" (First Report of the Commissioners appointed to Inquire as to the best Means of Establishing an Efficient constabulary Force in the Counties of England and Wales, pp. 160-1 [169], HC 1839, XIX, pp. 166-7, cited in Hawkins 1991:23).

[37]According to Hawkins, "the success of the Irish system validated further attempts on similar lines elsewhere; it served as a precedent to be cited...and a standard to be striven for; the extent to which it was actually copied in any given instance, however, remains to be established, and has too often been taken for granted" (1991: 24).

[38]The density of distribution of RIC was one to 400 or 500, more than twice as much as in Britain.

[39]Claims by contemporary administrators to have modeled their colonial police forces upon the RIC, are probably a sign that they understood the difference between a policing based on community consent, which protected a civilian (and civilized) population against external disorder, and a policing at the behest of the colonial state.

[40]Like the RIC, the North-West Mounted Police were organized in a military fashion, responsible to a central authority, and armed. The political nature of the force, as agents of Canada's sovereign claim and representatives of the State, was comparable to the function of the RIC in Ireland. Like the RIC, this force "not only enforced (and sometimes invented) the law, they also administered it" (Morrison 1991:101).

[41]Establishment of civic order correlates with police demilitarization, such that colonial forces more closely follow the London Metropolitan model (Killingray 1991:121). Internal security duties are transferred to the police who are recruited locally, although may maintain military elements. 'Settlement' occurs at different rates in different areas; in Egypt the Army was replaced by police only after W.W.I (Johnson 1991:158). On the other hand, if internal security responsibilities were shared by imperial troops and local police, withdrawal of troops from the colony may result in the militarization of the police: both Trinidad and the Bahamas the police force was reorganized along military lines and "by 1890, the Trinidad police force resembled the RIC" (Johnson 1991:85). "Increasingly the police took over the internal security role of the military, expanding to become the main coercive arm of the state in combating the growing weight of nationalist opposition" (Anderson and Killingray 1991:9).

[42]Between 1815 and 1914, Britain was at war only once with another imperial power, during the Crimean War of 1854-6.

460

[43]The post-Victorian era saw the development of two schools of strategic thought: the 'Continental' school which was represented by Godophin and Marlborough and the 'Maritime' which was backed mostly by land-owning Tories (see Jackson 1986).

[44]The 1908 Select Committee on The Employment of the Military in Cases of Disturbances, established that between 1878 and 1908 there were twenty-four cases when troops called out.

[45]To centralize or not to centralize command and control of troops and police is a recurrent issue in the history of UK domestic political disorder. Mockaitis' research suggests that command-issues regarding mixed forces contributed to the chaos during the Anglo-Irish war. War Office conferences were held to determine who, legally and bureaucratically, should assume control. Mockaitis (1990:71), contrary to the War Office's conclusion, asserts that police had the legal prerogative. The same dispute surrounding unification of command was reiterated in the 1970's. Vogler believes that the military interest in 'third force' issues reflected a contest in the 1972 National Security Committee over who would run the show during national emergencies. Judging by the current preference for police commando units (e.g., the Special Branch) rather than the SAS, funding and investment of police forces and police primacy in Northern Ireland, the police seem firmly entrenched as the stage-managers for any further riotous spectacles (Vogler 1991:91-94). The issue of legal control, however, remains very indeterminate in practice, as will be seen in Chapter Four.

[46]While the battlefield tactics of certain engagements resembled conventional warfare (the Zulu wars for example) small wars tactics were generally ad-hoc.

[47]Seventeen previous Coercion Acts had been in effect in Ireland in the forty six years since the Union. This new bill, however, as Lord Brougham remarked, "possessed a superior degree of severity" (cited in Woodham-Smith 1980:70).

[48]The Cardwell reforms also improved the training of soldiers and in 1881, the Army Discipline Act improved the conditions of soldiers by outlawing flogging for offenses, replacing it with the lashing of malefactors to gunwales for set periods of time. During this era, the Army officially replaced the 'red coat,' (which made its last appearance at the battle of Ginnis in the Sudan in late 1885) with khaki, an invention of the Indian garrisons who dyed their white dress uniforms with curry powder to conceal dirt (Featherstone 1973:19).

[49]The norm prior to the late 1800's was extended foreign posting of garrisons; not until 1865 were one third of the line battalions actually in Britain (Featherstone 1973:15).

[50]Actually, it would be plausible to argue that the counter-insurgency system functioned (or originated) during the period of eighteenth century revolutionary wars. Canny (1976), for example, points out the consistency in the ideology of colonization in the English experience of 'civilizing' Ireland and fighting first Indians and then rebels in America. MacLeod points out the consistency in frontier policy during the sixteenth and seventeenth centuries regarding both Celt and Indian (MacLeod 1928). It has also been argued that the second Boer War (1902) was the first guerrilla war (Dixon 1976:54), although it would seem that the Boer's were fighting in a rather unconventional fashion, while the British were not (see for example, Pemberton 1962). Even the American Civil War had aspects of guerrilla warfare, particularly the focus on the raid as a tactic in land warfare (Bellamy 1990:65-67). It might be interesting to compare this with Deleuze and Guattari's (1992) analysis of the raid as part of the nomad war machine.

[51]What had once been called a 'small war' was now, in alignment with the change in colonial status afforded by law, to be called 'imperial policing.' 'Small war' was now to be used to designate limited conventional conflicts. The same restraint which was supposed to be applied to civil unrest and allying internal disorders, was now to be applied in 'imperial policing', since, in fact, the empire was now to be considered as a single, unified political bloc. During this period, the Manual of Military Law distinguished between riot in which principle of minimum force applied and insurrection, where it did not. A pattern of response was being established (Mockaitis 1990:18) which would later become very important in the development of a full-blown counter insurgency doctrine.

[52]Despite the doctrine on minimum force adopted after Amritsar, atrocities continued to occur. During the Moplah Rebellion, two years after Amritsar, November 1921, 56 of 100 prisoners died of asphyxiation in a locked baggage car under police guard. The guards were tried and acquitted, the jury ruling that it was an accidental death. Decapitations of Moplah guerrillas by Chin and Karin irregulars of Burma Rifles are known to have occurred regularly (Mockaitis 1990:38). Evelyn Waugh satirizes decapitations of tribal opponents by members of the British Army in Men at Arms (1952).

[53]The Hunter Committee Report condemnation of Dyer, not for moral reasons, but because maximum use of force produces the opposite result to that desired. "Hearts and minds" counter-insurgency also sometimes seems to be more concerned with the instrumental rationality of ends than with means. Gwynn criticized Dyer on the grounds that Dyer set a dangerous precedent by firing without higher authorization from the civil power: "subordinates cannot be allowed to dictate policy" (1934:61) and "the military commander called in to replace the civil power should make the utmost use of the... judgment of the civil authority he replaces" (1934:62).

[54]Confusion about the legal status of troops on IS duties was a continual problem (see Evelegh 1978), exacerbated by the failure to properly codify IS rules in a timely fashion. British troops assigned to internal security duties in Palestine had to rely on Duties in Aid of the Civil Power, 1937 and Notes on Imperial Policing, 1934, neither of which referred to the political aspects of insurgency nor recommended minimum use of force in civilian engagements. Confusion about the status of the conflict combined with lack of application of the minimum force doctrine probably contributed to the atrocities committed under Wingate in 1938 (see Mockaitis 1991:34).

[55]General Sir Edward Bulfin, recently returned from controlling civil disorder in Egypt, refused the post of Inspector-General of Ireland in 1920, on the grounds that "it would be most distasteful to him to do any work...which was not of a purely military character" (Jeffery 1984:82).

[56]Churchill, for example, objected to the appointment of Sir Nevil Macready to replace Shaw, whose tenure as the General Officer Commanding-in-Chief was over. Lloyd George wanted to send Sir Nevil Macready to Dublin because had considerable experience of commanding troops 'in aid of the Civil Power' in Wales in 1910, and Belfast in 1914. As Commissioner of the London Metropolitan Police, he had dealt successfully with the police strike.

[57]This was countered by a suggestion from the Secretary of War that a 'Special Emergency Gendarmerie' of 8,000 old soldiers be raised to reinforce the RIC. A committee reported against a 'gendarmerie' and in favor of battalions, under the Army Act but limited to service only in the UK (see Jeffery 1984). For Churchill, declaring martial law was tantamount to proclaiming total fidelity to Union. Wilson "urged Winston...that the Government should govern, proclaim their fidelity to the Union and declare Martial Law" (cited in Jeffery 1984:87).

[58]This included sweeps and reprisal killings, which were common practice. General Tudor, in command of a reinforced RIC, was assured by the Prime Minister of 'full support'. According to Sir Henry Wilson, Lloyd George had a "theory that Tudor, or someone, was murdering 2 S[inn] F[einer]s to every loyalist the S.F.s murdered. I told him this was absolutely not so but he seemed to be satisfied that a counter-murder Association was the best answer..." (Wilson cited in Jeffery 1984: 85). Pressed for information, Lloyd George responded, "you must not ask me any questions but the thing is in operation already" (cited in Jeffery 1984:86).

[59]Because it seems to promise a fair fight, the British Army is still eager to have belligerent status declared for insurgents (see Simpkin 1985:319-321). According PIRA belligerent status, allowing them to procure proper weapons, clearing West Belfast of civilians, bringing in the ICRC to monitor the application of the laws of war, and letting the war rip would be an interesting way to end the conflict.

[60]Recognition of the IRA by the civil government in Ireland would have contradicted their own justifications for their continued presence in the country and created a legal conundrum under the two legal codes (the Restoration of Order Acts and Martial Law), where the IRA would have been a legal entity in some areas and not in others (See the Jeudwine papers 72/82/2).

[61]With police intelligence deteriorating, the Army assigned detachments of soldiers to protect besieged police posts. The Dublin district was divided into operational circles, which were seen as tactical areas: "In each circle a detachment of troops varying from a company to a platoon...will be or has been posted with the RIC at a central point" (directive dated 2 July 1920, cited in Mockaitis 1990:75). This type of arrangement, 'framework deployment,' was used quite successfully in Malaya and is still used in Northern Ireland. Framework operations, on the other hand, are planned overt operations by which the Army obtains detailed knowledge of an area, including patrolling, vehicle searches, and establishing a physical presence. Reactive operations are immediate reactions to terrorist incidents, preplanned reactions to specific types of information or intelligence and contingency plans. Specialist operations include bomb-disposal, air reconnaissance, etc. Covert operations include acquiring of evidence using specialized surveillance techniques and hands-on counterterrorist operations. Static tasks, such as guarding permanent check points and army bases, are probably the most boring. According to Urban (1992), "During the 1980's the term 'ambush' was replaced in SAS orders in Northern Ireland by 'OP/React', short for 'Observation Post/Reactive', according to an SAS man who served there. He says an OP/React order is 'to all intents and purposes an ambush'..." (164). Urban gives the following account of the interview:

> URBAN: What is the mission on an ambush?
> SAS MAN: You know what the mission is on an ambush, everybody knows what the mission is in an ambush.
> URBAN: Tell me what you think it is.
> SAS MAN: I know that when you do an ambush you kill people.

[62]According to Gwynn: "For although troops are at all times justified in taking ...the action which the immediate necessity of the situation...demands, their right to take action to prevent further outbreaks is very limited unless continuity of necessity is officially recognized" (1936:17).

[63]Statements of a minimum force directive became more overt in later pamphlets, eventually codified on 'Yellow Cards' in Northern Ireland. Imperial Policing and Duties in Aid of the Civil Power, 1949, rectified earlier omissions of minimum force, incorporated the experiences of Palestine and Malaya, advocated civilian-military cooperation and stressed the application of minimum force:

> The sole object of military intervention in civil disputes, or in dealing with the general unrest or even widespread insurrection or violence is the restoration of law and order by military means when other methods have failed...There is, however, one principle that must be observed in all action taken by the troops: no more force shall be applied than the situation demands (Imperial Policing and Duties in Aid of Civil Power, 1949:5)

[64]Jeffery writes that "colonial warfare...may be crudely described as conflict between white forces (or at least forces commanded by white men) and non-white groups..." (1988:24), the only exceptions being Great Britain in Ireland and Japan's involvement in Asia. Varieties of colonial war, according to Jeffery are imperial powers fighting indigenous forces, imperial powers fighting each other with indigenous forces over territory and pacification or peacekeeping activities in territories.

[65]Although Mockaitis argues that "[i]mplicit in such an approach was the concept of minimum force" (1990:186), Thompson, in fact, argues in favor of preemptive strikes (1966:50), a rather maximal sort of force.

[66]His work is apparently still used as instruction at Sandhurst (Mockaitis 1990:185 and fn. 190). Thompson, himself, is still working in this capacity to the U.S. government (see Proceedings of the Low-Intensity Warfare Conference 1986).

[67]Belfast is an excellent space for war, displaying a cross-section of military landscapes: fortress fortifications, fixed defensive positions, concealment and entrenchment of strategic points, the use of concrete and Nissen huts, structural prefabrication for maneuverability and easy assemblage, exploitation of natural cover for OP's, minimum dead ground in the field of fire, etc. (Mallory and Ottar 1973).

[68]The insurgency loop circulates in both directions: the IRA was apparently a model for Grivas' EOKA forces, which fought against the British in Cyprus (Beckett and Pimlott 1988:7). Grivas, of course, had also served with the British Army and his tactical guerrilla campaigns exhibit this odd double influence.

[69]'Economy of effort' replaced 'economy of force' in British military parlance after Montgomery became CIGS following WWII.

[70]World War II, for example, shifted military thinking from small operations back towards more conventional strategy. In Northern Ireland, the regimental system enhances unit cohesion, and created officers who are best at small unit operations, particularly emphasizing the junior NCO level of leadership: corporals and lance-corporals were typically in charge of a 'brick' (a four man patrol) on the street (see Dewar 1985:178).

[71]Describing the encampment/entrapment of the French Army at Verdun in 1916, Virilio writes: "the country is cut in two by a demarcation line," divided between a civilian rear guard with democratic government and economic and industrial activities; and a military zone, "fortified glacis where, Ferry notes: 'The supreme commander is no longer a chief of war, *but the administrator of territory*"— a territory in which civilian power hoped to crystallize the battle and enclose its military proletariat in an absolute war, a war 'without limitations in the use of violence,' but one that would not spread, could not be brought into the interior" (1986:52. Italics in original).

[72]Alves writes that "the character of the national security state can only be understood in relation to its interaction with...opposition movements in civil society" (1985:9). Anti-colonial movements use military discipline for "counter-hegemonic purposes" (Mitchell 1991: xi; Fergusson 1992; Crowder 1978); state armies derive their organizational forms from resistance movements (Leeman 1991).

[73]Some British internal security structures were developed with the Irish in mind. What is now known as the Special Branch first developed out of the CID at Scotland Yard in 1883 to combat Fenian bombings in London. Initially called the 'Political Branch' it was then renamed the 'Special Irish Branch', and finally the 'Special Branch' in 1888. Its particular task had always been, and is now, to cope with Irish political violence.

[74]While in the past, the execution of war was an exchange, in the present war means becoming the enemy. Tactical mimicry collapses all strategies; the state becomes indistinguishable from its enemy. Under these conditions, power is forced back onto itself, constituting itself as what Foucault calls 'peripheral subject' (1990:212) of its own effects.

[75]According to Faligot, for example, "as early as 1976, the IRA used electronic detectors to locate high frequency radios (ZB298) connecting the SAS ...with their HQ...consequently...IRA Volunteers adopted the same methods used by the SAS to successfully track down the SAS" (1983:45).

[76]A World War II era example of this was the Future Operation (Enemy) Section (FOES). Established in 1940 to assess the course of the war from the enemies point of view, the FOES was replaced in March 1941 by the Axis Planning Section (APS), which in turn evolved into the Joint Intelligence Staff (JIS), an inner committee of the JIC (Gudgin 1989:58). A more modern example is war-games predictive modeling: "...to the extent that the insights generated by watching wars between Red and Blue automata find their way into public policy and contingency plans, there is a sense in which our future is becoming increasingly dependent on correctly thinking Red" (De Landa 1991:100).

[77]Interestingly, Calvert was SOE trained, had commanded an SAS Brigade during WWII, and was posted as a staff officer in Hong Kong in 1950. Summoned by CLF Harding to make recommendations, he proposed the formation of the Malayan Scouts which later became the nucleus of 22 SAS (see Jackson 1991:36-7), strategic hamlets called kampongs, and food denial programs, which were incorporated into the Brigg's plan (Seymour 1985:275). The Malayan Scouts were an excellent example of environmentally adaptive warfare, and once spent 103 days in the jungle (the average was seven days for British soldiers).

[78]A similar 'peace line' was established in Nicosia on Cyprus between Turkish and Greek Cypriot communities (see Kitson 1977:250-1).

[79]Unionists seek to maintain the status Northern Ireland within the United Kingdom; their loyalty based on continuing defense of their interests by Crown. According to a 1978 survey, 85 % of Protestants stated that a "loyalist is loyal to Ulster before the British government" (Moxon-Browne 1983:86). Groups like the Ulster Volunteer Force, formed by Edward Carson in order to resist Dublin rule, willingly bite the hand that feeds. The British Army has been unwilling to act against Protestants: the Curragh Mutiny in 1914 was nearly replicated during the UWC strike, when soldiers refused to remove barricades. Police in Ulster are equally unpredictable: if the UVF had attempted by coup to establish an Ulster Provisional Government under Protestant control, the RIC might have offered assistance.

[80]Although the RUC favored de-militarization (Weitzer 1990:123), disarmament did not occur until 1970 (and then only temporarily). If the RUC refused to co-operate with the implementation of policy, it was feared that Britain would be unable to provide an imported substitute force; the British Police Federation, in any case, had objections to serving under the Unionist Minister of Home Affairs (Callaghan 1973:19, 22). If the Unionist government protested seriously against the way the Army was being used, "it would, in effect, be making a Unilateral Declaration of Independence along the lines that Ian Smith's white Government in Rhodesia had done" (Hamill 1985:16).

[81]Following a joint Army/police study day, the MOD appointed a Commander Land Forces (CLF) to deal with daily operations and left the GOC to control political business. GOC Sir Ian Freeland announced that political direction had recently been given for the RUC to resume control of maintaining law and order (Hamill 1986:39), so that the Army could concentrate on other tasks. Although the RUC did not resume control of policing until almost two years later, police primacy sharpened the British Army focus. In 1971, the Joint Security Committee, comprised of the Prime Minister of Northern Ireland, the GOC, Chief Constable, and others, was formed to coordinate security, while operational decisions were made at Lisburn (Hamill 1985:46).

[82]The Army and the Ministry of Defense favored direct rule, and believed that if Westminster took over security matters, an overall policy would be developed.

[83] "Imprisonment without trial is better than murder without trial," said one former GOC (Dewar 1985:114).

[84] Normally, internment has a war-time legal basis: the declaration of a state of emergency (as in Kenya and Ireland in 1916) allowed the government to issue detention orders (Clayton 1976:14). By the end of the Mau Mau Emergency, over ninety thousand people had been detained and interrogated by teams of police. An investigation of Captain Brian Hayward's team, who screened over two-hundred Kikuyu from October 14-16, 1953, revealed the systematic use of torture on detainees who had been whipped on the soles of their feet, burned with cigarettes, dragged on ground by leather straps tied to their necks. In other cases, suspicious detainees were often spared interrogation and executed outright. A policeman from London, H.A. Cross, posted in South Nyeri Reserve wrote that "Since I've been here, I inspect all prisoners brought in, and if they are a bit dubious, I refuse to have 'em — get called to a dead body the next day — and proceed Normally" (cited in Mockaitis 1990:46).

[85]It has been argued that the Army's minimum force doctrine constituted, in practice, a de facto form of political control (Hamill 1986:26).

[86]Other counter-insurgencies have been framed as tribal wars, as for example the war in Kenya (1952-60) with the Land and Freedom Army (Mau Mau). According to Paget (1967), in Kenya there was "no evidence that it was communist inspired or received material Communist support (86), but it was seen as the result of tribal in-fighting. The European plantation in the fertile White Highlands dispossessed Kikuyu tenant farmers. Although the Kikuyu grievances were of a political nature, the high number of Africans killed (1,819) by the Land and Freedom Army, suggested that the Mau Mau were conducting 'tribal warfare'. The killings of whites (32) allowed the British to represent the conflict as a racial struggle (Mockaitis 1990:126). The combined British and home-guard security forces killed over ten thousand Kikuyu. "When the enemy was a poorly armed Africa operating in thick forest, the boundary between war and hunting was too easily crossed" (Mockaitis 1990:50). Kitson described it as "an outlandish campaign left over from the Victorian era" (Kitson 1987:14).

[87]Small wars also included "campaigns of conquest when a Great Power adds the territory of barbarous races to its possessions..." (Callwell 1906:22). While annexation incorporated foreign territory into the boundaries of empire (for example, the campaign of the Punjab) subjugation of insurrection occurred within the boundaries of the imperial state (like the French pacification of Algeria). Small wars were also wars of expediency, undertaken to avenge a wrong, such as the Abyssinian expedition of 1868.

[88]"Beating of the hostile armies is not necessarily the main object...moral effect is often far more important than material success..." (Callwell 1906:42).

[89]In the Quteibi area, by May 1934, the blockade lasted sixty-one days, twenty-eight and half tons of bombs had been dropped, and forty thousand rounds of ammunition had been fired. Four villages were heavily damaged and seven people were killed as a result of tampering with UXBs (Mockaitis 1990:32).

[90]Stocking describes this process in a somewhat more hyperbolic fashion as "the advancement of native welfare through the systematic confiscation of tribal lands and the destruction of tribal institutions" (1987:81). The "biocultural conception of race" (1987:106) within Victorian ethnology tautologically legitimated the colonial conditions which produced the signs of 'savagery' (illiteracy, poverty, alcoholism).

[91]While the urban primitivism of pre-industrial London was increasingly disciplined by policing (institutionalized with the formation of London Metropolitan Police in 1829 and spread to provinces after Municipal Reform Act of 1835) the rural primitivism flourishing on the Celtic fringe (Stocking 1987:213) provided "a continuing 'primitive' base point for the 'civilizing' processes that were transforming them" (Stocking 1987: 210).

[92] Tribal models mistake the consequence for the cause: rather than interpreting conflict between a political minority and the government of settler state as *a result* of intervention, ethnic oppositions are posited as the cause of metropolitan politico-military intervention. The Mau Mau rebellion, for example, can either be interpreted as an anti-colonial conflict over the appropriation of tribal lands (see Barnett and Njama 1966), or as the product of 'ethnic tension' between different groups of Kikuyu (see Paget 1967). Ethnic conflicts in divided states generally do not reflect ancient, implacable antagonisms, but conditions imposed by powerful nation-states (Schaeffer 1990). A tribal model, by essentializing violence as 'natural' animosity, absolves the state of responsibility for the social and political conditions which may cause and sustain violence, leaving violence appearing as a pathological conundrum (Burton 1978).

[93]Humprey Atkins, on a visit to Northern Ireland, was shown a 'tribal map,' of Belfast, and failing to fully appreciate the significance of the color symbolism, asked "what's this bit of blue?" To which an Army officer relied, "that, sir, is Belfast Lough."

[94]Tuzo replaced Freeland as GOC.

[95]Using an Army for 'peace-keeping' and internal security operations contradicts the fundamental logic of war. The Widgery Report concluded that "if the Army had persisted in its 'low key' attitude and had not launched a large scale operation to arrest the hooligans the day might have passed off without serious incident" (point four from the Summary of Conclusions). Armies, however, cannot be expected to 'play nice'.

[96]A similar split of the Republican movement occurred in 1914 over the issue of participation in the British war effort. The constitutional nationalists favored joining, while the 'provisional' Fenian elements refused to join except as an independent state (see Townshend 1986:52).

[97]Protestant factions were actually the first to fire. The IRA was, in fact, much maligned within Republican communities for their inability to counter the British Army; graffiti covered Belfast and Derry until 1971 which read: "I.R.A: I ran away".

[98]By calling the conflict 'war', Chichester-Clark probably hoped to have more troops deployed and Faulkner to justify large scale actions against civilian communities. The rhetoric failed in both cases; no additional troops were granted to Chichester-Clark and public protest did accompany the beginning of internment under Faulkner.

[99]Political independence was preferable to having communists in power. The British did not resist Malayan independence, but used it as a weapon against Communism.

[100]Mao assumed that there would be an increase in civilian support for the guerrillas. Following the initial 'hit and run' military tactics, small-scale conventional engagements of short duration would be fought at insurgents choosing, followed by increasingly conventional large engagements. Guerrillas and official army forces do not meet on an open battle field, but engage in armed combat within civil communities. At the early stages, guerrillas in 'low-intensity conflict' tend not to wear uniforms, but are indistinguishable from the general population, blurring of distinction between soldier and noncombatant (Galvin 1991:13).

[101]Bunch of Five and Low Intensity Operations were used as teaching material by the Army, at least until 1985, the former being required reading for the Staff/Promotion exam (Mockaitis 1990:190fn.).

[102]Despite the fact that PIRA derived certain analytical devices from Marxism, the hype which surrounded their alleged 'communism' was grossly exaggerated.

[103]PIRA as 'communist agitators' may also have been an idea which Airey Neave, at that time the Conservative opposition spokesman on Northern Ireland, derived from PSYOP agitprop and disseminated in public debates (Foot 1990:173-4).

[104]After the institution of direct rule, a unified political and security policy was passed through the Secretary of State for Northern Ireland and the Northern Ireland Office. In the chain of command, the GOC served not the Secretary of State, but the Defense Secretary (Hamill 1985:160).

[105]These ideas had been floating around the halls of power for quite a while. In 1974, Rees said: "I believe that the cornerstone of security policy should be a progressive increase in the role of the civilian law enforcement agencies in Northern Ireland....in the long term it must be the community itself and normal police activities not military operation alone, which would finally defeat the terrorist" (HC vol. 871 col. 1466 4/4/74).

[106]The term 'ulsterization' was probably derived from 'vietnamization,' originally coined by Sir Robert Thompson to describe the US Army's policy mishaps.

[107]The force level of the Army fell from 22,000 in 1972 to 10, 123 by May 1988, while the force level of the UDR and RUC rose from 15,000 to 19,237 (Weitzer 1990:207).

[108]According to Foucault, power is tolerable only on condition that it mask a substantial part of itself. Its success is proportional to its ability to hide its own mechanisms....For it, secrecy is not in the nature of an abuse; it is indispensable to its operation (1990:86).

[109]Geraghty (1980) says 1974, Dewar (1985) says 1976, and Faligot (1983) says 1970. MacStiofain (1979) claimed in his memoirs to have identified them in Northern Ireland as early as May, 1971.

[110]Terrorism for Wilkinson is a criminal act perpetrated for personal gain by "desperate people bitterly opposed to the prevailing regime, alienated from all liberal democratic values" who read Sartre, and commit macabre crimes in "sacrificial ecstasy" (1986:94, 72, 67).

[111]Rational politics cannot be attributed to terrorism: it is polluted as it is formulated. 'Terrorism' is typically represented not only as a crime against the state, but against civil society: "a form of warfare in which violence is directed primarily against noncombatants (usually unarmed civilians), rather than operational military and police forces or economic assets (public or private)" (O'Neill 1990:25). 'Terrorism' is defined by its aims: "the threat of force or the use of violence for political ends" (Crozier 1960:158, see also Wilkinson's typology of terrorism, 1986:32-33). In a deft logical maneuver, the criminal methods of 'terrorism' make any ends illegitimate: "it is precisely because terrorists, by definition follow a systematic policy of terror, that their acts are analogous to crime" (Wilkinson 1986:65).

[112]Townshend subscribed to the view that "the belief that terrorism is fueled by publicity has become an *idee fixe*, although it has no historical basis" (1986:36).Whether violence had an inherent meaning was a nineteenth century anarchist debate, opposing theoretical propaganda to the propaganda of the deed (Lacquer 1987:48-51). 'Propaganda of the deed' attributed communicative power to the performative aspects of violence. Symbolic (rather than instrumental goals) are similarly said to distinguish "terrorism" from war (see Thornton 1964; Priestland 1974). Terror, "a symbolic act designed to influence political behavior by extra normal means, entailing the use of threat of violence" (Thornton 1964:73), resembles a speech act, rather than an act of war. Wilkinson, paraphrasing the influential Jenkins, writes, "the terrorist wants a lot of people watching rather than a lot of people dead" (Wilkinson 1986:51). Terror in this model would seem to be a theatrical spectacle, targeting not victims but spectators. (For a critique, see Schlesinger et al. 1983). What terrorism appears to do is often as important as what it does do: "terror is most effective when it is indiscriminate in appearance but highly discriminate in fact" (Thornton 1964:72). Wilkinson writes that "although it is often claimed that an act of violence is trying to 'say something,' violence is a singularly clumsy and ambiguous mode of communication" (1986:31). Leeman, in <u>The Rhetoric of Terrorism and Counter-Terrorism</u> (1991), argues that mutual interpretation by terrorist and counter-terrorist of each other's violence is a 'natural' result of 'dialogic' interaction (1991:13). "The conservative fallacy is not merely, or not at all, an argument that violence can only be restrained by violence, but that, in the hackneyed phrase, 'the only language the terrorist understands is the language of force'....It is to say that a moral or constitutional message can be transmitted by such means, which will in some sense evoke recognition and acceptance from those against whom it is directed. Such a response has always characterized some at lease of the British policy toward nationalist violence in Northern Ireland..." (Barker 1990:134).

[113]Liz Curtis, in The Propaganda War, demonstrates how media representation replaces political representation as the central goal of both 'terrorist' and 'counter-terrorist.' (see also Herman and O'Sullivan 1989; Chomsky and Herman 1988). The terms 'terrorist' and 'terrorism' which "have generally been confined to the use of violence by individuals and marginal groups" while "official violence which is far more extensive in scale and destructiveness is placed in a different category" (Chompsky and Herman, cited in Schlesinger 1981:80). Censoring IRA speech leaves the rationale for violence explored. "To say that the IRA is worth hearing, and is a real force in the conflict, is also to undermine the notion-- central to security policy polity from the mid-seventies onwards--that they are reducible to mere 'murderers and thugs' and represent only a tiny minority" (Curtis 1984:151). Deaths due to 'terrorism' are reported as human interest stories, rather than as news (Curtis 1984:112), which creates compassion for the victim rather than the perpetrators.

[114]The strategic value of PSYOP was increased by the formation of an Information Policy Department in September, 1971, headed by Colonel Maurice Tugwell, Colonel General Staff (Information Policy). Tugwell was previously an Intelligence Officer in Palestine, who had also served in Malaya, Cyprus, Arabia and Kenya. After his command in Northern Ireland, Tugwell was posted to Iran. The Army, after allegations were made of indiscriminate killings, began having spokesmen give their side of the story. Dis-information may have been leaked to the press. In February, 1977, BBC Northern Ireland Controller, Richard Francis, speaking to the Royal Institute of International Affairs said: "Sometimes...the Army's initial version of events turns out to be further away from the truth than that of the Provos" (cited in Curtis 1984:75). Soon afterwards, the Northern Ireland Office agreed to refrain from deliberately planting false stories, nor "any form of counter-propaganda not based strictly on the truth" (NIO statement cited in Curtis 1984:75).

[115]According to the British Army manual, <u>Land Operations:</u> <u>Volume III —Counterrevolutionary Operations, 1969,</u> the institutionalized form of this 'persuasion' is "the planned use of propaganda...to influence to our advantage the opinions ...and behavior of the enemy" (para 447). "The press, properly handled, is potentially one of the government's strongest weapons" (chapter 3, para 160). "Every local mention of the army affects its image. There is no neutral publicity, and the scope for favorable unit publicity includes the fostering of good community relations..." (chapter 11, para 532). By the end of 71, more than 200 officers had been put through course on television interviewing (Charters 1989:226). Effective interviewing for soldiers became Sandhurst curricula in 1972 (and still is), by 1982, four hundred soldiers a year were attending courses. Banal reports of Army activities contributed to the normalization of a military presence. Media attention often focused on stray dogs adopted by Army units: one Crossmaglen stray adopted by a British army unit became a celebrity. Although Rats was not the only dog living in the barracks, "a 'big and really dirty' black Labrador called Fleabus, was evidently not though suitable for the limelight" (Curtis 1984:85). Named 'Dog of War' by the popular press, Rats was awarded military honors for valued service, appeared on TV with the Queen Mother as she was saluted during the Royal Tournament at Earl Court, and received more Christmas cards than the entire Battalion combined. In October 1981, this dog had its biography published, entitled <u>Rats: the Story of a Dog Soldier</u> (Halstock 1981).

[116]As Der Derian points out, the discourse of terrorism produces a hyperreal threat of violence which distracts us from identifying actual threats (1993:81). Despite the fact that one is more likely to be stung by a bee, bee stings do not make the news. 'Criminalization' as a discursive strategy does not replace counter-insurgency, rather the discourse of 'terrorism' is itself, a product of counter-insurgency. 'Criminalization', 'normalization', and 'ulsterization' are *products* of a counter-insurgency program.

[117]The "ahistorical conspiracy theory" of counter-insurgency theorists "omits the possibility of the community sharing the IRA's conception of the police and the British Army as armed aggressors" and reduces the violence of the IRA to "a pathological conundrum" (Burton 1978:119-120).

[118]Burton's later work with Pat Carlen (1979) is an astute and penetrating piece of theory on state discourse of jurisprudence.

[119]The transcripts of the communiqués between PSF and Whitehall during the 1993 negotiations for a cease-fire/truce were reproduced by the Northern Ireland Information Service. Bodenstown speeches are apparently still a means of communication between the Republican movement and the British government. According to a British message sent on 17 July 1993: "There is one very important point which needs to be answered to remove possible misunderstandings. Recent pronouncements, including your Bodenstown speech, seem to imply that unless your analysis of the way forward is accepted within a set time, the halt in violence will only be temporary. This is not acceptable" (Northern Ireland Information Service document, 29 November 1993).

[120]A child sent the Army an Easter card in 1974, which read: "To the Army. Up IRA. Ha Ha Ha. Boo Hoo Hoo. Up the IRA. Do not take off this cover up IRA. Vote Bernadette with X. Signed Anonymus. Do not smoke this is Lent" (Army Ephemera Collection, Linen Hall Library).

[121]Terrorism, although prohibited by the Geneva Convention and the Protocol Addendums, is not defined within these texts. Key elements of the prohibition of terrorism in article 3 of 1949 Geneva Convention and article 4(2)(d) and article 13 of Protocol II are the intent to use terror, the indiscriminate effect of terror and the absence of proportionate military necessity. Walzer has argued in addition to these qualities, that "randomness is the crucial feature of terrorist activity" (1977:197). Furthermore, this 'randomness' must create terror through symbolic, unpredictable use of force or threat of the use of force for the purpose of influencing a target group or individual (Hanle 1989:104). Rather than interpreting terrorism as an indirect "continuation of war by political means" (Walzer 1977:198), terrorism can be more easily defined as a continuation of politics by martial means, using indiscriminate force for political purposes without combatant distinctions.

Whereas terrorism fails to recognize non-combatant distinctions, guerrilla warfare, is a type of levee en masse, "authorized from below" (Walzer 1977:180). Luttwak distinguishes two types of guerrilla warfare: first as a "theater scale enlargement of agile, light-infantry tactics" and second within the context of an internal armed struggle as "only the military component of a grant strategy-revolutionary war, whose political component is subversion" (1987:132n.). The primary conceptual distinction between terrorism and guerrilla warfare is the level of reliance on civilians. Whereas terrorists operate outside of the confines of a social community, with political aims disassociated from the political goals of a specific subgroup, guerrilla warfare necessitates an immediate and direct dependence on civilians.

[122]The Provisional Irish Republican Army is the center of legal power and the central cultural, political and social fact of the Republican community of which they are members. The organization is highly secretive, tightly-knit, and omnipresent, the center of legal power exercises social control over the community from which it springs, and over the military behavior of its members. Behavior of the members of PIRA must be seen as part of the context of Republican culture (see for example, Burton 1978; Feldman 1991).

[123]Similarly, the announcement of the 1980 hungerstrike was framed in the language of the Protocols, demanding "as of right, political recognition and that we be accorded the status of political prisoners". Furthermore, "we claim this right as captured combatants in the continuing struggle for national liberation and self-determination" (10 October communiqué from the blanketmen in Long Kesh, cited in Clarke 1987:123).

[124]While PIRAs command structure and internal disciplinary system are probably developed enough to allow it to be included in either Protocol, there is disagreement as to whether PIRA 'holds territory' in any formal sense. During the 1970's and 80's PIRA established no-go areas in Belfast and [London]Derry, setting up road blocks and controlling access. With no ground patrols in South Armagh (British Army troop movements are accomplished by air), the Army can hardly be said to control this territory. PIRA also controls West Belfast to the degree that it can effectively police the population and expel people from communities.

[125]Early counterinsurgency theorists advocated imposing martial law and using the army (rather than the police) to enforce control in the colonies. According to Gwynn, "the principle police task of the Army is no longer to prepare the way for civil control, but to restore it when it collapses" (1934:5). Gwynn's statement recalls Derrida's distinction regarding the uses of violence:

> first, there is the distinction between two kinds of violence in law, in relation to law (*droit*): the founding violence, the one that institutes and positions law (*die rechtsetzende Gewalt*, 'law making violence') and the violence that conserves, the one that maintains, confirms, insures the permanence and enforceability of law (*die rechtsverhaltende Gewalt*, 'law preserving violence') (1992:31).

[126] Developed from Anglo-Irish war jurisprudence, the British common law concept of martial law within the United Kingdom is based on the idea of a *de facto* suppression of civil law. According to Sir James Mackintosh's 1824 legal judgment, "When it is impossible for Courts of Law to sit or to enforce their Judgments, then it becomes necessary to find some rude substitute for them, and to employ for that purpose the Military, which is the only remaining Force in the Community" (cited in Townshend 1986:21. See also Townshend 1982:127). The common law, thus, allows for the exercise of martial law by the military, in effect, within the United Kingdom (Evelegh 1978:33).

[127]From 1600-1855, the British Army was under civilian control through the Horse Guards; only after the Crimean War was the War Office established (Vogler 1991:86). Until 1910, the military could not operate domestically unless accompanied by a representative of the civil power (Vogler 1991:84). To this day, when the Army is called out 'in aid of the civil power', soldiers are answerable to the common law rather than a military code. Military command during domestic civil disorder only became common following the Tonypandy Riots of 1910. Troop deployment as the first response (Geary 1985:47; Vogler 1991:89) to civil disorder demonstrates the increased military interest in the civil arena during the Macready (1910-26) and the Kitson eras (1970's) and can be directly correlated with decreasing levels of military commitment overseas (Vogler 1991:91), and the presence of a standing army in Britain (Howard 1957:14).

[128]Reith (1943), for example, argued that the military during civil disorder "are no longer soldiers but instruments of law functioning in accordance with police principles" (1943:9).

[129]Under the Defense of the Realm Acts and Regulation, what Hadden, et al. refer to as "statutory martial law" (1988:15), competent military authorities were empowered to order trial by court-martial. The Army Council was given power to designate 'Special Military Areas', in which special restrictions could be imposed. In the 1920's, official reprisal killings were instigated in areas designated as under martial law. During the Irish Civil War, a non-statutory system of martial law allowed for internment of suspects and the executions carried out by military courts.

[130]Such as the Civil Authorities (Special Powers) Acts (Northern Ireland) 1922-23 and the Criminal Law Act 1967 and the Army Act 1955.

[131]The need for a 'third force' reflected a new style of insurgency could "paralyze the legal system without producing the traditional symptoms of a 'state of war'" (Townshend 1986:21).

[132]Evelegh calls for the abolition of "the particular common law doctrine of martial law existing as a state of fact deriving from the inability of the civil courts to function" (1978:106) and advocates civilian-systems of control (1978:107). The common law concept of martial law does not protect the military from the consequences of unreasonable acts committed by suppressing rebellion, since the military is still answerable to civil law (see <u>Higgins v. Willis</u> in 1921 and <u>Rex (Ronayne and Mulcahy) v. Strickland</u>, ([1921] 21R-333)). The common law doctrine of martial law does not protect, and is not designed to protect, soldiers.

[133]This passage is usually cited out of context; Kitson, if fact, rejected the use of law as a weapon in favor of counter-insurgency operations within the law, no matter how flexible.

[134]Laws guiding the application of lethal force were usually clarified after the fact. For example, after the first petrol bomber was shot, the rules for firing were outlined on a 'yellow card', carried by all soldiers. While responsibility for firing orders normally rests with the commander, the "Instructions by the Director of Operations for opening fire in Northern Ireland" allow a soldier to open fire when someone with a clearly identifiable firearm is about to use it offensively and refuses to halt and desist when called on to do so. Firing may commence without warning in self-defense, if there is no other way a soldier can prevent himself from being killed or seriously injured. If life is endangered, petrol bombers may be shot after a warning is given (Dewar 1985:220).

135Power relations are ultimately capable of generating the evidence needed for their own legitimation (Beetham 1991:60). Law, in this sense, self-authorizes. According to Kerruish, it is the ideological project of law to persuade those who experience it of its truth. "Jurisprudence has the task of imposing meaning on legal practices; of making the liberal lawyer's convictions that the rule of law legitimates the coercive, exercise of power by the state, truth" (Kerruish 1991:118; Cornell 1992:155-169). Martial law has no pretense of neutrality, and therefore no need for legitimation. In a Kitsonian universe of expedient force, the coercive element of law remains transparent.

136The Coercion Acts established special courts for the prosecution of rebels, enacted devices to aid criminal prosecution (especially the discretion to admit evidence), stipulated that convictions for political violence carry special, increased penalties, allowed for detention without trial, proscribed certain groups, and restrained political expression.

[137]In the eighteenth and nineteenth centuries, political violence led to the suspension of trial by jury. Under the Criminal Law and Procedure (Ireland) Act of 1887, permanent special provisions were introduced. Government fears about the reliability of juries led to the enactment of the Public Safety Act 1927 which authorized the creation of a non-jury military Special Court (never implemented). During the 1930's political offenders were tried in accordance with the Constitution (Amendment No. 17) Act 1931 by a constitutionally established tribunal of Army officers operating under special powers. Following the abolition of the Tribunal 1936, the Offenses against the State Act 1939 established permanent provisions for non-jury Special Criminal Courts composed of Army officers or civilian lawyers appointed by the government. The court functioned between 1939-46 and 1961-62. Another non-jury military court operated under the wartime Emergency Powers Acts from 1940-43. In Northern Ireland, the Criminal Procedure Act (Northern Ireland) 1922 provided for special non-jury courts. The Civil Authorities (Special Powers) Acts (Northern Ireland) 1922-23 created special courts to try offenses against regulations made under the Act. "Despite the precedents, however, it was not until the introduction of the Diplock courts in 1973 that non-jury trial on indictment became a standard procedure in Northern Ireland" (Hadden, et al. 1988:14). The Diplock Report, upon which the changes in legal structure were based, altered the way in which a 'normal' level of violence could be perceived: "The Diplock text's radical transformation of due process leans heavily upon the invocation of the abnormal conditions in Northern Ireland which render the technical guarantees of judicial objectivity 'impracticable'. These abnormalities bestow upon the text a discursive confidence that ...exposes the pragmatist nature of juridical precedent...Its intra-discursive logic is as incoherent as its epistemological justification" (Burton and Carlen 1979:83).

[138]An interesting aspect to this legal posturing by the British government, and their Republican opponents, is the way test cases poke and prod at weak points in British and international law. Beetham points out that "the legitimating ideas...that underpin the given institutions of power define which challenges the ruler has to take most seriously, because they strike at the basis of the system of rule itself" (1991:36). Since the legitimating fiction of British policy is governance not by 'force' but by the rule of law, the legitimacy of law is a primary Republican target. The Republican refusal of British law seeks to expose the illegitimacy of the legal order itself and to offer what Derrida, after Benjamin, has called a 'critique of violence' (1921):

The state is afraid of fundamental, founding violence, that is, violence able to justify, to legitimate,....or to transform the relations of law, and so to present itself as having a rights to law....Only this violence calls for and makes possible a 'critique of violence' that determines it to be something other than the natural exercise of force (1992:34-35).

Test cases are at the center of the Republican challenge to British sovereignty: the Republican movement has consistently tried to demonstrate the illegality of law, to offer a 'critique' of legal violence *through legal means*. In defying the law, the Republican movement has tried to lay "bare the violence of the legal system, the juridical order itself" (Derrida 1992:33). Test cases use British law as a weapon against Britain, forcing the state to define its legal position *vis-a-vis* its own law.

[139]Under the EPA, all statements are considered admissible for scheduled offenses unless "prima facie evidence was adduced that the accused was subjected to torture or to inhuman or degrading treatment in order to induce him to make the statement" (section 6), in which case the court was to exclude the statement.

490

[140]Written statements obtained from dead, crazy or unavailable authors would now be admissible as evidence. Such statements would have been inadmissible under the hearsay rule of the common law (Boyle, et al. 1975:104).

[141]The Compton Committee (1971) found that 'interrogation in depth' had occurred in a number of cases (Taylor 1980). These techniques "had been developed by the British Army in the course of 'emergency' operations in a number of British colonial territories ..., and were passed on to RUC Special Branch officers in the course of special training sessions arranged with a view to securing the maximum intelligence results from the internment operation..." (Boyle, et al. 1975:49). The majority view of the Parker committee, investigating whether interrogation in depth should continue, was that since "nothing unlawful had been or could be authorized under the prevailing military directive the government should...ensure protection for those taking part in the operation" (para. 38, Parker Committee Report). In other words, since torture was illegal, it had not occurred.

[142]The sheer number of people convicted in this manner would seem to support referring to it as a 'system'. Between November 1981 and November 1983 seven Loyalist and eighteen Republican supergrasses caused nearly six hundred people to be arrested and charged with offenses connected to paramilitary violence (The figure 593 for 1982-5 was given in a Commons written reply by Mr. Nicholas Schoot, Northern Ireland Junior Minister. H.C. Debs., vol. 73, col. 100. cf., Greer 1988:73).

[143]The Diplock Report justified the shift from internment under the Unionist regime to detention without trial under Westminster thus: "The only hope of restoring the efficiency of criminal courts of law in Northern Ireland to deal with terrorist crimes is by using an extra-judicial process to deprive of their ability to operate in Northern Ireland those terrorists whose activities result in the intimidation of witnesses" (para. 27). This new approach was sanctioned by a Westminster Order in Council in November 1972 before the final Diplock report was produced. The Detention of Terrorists Order was later incorporated into the EPA 1973.

[144]The Prevention of Terrorism Act empowers the police to detain suspects for up to forty-eight hours, which can be extended for an additional five days. In 1987, the European Commission found that seven day detentions violated the European Convention on Human Rights, which stipulates that detention cannot exceed five days. The British government responded to this finding by requesting, in March 1989, a temporary derogation from European Convention.

[145]This absence of control resulted in two distinct variations in the interpretation of the EPA. The Army's "military security" approach to internal security operations involved under-cover and screening operations (formerly the role of the police 'Special Branch') to identify, rather than arrest, suspects. The Army's strategy for controlling terrorism was to put behind bars as many men as possible, removing opponents from the field of battle. The "police prosecution" approach, on the other hand, emphasized arresting those actually suspected of committing specific criminal offenses, and to proving cases in criminal courts (Boyle, et al. 1975:43-46). There was a "substantial difference in the treatment of those suspects who were processed as a result of military and of police activity" (Boyle, et al. 1975:46). Some of these differences were due to jurisdictional organization: Republican areas where the RUC could not penetrate were under Army jurisdiction, and the population was thus exposed to 'command screening'. With the RUC acting mainly in Protestant areas, specific incidents were the focus of investigation. This resulted, according to Boyle, et al. in a "differential flow of suspects into the twin system of court trial and administrative detention" In 1974, for example, although there were five hundred Republican detainees and only fifty Protestant/Loyalist detainees, roughly the same number of suspects appeared in court (Boyle, et al. 1975:48).

[146]Because the conflict in Northern Ireland could not be identified as a war, the Widgery Report exonerated the soldiers from the charge of indiscriminate firing during Bloody Sunday on the basis that they had acted as *individuals*: in essence, like good police. When firing orders cannot be controlled in battle, British soldiers are allowed to act individually without direct orders. In the events which took place on 30 January the soldiers were entitled to regard themselves as acting individually and thus entitled to fire under the terms of Rule 13 without waiting for orders. ...in the prevailing noise and confusion it was not practicable for Officers or NCOs always to control the fire of individual soldiers. The soldiers' training certainly required them to act individually in such circumstances and no breach of discipline was thereby involved (Widgery Report, para. 95).

British soldiers are required because of the way the law is framed to respond as if they were police. Police, trained to act individually, are distinguished from soldiers, trained to function as an obedient group. This amounts to a "direct clash between civil and military logic. The clash centers on the issue of concentration versus dispersion...the civilian priority is protection and dispersion, the military aggression and concentration" (Townshend 1986:31). While soldiers conceptualize guerrillas as a military force to be defeated, policemen see them as criminals to be apprehended through investigative process and punished judicially (Hutchinson 1969; Charters 1989).

494

[147]A 'third force' is essentially a response to the conditions of internal conflict, where " a force designed to maximize shock faces a situation in which minimum force may be necessary....once insurgency passes beyond the stage of episodic violence, which the police can deal with, it definitely becomes a form of warfare, fought by distinctly military techniques which it is neither easy nor, perhaps, desirable for the police to master. As a hybrid form of conflict it calls for a synthesis of police and military skills (Townshend 1986:30). 'Third force' units, usually conglomerates of police constables imported from the metropolis to the colony, members of the settler community, or members from the insurgent 'community' monitoring their own population. During the Mau Mau rebellion, in addition to a police presence of imported constables, the British established a Home guard of local Kikuyu, the tribe from which most of the Mau Mau came. Local, indigenous peoples were thus hired to patrol and monitor their own tribe. The third police presence during the uprising in Kenya was the 'Kenya Police Reserve', a group of white highlanders with a vested interest in seeing the rebellion suppressed, similar to the UVF established in Ireland in 1914. Their tendency towards violence led Commander-in-Chief of East Africa, General George Erskin, to write "...I most strongly disapprove of 'beating up' the inhabitants of this country just because they are the inhabitants...Any indiscipline of this kind would do great damage to the reputation of the security forces..." (cited in Mockaitis 1990:49).

Third Force units are often prone to use of excessive force. Police given military equipment but allowed individual initiative tend to ignore the minimum force directive. Latitude in the use of force may also result from deploying police constables into an unfamiliar, foreign culture, as in Kenya and Ireland of 1921. Third force units composed of members of the settler community often end up representing sectarian political interests. In 1920, Lloyd George approved the UVF to act in an official capacity as the USC (Dewar 1985:99). A former Chief Constable said in relation to the USC "they were not trained in normal policing, were not subject to disciple, and tended to be a law unto themselves" (cited in Weitzer

[148]Whether legal decisions can ever redress political problems is, of course, debatable.

[149]Hogan and Walker argue that special provisions are superfluous in relation to 'arrestable offenses' under section 2 of the Criminal Law Act (Northern Ireland) 1962, although section 2 is rarely used. They believe that soldiers are handicapped under section 2 because it offers "no power of arrest for future offenses and past offenses must really have been committed" (1989:57). Additionally, they point out that the requirement of reasonable suspicion is too precise to allow pre-emptive intervention.

[150]Most acts enacted from 1761-1922 were repealed, but the Acts of 1775 and 1787 became permanent, and some parts, according to Hogan and Walker (1989:12) survive today (As amended by the New Whiteboy Act 1931 (1 &2 Will. 4 c. 44) and the Capital Punishment (Ireland) Act 1842 15 & 6 Vict. c. 28)), though there are few recorded prosecutions since the middle of last century (see The State (O'Connor) v. O'Caomhanaigh [1963] I.R. 112 (prosecution for threatening letters contrary to the Tumultuous Risings (Ireland) Act 1831, section 3).

[151]These are the rights established by Article 3 (prohibition on torture), article 4 (forced labor), article 7 (retroactive penal law), article 41 (discrimination), and article 2 (the right to life except 'in respect of deaths resulting from lawful acts of war').

[152]Normative codes generated by paramilitary groups are rarely perceived as full-blown legal systems, but more commonly as artifacts of custom. The discursive link between terrorism and primitives again becomes relevant. While custom is attributed to primitive societies, law appears as the prerogative of civilization. Terrorists and tribes exist outside of the hegemonic legal structures of Western nation-states and have a fuzzy position *vis-a-vis* international humanitarian law of armed conflict. The law of war may become the symbolic battlefield for defining the limits of cultural interaction, and for refusing to share or acknowledge codes between states and subaltern groups. The propensity to use international law as a theater of justice is especially relevant for Republicans who have traditionally framed political claims in terms of law.

[153]Low-intensity conflict is typically defined in British military usage as: "a situation between peace (when no arms are used) and the full-scale use of weapons in a conventional war. During such a situation there is extensive diplomatic economic and psychological activity, but its primary feature is the covert or overt use of military force short of war" (Vetschera 1993:1575).

[154]Simultaneously, the terror-threat of the Provisional Irish Republican Army (PIRA) is already a cyberspatial simulation of war, a hyper-paranoid diffusion of a temporal form of violence by the global communication network (Der Derian 1992:116).

[155]Der Derian defines intelligence as "the continuation of war by the clandestine interference of one power into the affairs of another" (1992:21).

[156]Despite the fact that my argument emphasizes the importance of HUMINT in LIC, obviously other sorts of intelligence sources are very important. Surveillance by photo reconnaissance, for example, provides a "mosaic" image of the battleground in Northern Ireland (Unattributed source, 1994) and electronic counter-measures are in common use. Furthermore, this argument is not intended to suggest that Northern Ireland is a 'total surveillance society'; despite the widespread introduction of computers to war, intelligence systems in LIC still rely on the "helter-skelter pursuit of information" (Kitson 1977:23) and lack of basic intelligence is a constant complaint of soldiers.

[157]During WW I, for example, agents never provided more than five percent of the information received by the War Office (Gudgin 1989:62).

[158]In 1906, Sir Charles Callwell advised that "the ordinary native found in theaters of war peopled by coloured races.... [is] far more observant than the dweller in civilized lands. By a kind of instinct they interpret military portents" (50, 54).

[159]Military intelligence includes: orders of battle, organizational details, unit and formation identification, personality and leadership details, readiness for war, weapon and equipment details, mobilization plans, details of land defenses and topographical features (geography, climate, endemic diseases, etc.)

[160]Obtained through overt reconnaissance, tactical information was considered 'public' information and was not tainted by the British aversion to gathering information by stealth. Covert intelligence (spying) was reviled (Gudgin 1989:22). Colonel GA Furse wrote in Information in War, 1895: "The very term "spy" conveys to our mind something dishonourable and disloyal...we would blush at the very idea of having to avail ourselves of any information obtained through such an agency" (cited in Gudgin 1989:12). Clandestine gathering of knowledge was repulsive to the gentlemanly ethos of war.

[161]To prevent the autonomy of spies, information was centralized and made accessible to the 'brain' of the army. The Depot of Military Knowledge, formed in 1803 on the model of the French <u>Depot de la Guerre,</u> was an information storage system designed to increase military effectiveness, and to gather military intelligence during peacetime. In 1855, its replacement by the Statistical and Topographical Office (War Department) constituted the first step toward the creation of a British General Staff (Gudgin 1989:22) which prevented intelligence operations being conducted under separate chain of command. The degree to which Intelligence officers have the authority to assign tasks to soldiers is still an issue. Modern war staffs are "central military organs assisting the supreme military authority of the state...in determining and implementing the higher directives which are to govern military activity" (Irvine 1938:162; Dandaker 1990:87). The conduct of large-scale European wars between nation-states after the eighteenth century depended on the administrative system of the general military staff. War staffs developed as a result of an overall increase in administrative tasks within armies, improvements in military technology, cartography and transportation logistics, and the institutionalization of the divisional system (Irvine 1938). War staffs relied on the collective performance of administrative duties: "the brain' of military organizations became collectivized" (Dandaker 1990:87). The general military staff, the penultimate example a hierarchical control system, depended a military bureaucracy to efficiently store and retrieve information about the enemy. Collective decision making at the top of the hierarchy relied on centralized intelligence processing.

[162]The US Army "met defeat because it failed to 'know the enemy' and therefore could not adopt the strategy and tactics that were specific to the particular enemy it faced..." (Cincinnatus 1981:9). MacNamara's quantitative scientism of body counts, kill ratios, ergonomics, and crisis management has been identified as a "technobureaucatic or production logic" (Gibson 1986) of war. Wars always 'produce' death (see Chapter Five); Vietnam merely shifted the rhetoric. General Westmoreland, translating McNamara's technowar logic into military operations, persisted with a high-cost search-and-destroy strategy of using helicopters for recon, which gave LZ positions away and alienated local Vietnamese (Cincinnatus 1981:79). Westmoreland's war of attrition was far closer to a conventional war than a counter-insurgency approach (Krepenevich 1986). Despite the availability of analytic approaches to guerrilla combat such as Robert Taber 's War of the Flea (1965), the voice of sympathetic journalist Bernard Fall (1966), and the stern admonishments of British military advisor Sir Robert Townshend, the US Army continued its "technowar" approach, and failed to incorporate British COIN strategy. In 1961, the British Advisory Mission suggested the strategic hamlet approach to the South Vietnamese government, and as Thompson (1966, 1969) has pointed out, it was horribly botched. Despite the US Army's subaltern history of guerrilla warfare (e.g., the War of Independence, the Indian wars, the Spanish-American Wars, and the Philippine Insurrection), Vietnam was fought as a technowar where "machine-system meets machine-system" (Gibson 1986:23). This war-paradigm certainly has contributed one strain of strategic thinking to the cyberwar concept, and British distillation of maneuver theory, as I will argue, contributes the other.

[163]An excellent example of how netwar/cyberwar attacks information nodes (resembling something like 'information war') the first 'Bloody Sunday' on 21 November 1920. British intelligence operations, which during the early part of the war were quite amateur, improved with the introduction of a group of operatives called the 'Cairo Gang' . These intelligence officers, headed by Colonel Ormonde d'Epee Winter (code-name: "O"), had been transferred from the Middle East to Dublin. The 'Cairo Gang' was trying to flush out the intelligence network of the IRA, who became aware of the plan, identified the members of the British Intelligence operation, placed them under surveillance, and carried out simultaneous, timed executions across the city.

[164]The concept of a 'net' is not unheard of even in conventional air-land battles. As a tactical defence concept, Simpkin proposes a "universal net" composed of locally recruited militias who would function as a local defence force and as observers on the ground during combat. "The problem in NATO terms is that a universal net of this kind should — in fact must — be found by indigenous troops'" (1985:303). Local knowledge, as Simpkin insists, more than offsets any limitations in training in third force troops.

[165]According to Arquilla and Ronfeldt, "later, when it was clear that the world was not ending, the Mongols willingly adopted both Christianity and Islam, whichever eased the burden of captivity for particular peoples" (1993:150). Adaptability and flexibility regarding the social organization of defeated peoples should also be attended to as part of doctrine.

[166]A classic example of a state imitating a nomad war machine can be found in Chapter 33 of T.E. Lawrence's <u>Seven Pillars of Wisdom</u> (1933) (thanks to John Pimlott for this reference). Lawrence, in this chapter outlines the basis of a strategy for guerrilla warfare, which predates Mao's by about twenty years. Arquilla and Ronfeldt also link Mao and Lawrence theoretically in the context of a discussion of decentralization vs. top sight in communications systems. Mao's doctrine of strategic centralization of command and tactical decentralization is, according to Arquilla and Ronfeldt, confirmed by TE Lawrence's analysis of the Desert Revolt (1992:163fn.).

[167]If PIRA can be said to have a legal code, or complex political organization and military structures, these are a product of their interaction with the British Army. The mirror modeling of PIRA on the British Army discussed at the end of Chapter Three substantiates this principle.

[168]Although all military structures aim towards conservation of energy, generally true of most open systems, guerrilla warfare aims explicitly for non-battle, since in open confrontation with the forces of the state, firepower is not in their favor.

[169]"Among the many levels of interpretation to which the Chinese master's 'ordinary force' (<u>cheng</u>) and 'extraordinary force' (<u>ch'i</u>) lend themselves is the physical one which equates 'ordinary force' (with which one engages the enemy) to 'holding force', and 'extraordinary force' (with which one wins the battle) to the 'mobile force'. Again Sun Tzu's analogy of a torrent of water ('Now the shape of an army resembles water') perfectly expresses the dynamism of manoeuvre theory..." (Simpkin 1985:37)

[170]Certain aspects of cyberwar are nevertheless consistent even in low-intensity conflicts. Simulation, for example, though much emphasized in critiques of cyberwar in the Gulf (Der Derian 1993) has always been a component of warfighting. Simulation, characteristic of cyberwar, has a place in low-intensity conflict. Before deployment to Northern Ireland, all soldiers undergo training at the Northern Ireland Training and Advisory Team (NITAT), which provides classroom training, role playing sessions, and a full-scale simulated combat environment to prepare soldiers for actual on-the-ground conditions in Northern Ireland. NITAT staff provides commanders with intelligence updates on PIRA weapons, tactics and operating procedures. The simulated combat sequences are conducted in Kent, with some soldiers assuming the role of rioting crowds, fleeing terrorists, and angry housewives. Counter-interrogation training also takes place:

> In one exercise, on Dartmoor, you are captured and interrogated.... When you are captured...you are slung into the back of a dustcart in patch darkness, driven around for 1-1 1/2 hours then dragged out into spotlights, beaten up, stripped or lined up against a wall. Someone was stood in a stream, in the middle of the night....In another exercise we had to interrogate RAF officers as part of their officer training. We were to give them a real kicking to make it realistic....That was in 1973... (Royal Marine Commando, cited in Information on Ireland Pamphlet No. 1:24).

[171]In fact, the discursive formation of fighting Indians along a frontier was derived from Ireland: during the 16th and 17th centuries, Britain was developing frontier policies on the Celtic frontier, as well as in the Far East and in America (MacLeod 1967:25). While Britain was colonizing North America, Celts still held 50% of the British Isles: "the frontier problem at home was more serious in some ways than frontier problems abroad" (1967:27). The reservation system was developed in Ulster, and transplanted to the New World. In 1609, Ulster, was divided into six counties, and natives were ordered to gather together onto reservations or be exterminated. Any native found outside of Connaught by May 1, 1654 was to be put to death. At one time, reservation land comprised one-fifth of the total land area of Ulster, the native eventually gaining acceptance as tenant farmers outside of the reservation by 1660. In Virginia, the London Company and Indian tribes were signing treaties which did not require fealty; the Celtic tribes, on the other hand, were required to submit to the Crown (MacLeod 1967:34-36).

[172]The American Joint Chiefs of Staff's Dictionary of Military and Associated Terms defines C^2 as "the facilities, equipment, communications, procedures, and personnel essential to a commander for planning, directing, and controlling operations of assigned forces pursuant to the missions assigned" (1987:77). C^3I includes information, as well.

[173]This is hardly a new military skill: "Simulated disorder postulates perfect discipline...Hiding order beneath the cloak of disorder is simply a question of subdivision..." (Sun Tzu 1983:22).

[174]As Kingsley Amis writes in The Anti-Death League (1966), "a system that runs itself is still a system. You don't have to believe in a weather god to find a climate unbearable" (207).

[175]Richard Bissell, formerly the CIA's Deputy Director, Plans, listed eight categories of covert operations: political advice and counsel, subsidies to an individual, financial and technical support to political parties, support of private organizations, covert propaganda, private training and exchange of individuals, economic operations and paramilitary operations (Marks and Marchetti 1974:41)

[176]The Special Air Service (SAS) has "a unique and inherently political function beyond that of the British armed forces as a whole" (Bloch 1983:40; see also Geraghty 1980:1) The SAS, formed in 1941 by Lt.-Col. David Stirling to carry out sabotage and reconnaissance missions behind enemy lines, was first deployed in North Africa in 1941. In 1947, the SAS reformed as a territorial and volunteer unit, the 21st SAS (Artists). A regular (full-time paid) regiment was created in 1952, the 22nd SAS, of Malayan Scouts. In 1952, 22 SAS left Malaya and returned to UK. Recruits are drawn only from within the armed forces, mainly from the Paratroopers.

[177] According to the Franks Committee Report 1972, the Official Secrets Act 1911 "catches all the official documents and information. It makes no distinction of kind, and no distinction of degree. All information which a Crown servant learns in the course of his duty is 'official' for the purposes of section two, whatever its nature, whatever its importance, whatever its original source. A blanket is thrown over everything: nothing escapes (Cmnd 5104, para. 17). The various sub-sections of the Act have been estimated to produce over 2,314 possible offenses (Ponting 1990:12). The Official Secrets Act 1920 carried over a number of criminal offenses from wartime legislation (such as the wearing of a military uniform) and "blurred the distinction between espionage offenses and breaches of official trust" (Ponting 1990: 15). The onus for proving innocence was clearly placed on the defense, where "an act shall be deemed ...prejudicial to the safety and interests of the State unless the contrary is proved." (section 1, para 2) The 'interests' of the State were defined by the government, rather than by the courts.

Because of the general nature of section 2 of the 1911 Official Secrets Act, technical breaches occurred constantly, while the relatively rare legal proceedings were "remarkable only for their triviality, at times verging on the ludicrous. They served as a periodic reminder that *any* unauthorized disclosure could attract criminal charges--perhaps this was their *raison d'être*." (Ponting 1990:56). Absurd prosecutions included Compton Mackenzie, a captain in the Royal Marines during WW I who upset everyone by publishing his memoirs of military intelligence in the Balkans. He later wrote a satire of military intelligence called <u>Water On the Brain</u> (1934). Peter Wright's book <u>Spycatcher</u> (1987) was prohibited from UK publication, and the government attempted to stop publication in Australia for a time under the Official Secrets Act.

[178]The Oath of the Privy Councilors (to keep all advice given to the monarch secret), lays the foundation of a doctrine of executive secrecy. The oath assured collective responsibility for Cabinet decisions and protected individual ministers from arbitrary interference by the monarch. Dating back to the thirteenth century, the Oath possibly Britain's oldest secrecy provision.

[179]Weber (1984) writes that "everywhere that the power interests of the domination structure towards the outside are at stake...we find secrecy" (1984:48). Information is very largely the property of the state in Britain. Withholding information is a crime under the PTA, Section 18(1): "A person is guilty of an offence if he has information which he knows or believes might be of material assistance..." in the eradication of terrorism by security forces. Unlawful collection of information under Section 22 of the EPA 1978 is also prohibited. Section 22 reads:

> (1) No person shall, without lawful authority or reasonable excuse (the proof of which lies on him):
> (a) collect, record, publish, communicate or attempt to elicit any information with respect to any person to whom this paragraph applies which is of such a nature as is likely to be useful to terrorists.
> (b) collect or record any information which is of such a nature as is likely to be useful to terrorists in planning or carrying out any act of violence; or
> (c) have in his possession any record of or document containing any such information as is mentioned in paragraph (a) or (b) above.

These combined sections pretty much rule out the possibility of doing anthropology in Northern Ireland!

[180]Ministers in each department are supposed to control their agencies: the Home Secretary is responsible for MI5 and the Defence Secretary for DIS, while MI6 and GCHQ are under the Foreign Secretary. This division of responsibilities is quite flexible, since the Prime Minister may take a strong interest in intelligence and appoint, as Harold Wilson did with George Wigg and Thatcher did with Cranley Onslow (Bloch 1983:52), a minister to oversee the entire bureaucracy. All the directors have direct access to the Prime Minister.

[181]Authority rests with a permanent secretaries' steering group: Permanent Secretaries Committee on Intelligence Services controls the budget and, in consultation, approves major covert operations.

[182]Collating and analyzing different agencies reports belongs to the Joint Intelligence Committee (JIC) and to departments of the Cabinet Office Secretariat (composed of the Assessments Staff and the Current Intelligence Group). The JIC is composed of the heads of the four intelligence agencies, the chair and deputy chair of the Assessments Staff, the head of the Permanent Undersecretary's Dept. in the FCO and the Co-ordinator of Intelligence and Security. While the JIC is the formal end of the line in terms of bureaucratic organization, it does not direct or supervise the work of the intelligence agencies.

[183]MI6 both collects foreign intelligence using human sources and conducts political operations. It evolved from a group of organizations set up by the Foreign Office, the Colonial office and the India Office, which became the Secret Service Bureau in 1909. The Bureau's Home Department was the MI5 and its foreign office was the MI6. In the 1930's, it was renamed the Secret Intelligence Service (SIS). MI5 and MI6 are now known respectively as DI5 and DI6, for Defence Intelligence.

[184]The covert intelligence war between military intelligence organizations resulted from organizational chaos existing in Northern Ireland following the collapse of Stormont. The introduction of internment, and the interrogation of PIRA suspects, represents the failure of an outdated police intelligence system. The internment operation begun on August 9, 1971 by the British Army and the RUC, acting under authorization from Stormont, was based on outdated Special Branch files dating from 1956-62. Most active Republicans had foreknowledge of the predawn raids, and evaded arrest. Increased political pressure to halt the escalating war, led to the attempt to rationalize the intelligence structure in Northern Ireland. The internecine war was the eventual result of the fact that all of these various agencies tried simultaneously to establish their own independent networks.

[185]Colin Wallace, an ex-Intelligence officer in Northern Ireland, asserts that Oldfield while Security Supremo was also the target of an MI5 campaign and may have been murdered by MI5 (Cavendish 1990:173). The source of MI5 disaffection was their apparent jealousy of MI6's move into Northern Ireland in 1971, when Prime Minister Edward Heath invited Sir John Rennie (Chief of MI6) to assist with Intelligence matters in Northern Ireland. The initial appointment of MI6 over MI5 reflected Prime Minister Harold Wilson's mistrust of MI5, and political jurisdictions controlling the exchange of security information between the UK and the Irish Republic, which was considered an external affair and controlled therefore by the Foreign Office, rather than the Home Office (Geraghty 1980:191). The Chief Intelligence post was given to an MI5 officer, Dennis Payne. When MI5 finally gained control over intelligence they attempted to get MI6 members removed from their posts and install MI5 members in their stead. The fur was flying, thick and fast.

[186] The MI5, engaged primarily in internal intelligence operations, has always had close connections with the police. In 1950, MI5 and the Special Branch offered courses to colonial Special Branch. From 1956-62, MI5 seconded Security Intelligence Advisors to the Colonial Office to advise the Colonial Secretary (Bloch 1983:30). MI5 has rarely engaged in covert operations overseas: the last documented was in collusion with the CIA in British Guyana during 1963-4.

[187]Because the conflict in Northern Ireland is officially classified as an internal security operation within the borders of the United Kingdom, the doctrine of minimum force is strictly enforced. Soldiers are restrained by law from firing except under very specific conditions, to their great chagrin. Despite various public scandals (Stalker's investigation into police practices, the reports of Amnesty International and the International Lawyer's Inquiry), British security forces have never operated with a shoot-to-kill policy in the sense of a blanket order. While killing an unarmed PIRA member may be politically unacceptable, killing an armed man is considered, even by Republicans, to be fair. Lethal force used in a fair manner, within the law, is called a 'clean kill' (Urban 1992:164). In one sense, clean kills are a form of nomad law by which the states must abide when dealing with insurgents.

[188]The difference of course is that military independence is institutionalized in Republican ideology, and in PIRA's chain of command.

510

[189]It seems quite likely that these techniques were used by the Provisional IRA on a British soldier. Robert Nairac was a Grenadier Guards officer who acted as a liaison officer for the 3 Brigade Detachment of 14 Intelligence Company, also known as 3 Brigade Survey Troop. Although certain allegations made by Holroyd regarding Nairac's collusion with Loyalist terrorists in assassinations of Republican suspects (1989), the more interesting aspects of Nairac's story are the less obvious ones. In May 1976, Major Julian Ball of the SAS apparently requested that Nairac act as a liaison with the RUC for the SAS squadron in south Armagh (Urban 1993:54). In May 1977, Nairac was kidnapped from a pub in Crossmaglen. Nairac apparently enjoyed 'passing' as a Republican; he knew many Republican drinking songs by heart and was known for buying rounds of drinks for the 'boys'. On the evening of Nairac's disappearance, two Provisionals in the pub suspected that he was not the out-of-town Republican that he claimed to be, but was a British soldier. As he left the pub, he was followed, abducted, interrogated, and tortured by members of PIRA. Before executing Nairac, the Provisionals waited for authority from higher up their chain of command, and sent for another weapon since it would have been disrespectful to execute Nairac with his own gun. I am under the impression that they had some respect for him as a soldier, and Nairac's death certainly raises issues of codes of honor between soldiers and paramilitaries. In a deeper sense, his posturing as a Republican (for whom he allegedly had a great deal of political sympathy), shows what it means to 'know the enemy', and to use 'local knowledge' to the degree that one becomes just like them. Nairac also was doing these hi-jinx on his own time: no one authorized his activities and this absent-without-leave behavior on the night of his death shows the degree to which special forces do establish their own autonomy from parent organizations. Nairac, in a sense, was conducting 'netwar' by disrupting what the enemy knows (or thinks they know) about themselves. Nairac had a certain kind of 'background' which gave him access to power and authority the military establishment which may have been premature. Finally, Nairac's death suggests a limit to cultural mimesis: he

[190]Following allegations that the internees had been tortured, the Compton Report concluded that there had been no brutality, because the interrogators did not intend to inflict pain: "We consider that brutality is an inhuman or savage form of cruelty, and that cruelty implies a disposition to inflict suffering, coupled with an indifference to, or pleasure in, the victim's pain. We do not think that happened here." The Attorney General announced in May 1972 that interrogators would not be prosecuted, but the government paid out 200,000 pounds for fourteen men by 1978 (Curtis 1984:35).

[191]"By the late 1970's, this had become so widely known as the standard method of recruitment that anybody who had gone free after an arrest was liable to fall under suspicion within republican communities — it became common for people to announce the fact that they had been approached by the police while in custody in the pages of nationalist newspapers as a means of trying to dispel suspicious about themselves" (Urban 1992:105).

[192]The body uses up all reserves of glucose in the fatty tissues and begins to manufacture glucose from proteins in the muscles. The hunger stops after 4 or 5 days, as the metabolic rate falls, pulse slows and blood pressure drops. The body experiences severe coldness in the extremities between about the ninth and thirty-second day. After about forty-five days, the body consumes the tissues of the eyes, which collapse, causing irreparable blindness. After about 50 days, hearing starts to fail. Pain in the stomach and chest is constant, and is accompanied by acute vomiting. When the muscles are gone, the brain digests itself, causing convulsions and dementia. A few days before his death, Bobby Sands "was lying on the waterbed, his left eye was black and closed, the right eye nearly closed and his mouth twisted as if he had suffered a stroke. He had no feeling in his legs and could only whisper. Every now and then he started dry retching" (Beresford 1987:97).

[193]Sands had been convicted of possessing a revolver and was sentenced to fourteen years. Francis Hughes from South Derry was convicted of murder of Lance-Corporal David Jones and sentenced to life in prison. Raymond McCresh, serving fourteen years, was convicted of attempted murder. Patsy O'Hara was serving eighteen years for possession of a hand grenade. Joe McDonnell, from Belfast, was convicted of possessing the same revolver as Bobby Sands, and sentenced to fourteen years. Martin Hurson from East Tyrone was convicted of conspiracy. Kevin Lynch, twenty five years old, died after seventy one days. Kieran Doherty, from Belfast, was sentenced to eighteen years for possession of arms and explosives. Thomas McElwee was convicted of manslaughter and possession of explosives and sentenced to twenty years. Mickey Devine from Derry took part in a raid on a police armory and was sentenced to twelve years.

[194]The relationships can be schematically represented in the following manner:

WarTerrorism
legal (intra-state)criminal (extra-state)
rationalirrational
technicalcounter-technical
public neg. of meaningprivate (secret) neg. of meaning
sacrificialmurderous
contractualnon-contractual
conventionalnon-conventional

War, for example can be characterized as territorially oriented, legal or governed by law, representing a popular mandate, and discriminating between combatants and non-combatants. The application of the war/terrorism categories rather than the internal terms of reference, is the issue.

[195]Although the purpose of the 1981 hungerstrikes was overtly political, the spectacular, public manifestation of a political 'text' on the starving bodies of political prisoners could be considered as a ritual performance. Turner defines ritual as "prescribed formal behavior for occasions not given over to technological routine, having reference to beliefs in mystical beings or powers" (1967:19). Such rituals are goal oriented, that is they set out to do something within the psycho-social realm. Cannadine points out that "politics and ceremonial are not separate subjects, the one serious, the other superficial. Ritual is not the mask of force, but is itself a type of power" (1987:19). Ritual also makes politics 'real': Ritual "displays symbols of their [ideologies] existence and by implicit reference postulates and enacts their 'reality'" (Moore and Meyerhoff 1977:14).

The explicit goals and instrumental purposes of the ritual — in this case, asserting the military legitimacy of PIRA — determine how specific sets of interrelating symbols will be used and manipulated within the ritual, thus "each ritual has its own teleology" (1967:32). In other words, "the goals of the ritual will have overt and implicit reference to the antecedent circumstances and will in turn help to determine the meaning of the symbols" (Turner 1967:45).

[196]The history of Irish hungerstrikes against perceived British injustice is a long one, to which this paper cannot do justice. Republicans have also gone on hungerstrike against the Irish Republic: in 1923, eight thousand anti-Anglo Irish Treaty prisoners went on hungerstrike against the Irish Free State. For an excellent summary, see Coogan 1980.

[197]Sands described this state as being "crouched naked upon the floor in a corner, freezing cold amid the lingering stench of putrefying rubbish, with crawling, wriggling white maggots all around you, fat bloated flies pestering your naked body, the silence is nervewracking, your mind is in turmoil" (cited in O'Malley 1990:50).

[198]Saussure writes that, "what can be chosen is already determined in advance" (1983:71). Sahlins, in Islands of History, argues via Saussure, that culture exists in relation to conceptions of history embedded in current practice. According to Sahlins, "from the arbitrary nature of the sign it follows that culture is, by its own nature, an historical object" (1985:148).

[199]Under the Brigg's Plan squatter settlement were moved to new villages where the food supply could be strictly controlled through rationing. This guaranteed that collaborators would have to deprive themselves of food in order to pass it to guerrillas, a strategy which forced guerrillas to spend more time securing food. Rebel sympathizers were presented with a choice: to continue supplying food to the rebels and go to prison, to stop supplying food and be murdered for cooperation with the security forces, or cooperate with the security forces, and be relocated to Hong Kong (Mockaitis 1990:117).

[200]According to a 1979 pamphlet, the H-blocks were also known as the 'hunger blocks': "of all the despicable forms of torture tried in Northern Ireland...hunger is the most despicable."

[201]Ellmann identifies circulation and exchange as central elements of the hungerstrike: "Since food and words are circulating currencies, the faster by refusing one such form of interchange seems to be impelled to look for satisfaction in the other. It is circulation, therefore, that underlies the art of hunger, and it is necessary to investigate this ruinous economy" (1993:70). Hungerstriking, as I will argue, is not just an exchange of words and food, but a sacrificial exchange of warfare.

[202]1607 is the date normally considered to mark the ascension of British law over Brehon law, when Sir John Davis, Attorney General for Ireland (1603-1619) judged that Irish land inheritance patterns were invalid. Since then, land inheritance has followed the British pattern.

[203]A Belfast joke pokes fun of the refusal to recognize: Q: How many Provos does it take to screw in a light bulb? A: We refuse to recognize the authority of this court to ask the question.

[204]Similarly, the effect of judicial exclusion orders against Sinn Fein "is to shield the people of Britain not so much from violence as from unpalatable political viewpoints (Walker 1988:86).

[205]PIRA and the INLA carefully screened hungerstrike volunteers, before they were allowed to begin the fast. The would not take any one convicted of a violent crime, since it was thought that prior violence of a certain sort would jeopardize the 'purity' of the action. According to Girard, "Therefore...whenever violence is inevitable, it is best that the victim be pure, untainted by any involvement in the dispute" (1972:28)

[206]The malleability of the human body as a ritual symbol (Van Gennep 1960:7) culminates in the decomposing corpse as a symbol of transitional liminality (Turner 1967). Human biological processes of death and decomposition "give an outward and visible form to an inward and conceptual process" (Turner 1967:98).

[207]The Phoblacht na n-Eireann, *the Proclamation of the Irish Republic*, 1916, signed by Thomas J. Clarke, Sean MacDiarmad, Thomas MacDonagh, P.H. Pearse, Eamonn Ceannt, James Connolly and Joseph Plunkett.

[208]In June 1981, hungerstrikers Paddy Agnew and Kieran Doherty were elected in Louth and Cavan/Monahan, respectively.

[209]Foucault perhaps derived this from Hobbes, who ascribed to the view that only in death did man's desire for power finally cease. In Chapter Eleven of Leviathan, Hobbes writes: "So that in the first place, I put a general inclination of all mankind, a perpetuall and restless desire of power after power, that ceaseth onely in Death" (1968).

[210]Baudrillard, after Foucault, argued that the code always reconstitutes itself, any disruptions merely revert to a prior norm. This is the essential pessimism of post-modern theory. Foucault's theory of power problematizes the interpretation of resistance. According to Foucault, discourse is unipolar; resistance is not outside of power, but is a capillary of power constantly flowing and dispersing. The case of political violence in Northern Ireland demonstrates that even if the code regenerates itself, acts within the symbolic order can not only disrupt the code, but reformulate it. The bullet in the ambush is no simulacra to its human victim: the praxis of violence is extremely concrete.

[211]Although Countess Markievicz may have longed for the peace of the Republican plot, in recent years, Republican burial plots have hardly been peaceful.

[212]Regulating the public production of death increased the reliance Security Forces on covert intelligence operations (see Chapter Five).

[213]For example, in the case of the burial of Dan McCann, Sean Savage and Mairead Farrell killed in Gibralter by the SAS. During the funeral in Milltown cemetery, a Loyalist paramilitary, Michael Stone, began firing on the crowd and threw at least one grenade. Security Forces standing at the cemetery gate allegedly did not make any attempt to intervene. This attack resulted in two deaths, and during the funeral the following week, two British soldiers drove into the Republican funeral cortege and were abducted and later executed by members of the crowd.

214The military logic of hungerstrikes brings us back to Francis Hughes, with whom this project began and who exemplifies the way in which the war was brought into the prisons. Hughes was destined to be a soldier: following the 1972 ceasefire of the Officials, Hughes left their organization and established an independent commando unit in South Derry which conducted operations against the British Army. During their reorganization, the Provisionals lassoed all the renegade units operating in the rural backwoods. Normally, only one or two paramilitaries would be accepted from such independent groups, but Hughes had so thoroughly trained his men that they were accepted en masse. Hughes preferred to wear combat gear while on operations, a fact which eventually allowed Lance Corporal David Jones to mistake Hughes for a British soldier. After the firefight, Hughes was taken to Musgrave Park military hospital, where the surgery performed on his shattered hip left him dependent on a crutch to walk. After being sentenced to 83 years in prison, Hughes reportedly told medical staff, "I'm dying, I'm dead, that's it" (cited in Clarke 1987:163).

215Gramsci defined 'hegemony' as "a conception of the world that is implicitly manifest in art, in law, in economic activity and in all manifestations of individual and collective life" (1971:328). In condition of political unrest, 'collective' life can at best, only apply to a fraction of the population.

[216]"When the suspect has committed an offence and remains silent, there is no infringement of section 18, but the Order (especially Article 3) may penalise silence" (Walker 1992:138). According to Walker, Article 3 "provides that if a person being questioned by the police about an offence or being charged with an offence fails to mention any fact which could reasonably have been mentioned at the time and was subsequently relied upon in his defence, the trial court may draw adverse inferences from the silence or may treat it as corroborating other evidence" (1992:137). These legislative articles, combined with the emphasis on individual confessions as evidence in Diplock trials, essentially repealed any right to silence deriving from the common law, allowing silence to be interpreted as evidence of guilt in court.

[217]While much has been made of Adam's widely reprinted comment (e.g., that Sinn Fein and PIRA nearly had an ideological divorce over the issue, that it signified weak political support within the communities, that the presaged a truce which PSF opposed), it must be remembered that the last hungerstrike in the Republican Movement which ended in death was that of Frank Stagg in 1976. Stagg's family was politically divided in their support for the movement, and some family members did not support Stagg's fast. The misapprehensions were such that Maire Drumm actually had to explain in public that PIRA had not ordered Stagg to die, that he had taken on a hungerstrike willingly as a protest. PIRA and Sinn Fein wanted to avoid at all costs the impression that they were giving direct orders to the prisoners to begin the fast. On the contrary, prisoners were carefully selected for the honor of hungerstriking and relatives action committees (RACs) were established to encourage support.

[218]In November, female PIRA prisoners in Armagh went on hunger-strike for political status "in spite of appeals from the Republican movement advising them against such action" (Reed 1984:340).

[219]MacSwiney's speech has been transformed into a graffiti slogan in urban areas in Northern Ireland: "inflict and endure". The shift in emphasis in this slogan is perhaps note-worthy.

[220]"But it is in the dialectical relationship between the body and a space structured according to the mythico-ritual oppositions that one finds the form par excellence of the structural apprenticeship which leads to the em-bodying of the structures of the world..." (Bourdieu 1977:89).

[221]Normally, prisoners or convicted felons were not allowed to hold public office. Sands had been elected via a legal loop-hole: the Criminal Law Act of 1967 had abolished the legal category of 'felon', making prisoners eligible for electoral office. After his election, there was a question about whether Sands, being unable to be sworn in or to attend sessions in the House, might resign, to which Sands replied, "What do I want to resign for? I only have two weeks to live" (cited in Newsweek, 20 April 1981:14). There was also doubt about whether the Commons might vote to expel Sands on the ground that he was a criminal. To this Sands said: "Your government is scurrying to unseat a dying man, which if you allow it, will shame you in the eyes of the world" (cited in Irish Times 13 April 1981:5). According to a group of MP's who opposed the attempt to expel Sands, any attempt to unseat him would be a violation of the democratically expressed wishes of 30,000 electors in Fermanagh and South Tyrone (statement from Clive Stoley, MP, Ernie Roberts, MP, Lord Gifford and Lord Milford, cited in Irish Times, 14 April 1981:1). Sand's victory was based on a voter turn-out of 86.8%, and constituted a larger popular mandate than Thatcher had received in her own home constituency.

[222]Some of the organizations which offered to intervene were the Irish Council for Justice and Peace, International Committee of the Red Cross (ICRC), European Committee for Human Rights, a special emissary sent by the Pope, Amnesty International, etc. Even concessions offered indirectly through covert sources, were rejected by the hungerstrikers. Negotiations with Whitehall were carried out through a highly placed contact called, mysteriously, 'the Mountain Climber'.

[223]During the first hungerstrike, although the government publicly refused to negotiate, Sands and Brendan Hughes sat in conference with NIO officials inside of the prison. They were presented with a 34 page document outlining acceptable compromises. Further negotiations took place with the Foreign Office, though a secret channel and it looked as if things would work out. McKenna was dying, so Hughes called off the hungerstrike. When final offer came through, it did not meet the specified qualifications and thus, the second hungerstrike was begun.

[224]Hungerstrikes also re-particularize death. We are no longer living in a Weberian world in which the state has a monopoly on the use of force. Neither does the state retain a monopoly on death, which is threatened by multi-nationals, tribal ethnocides, and insurgencies. Martyrdom is becoming increasingly obsolete as a revolutionary strategy; frankly, nobody cares all that much. After Auschwitz, no corpse is unique, and no particular death is particularly meaningful (see Adorno 1973; Feldman 1991). The hungerstrikes, by explicitly re-focusing on the individual deaths of individual hungerstrikers who were the bearers and symbols of Republican nationalism, forced a recognition of the means of production of death (Virilio and Lotringer 1983:131).

[225]In the act of exchange, is also death. The hungerstrikes are the final element in the exchange metaphor which has dominated this project: "This production and reproduction, this distribution and circulation of bodies formed and signified by violence and signifying through violence, constitute the vast economic enterprise and structure of exchange that is warfare in Northern Ireland" (Feldman 1991:146).

[226]What does it all mean? It could be argued that within the hungerstrike discourse, "the author's intention and the meaning of the text cease to coincide" (Ricoeur 1979:78). Just as "the text's career escapes the finite horizon lived by its author" (1979:78), so the narratives embodied in the hungerstrikers became disassociated from the event of the hungerstrikes. Like the author who inevitably says more than he wishes to, the hungerstrikers 'say' things which are unintended and can only be retrospectively interpreted. Persisting patterns inscribed in a temporal space produce an autonomous history, or a "sedimentation in social time" (1979:85), through which deeds become institutions. Meaning can be evaluated only through history itself, so that "human action is an open work" (1979:86), whose meaning is 'in suspense', awaiting interpretation under the auspices of present praxis. In other words, lots of people have different interpretations of the same event.

And what did the hungerstrikes do for the Republicans? As soon as the hungerstrike ended, the substance of the five demands was quietly granted. "The Rule of Law had prevailed, the identity of the prisoners as 'criminals' had been constituted, and the violence of the state had been legitimated...the juridical ritual had been consummated by death" (Devlin 1993:173). The British government vanquished its Republican enemies. Although the failure of the hungerstrikes to achieve political concessions has given the protest an aura of pathos, the hungerstrike was not mere fatalism in the political realm, or a "paralyzed gesture...of aestheticized powerlessness..." (Said 1989:223). For Republicans, the hungerstrikes vanquished *death* and reaffirmed the historical transcendence of Republicanism. Thus, "if we can speak of a reassertion of the social order at the time of death, this social order is a product of rituals...rather than their cause" (Bloch and Parry 1982:6).

[227]As the hungerstrikes progressed, it became apparent that there were no procedures to oversee the implementation of settlement, and confusion about what "special status" actually entailed. McGeown wrote to Gerry Adams: "the protest is now the principle and not how to solve it. I think what we need to do is to begin to say the solution is the principle, not the protest..." (cited in O'Malley 1990:79). "[E]ndurance became more important than the end for which it was employed" (O'Malley 1990:111).

> They maintained the line and over a period of weeks, they argued, they trashed it out and in the end we were just terribly immobile. We weren't changing our mind. Frank Hughes died and then you went into the whole process of having no way out. At that point the strategy actually collapsed, nobody knew who could take a decision to end the whole thing (cited in O'Malley 1990:75)

No one, including the Army Council of PIRA, could control the deterioration of the hungerstrike: the Army Council "may as well have been talking to the wall" (O'Malley 1990:80). Although the prisoners were answerable to formal military organization and were under the direct command of the Provisional Army Council, they blocked all external attempts to control the situation. Technically, they could have been subject to court-martial.

Acting in direct opposition to PIRA, with the encouragement of the Church, the families of the remaining hungerstrikers quickly removed them from the strike. This would seem to offer evidence for the primacy of genealogical networks, after all. Although some hungerstrikers drew up legal documents to prevent the intervention of their families when they became delirious or comatose, after ten deaths, the remaining hungerstrikers were quickly removed from the fast. On the day the tenth hungerstriker, Mickey Devine, died, Paddy Quinn was taken off the fast after forty-seven days by his mother, who 'could not bear the sight of him kicking and screaming in so much pain" (Buckley 1985:160). Adams responded to this 'betrayal' by stating that "While we must

[228]The Irish Republic was never legitimately manifest in the Irish Free State, thus the contempt in which complacent 'Free Staters' are held by Republicans, who from a Republican perspective, are in cahoots with the British.

[229]Collins, military pragmatist that he was, was pro-Treaty since he believed that the Treaty was the only concession which the British could offer in 1921. Collins was obsessed with the North, however, and determined to use the IRB to protect the Republicans there against Protestant anti-Nationalists. Collins by this time, despite the fact that he had become a minister in the Free State government, was carrying out a series of commando raids against British forces stationed along the border. At this time, a series of pogroms were being carried out in the North, and these were widely considered to be the work of Field Marshal Sir Henry Wilson, ex-Chief of the Imperial General Staff and current military advisor of the Unionist government in Belfast (mentioned in Chapter Two as the man who advocated the declarations of martial law in 1921). Wilson, a vehement anti-nationalist, was killed in London on the order of Michael Collins by the IRB. Wilson's "killing was the pretext that finally tumbled the two Sinn Fein factions into Civil War" (Cronin 1980:151). For a number of very complex reasons, Collins had no alternative but to begin the war against O'Connor in the Four Courts in order to save the treaty. Collins was assassinated in 1922.

[230]Casualty rate differentials in combat during wars can be explained as a result of the relation of dispersion to weapons lethality. According to Bellamy, "dispersion has in fact increased more than lethality, and this explains why, for example , casualty rates were lower in World War Two than in World War One, or the American Civil War" (1990:47).

[231]While it may seem that the point of penetrating tanks is to immobilize it by killing the men inside, in fact in the language of weapon systems design and combat-speak, the aim is explicitly to kill the tank. The men who die are incidental to the destruction of the technology, which is the primary target.

[232]Gabriel argues that the extremely lethal conditions of modern war privilege the already insane. "The soldier must be made abnormal in order to behave 'normally' on the battlefield. If he is to function efficiently, he must first be made insane" (1987:149).

[233]"The British believe that killing by a gun or bomb is right, while killing with a panga [machete] is evil" (member of Mau Mau, cited in Mockaitis 1991:5).

[234]Bunker architecture has probably influenced civilian design styles, especially concrete tower blocks (Mallory and Ottar 1973).

[235]Military intelligence resembles anthropology, and connections within the academic-military complex have been noted (Asad 1975); specifically, "the function of social scientists in helping the government to formulate, evaluate and *implement* counterinsurgency policy" (Wakin 1992:59). Sir Richard Burton, "one of the greatest Secret Service agents of the Victorian era," combined proto-ethnography and espionage: "There cannot be many spies who have submitted a report of pederasty following nights concealed in the male brothers of Karachi!" (Dixon1976:295). In Northern Ireland, where information collection gives everybody the heebie-jeebies, ethnography is hardly distinguishable from intelligence gathering. John Darby and John Burton, collecting information for projects in Belfast in the late 1970's, were held by the PIRA and would have been executed, except for the intervention of Sinn Fein counselor, Rauriagh O'Braidaigh.

Both types of observation have a particular relationship to space: field sites, like battlefields, are panoptic spaces. The lexicon of fieldwork and intelligence operations is governed by a judicial metaphors of 'informants' offering 'revealing' (impressionistic and unverified) 'evidence' to fieldworkers who have 'penetrated' local culture, gathering 'local knowledge', etc. The plausibility of ethnographic reporting is based on "having actually penetrated (or, if you prefer, been penetrated by) another form of life..." (Geertz 1988:4). The intelligence operative, likewise, must penetrate and report on foreign, hostile groups, "transgressing the integrity of [his] informants" in a symbolic act of violence (Rabinow 1977:129). The ethnographer collects 'fieldnotes', subjective impressions derived from human contact. The unstructured impressions of state intelligence agencies increasingly resemble fieldnotes: "these files [CID notes] are highly evanescent...This unstructured, discursive quality...makes this information all but inaccessible to other police agencies..." (Rule 1973:78).

[236]Clifford describes a "war of gazes" between Griaule and his informants. According to Clifford, "[t]he researchers feel themselves under surveillance: 'hundreds of eyes follow us. We're in full view of the village; in every crack in the wall, behind every granary, an eye is attentive" (cited in Clifford 1983:135-134).